AMPL

A Modeling Language
for Mathematical Programming

Second Edition

Duxbury Titles of Related Interest

AMPL

A Modeling Language
for Mathematical Programming

Second Edition

Robert Fourer

Northwestern University

David M. Gay

AMPL Optimization LLC

Brian W. Kernighan

Princeton University

THOMSON

BROOKS/COLE

Australia • Canada • Mexico • Singapore • Spain • United Kingdom • United States

THOMSON

™

BROOKS/COLE

Publisher: Curt Hinrichs
Assistant Editor: Ann Day
Editorial Assistant: Katherine Brayton
Technology Project Manager: Burke Taft
Marketing Manager: Joseph Rogove
Advertising Project Manager: Tami Strang

Project Manager, Editorial Production: Karen Haga
Print/Media Buyer: Jessica Reed
Permissions Editor: Bob Kauser
Cover Designer: Carole Lawson
Cover and Text Printer: Transcontinental Printing, Inc.

This book was typeset
(grap|pic|tbl|eqn|troff −mpm) in Times
and Courier by the authors.

Printed in Canada
3 4 5 6 7 06

For more information about our products,
contact us at:
Thomson Learning Academic Resource Center
1-800-423-0563
For permission to use material from this text,
contact us by:
Phone: 1-800-730-2214
Fax: 1-800-730-2215
Web: http://www.thomsonrights.com

Library of Congress Control Number: 2002112121

ISBN 0-534-38809-4

Brooks/Cole—Thomson Learning
511 Forest Lodge Road
Pacific Grove, CA 93950
USA

Asia
Thomson Learning
5 Shenton Way #01-01
UIC Building
Singapore 068808

Australia
Nelson Thomson Learning
102 Dodds Street
South Melbourne, Victoria 3205
Australia

Canada
Nelson Thomson Learning
1120 Birchmount Road
Toronto, Ontario M1K 5G4
Canada

Europe/Middle East/Africa
Thomson Learning
High Holborn House
50/51 Bedford Row
London WC1R 4LR
United Kingdom

Latin America
Thomson Learning
Seneca, 53
Colonia Polanco
11560 Mexico D.F.
Mexico

About the Authors

Robert Fourer received his Ph.D. in operations research from Stanford University in 1980 and is an active researcher in mathematical programming and modeling language design. He joined the Department of Industrial Engineering and Management Sciences at Northwestern University in 1979 and served as chair of the department from 1989 to 1995.

David M. Gay received his Ph.D. in computer science from Cornell University in 1975 and was in the Computing Science Research Center at Bell Laboratories from 1981 to 2001; he is now CEO of AMPL Optimization LLC. His research interests include numerical analysis, optimization, and scientific computing.

Brian Kernighan received his Ph.D. in electrical engineering from Princeton University in 1969. He was in the Computing Science Research Center at Bell Laboratories from 1969 to 2000 and now teaches in the Computer Science department at Princeton. He is the co-author of several computer science books, including *The C Programming Language* and *The UNIX Programming Environment*.

Contents

Introduction

As our title suggests, there are two aspects to the subject of this book. The first is mathematical programming, the optimization of a function of many variables subject to constraints. The second is the AMPL modeling language, which we designed and implemented to help people use computers to develop and apply mathematical programming models.

We intend this book as an introduction both to mathematical programming and to AMPL. For readers already familiar with mathematical programming, it can serve as a user's guide and reference manual for the AMPL software. We assume no previous knowledge of the subject, however, and hope that this book will also encourage the use of mathematical programming models by those who are new to the field.

Mathematical programming

The term "programming" was in use by 1940 to describe the planning or scheduling of activities within a large organization. "Programmers" found that they could represent the amount or level of each activity as a *variable* whose value was to be determined. Then they could mathematically describe the restrictions inherent in the planning or scheduling problem as a set of equations or inequalities involving the variables. A solution to all of these *constraints* would be considered an acceptable plan or schedule.

Experience soon showed that it was hard to model a complex operation simply by specifying constraints. If there were too few constraints, many inferior solutions could satisfy them; if there were too many constraints, desirable solutions were ruled out, or in the worst case no solutions were possible. The success of programming ultimately depended on a key insight that provided a way around this difficulty. One could specify, in addition to the constraints, an *objective*: a function of the variables, such as cost or profit, that could be used to decide whether one solution was better than another. Then it didn't matter that many different solutions satisfied the constraints — it was sufficient to find one such solution that minimized or maximized the objective. The term *mathematical programming* came to be used to describe the minimization or maximization of an objective function of many variables, subject to constraints on the variables.

In the development and application of mathematical programming, one special case stands out: that in which all the costs, requirements and other quantities of interest are terms strictly proportional to the levels of the activities, or sums of such terms. In mathematical terminology, the objective is a linear function, and the constraints are linear equations and inequalities. Such a problem is called a *linear program*, and the process of setting up such a problem and solving it is called *linear programming*. Linear programming is particularly important because a wide variety of problems can be modeled as linear programs, and because there are fast and reliable methods for solving linear programs even with thousands of variables and constraints. The ideas of linear programming are also important for analyzing and solving mathematical programming problems that are not linear.

All useful methods for solving linear programs require a computer. Thus most of the study of linear programming has taken place since the late 1940's, when it became clear that computers would be available for scientific computing. The first successful computational method for linear programming, the simplex algorithm, was proposed at this time, and was the subject of increasingly effective implementations over the next decade. Coincidentally, the development of computers gave rise to a now much more familiar meaning for the term "programming."

In spite of the broad applicability of linear programming, the linearity assumption is sometimes too unrealistic. If instead some smooth nonlinear functions of the variables are used in the objective or constraints, the problem is called a *nonlinear program*. Solving such a problem is harder, though in practice not impossibly so. Although the optimal values of nonlinear functions have been a subject of study for over two centuries, computational methods for solving nonlinear programs in many variables were developed only in recent decades, after the success of methods for linear programming. The field of mathematical programming is thus also known as *large scale optimization,* to distinguish it from the classical topics of optimization in mathematical analysis.

The assumptions of linear programming also break down if some variables must take on whole number, or integral, values. Then the problem is called *integer programming,* and in general becomes much harder. Nevertheless, a combination of faster computers and more sophisticated methods have made large integer programs increasingly tractable in recent years.

The AMPL modeling language

Practical mathematical programming is seldom as simple as running some algorithmic method on a computer and printing the optimal solution. The full sequence of events is more like this:

- Formulate a model, the abstract system of variables, objectives, and constraints that represent the general form of the problem to be solved.
- Collect data that define a specific problem instance.
- Generate a specific objective function and constraint equations from the model and data.

- Solve the problem instance by running a program, or *solver*, to apply an algorithm that finds optimal values of the variables.
- Analyze the results.
- Refine the model and data as necessary, and repeat.

If people could deal with mathematical programs in the same way that solvers do, the formulation and generation phases of modeling might be relatively straightforward. In reality, however, there are many differences between the form in which human modelers understand a problem and the form in which solver algorithms work with it. Conversion from the ''modeler's form'' to the ''algorithm's form'' is consequently a time-consuming, costly, and often error-prone procedure.

In the special case of linear programming, the largest part of the algorithm's form is the constraint coefficient matrix, which is the table of numbers that multiply all the variables in all the constraints. Typically this is a very sparse (mostly zero) matrix with anywhere from hundreds to hundreds of thousands of rows and columns, whose nonzero elements appear in intricate patterns. A computer program that produces a compact representation of the coefficients is called a matrix generator. Several programming languages have been designed specifically for writing matrix generators, and standard computer programming languages are also often used.

Although matrix generators can successfully automate some of the work of translation from modeler's form to algorithm's form, they remain difficult to debug and maintain. One way around much of this difficulty lies in the use of a *modeling language* for mathematical programming. A modeling language is designed to express the modeler's form in a way that can serve as direct input to a computer system. Then the translation to the algorithm's form can be performed entirely by computer, without the intermediate stage of computer programming. Modeling languages can help to make mathematical programming more economical and reliable; they are particularly advantageous for development of new models and for documentation of models that are subject to change.

Since there is more than one form that modelers use to express mathematical programs, there is more than one kind of modeling language. An *algebraic* modeling language is a popular variety based on the use of traditional mathematical notation to describe objective and constraint functions. An algebraic language provides computer-readable equivalents of notations such as $x_j + y_j$, $\sum_{j=1}^{n} a_{ij} x_j$, $x_j \geq 0$, and $j \in S$ that would be familiar to anyone who has studied algebra or calculus. Familiarity is one of the major advantages of algebraic modeling languages; another is their applicability to a particularly wide variety of linear, nonlinear and integer programming models.

While successful algorithms for mathematical programming first came into use in the 1950's, the development and distribution of algebraic modeling languages only began in the 1970's. Since then, advances in computing and computer science have enabled such languages to become steadily more efficient and general.

This book describes AMPL, an algebraic modeling language for mathematical programming; it was designed and implemented by the authors around 1985, and has been evolving ever since. AMPL is notable for the similarity of its arithmetic expressions to customary algebraic notation, and for the generality and power of its set and subscripting expressions. AMPL also extends algebraic notation to express common mathematical programming structures such as network flow constraints and piecewise linearities.

AMPL offers an interactive command environment for setting up and solving mathematical programming problems. A flexible interface enables several solvers to be available at once so a user can switch among solvers and select options that may improve solver performance. Once optimal solutions have been found, they are automatically translated back to the modeler's form so that people can view and analyze them. All of the general set and arithmetic expressions of the AMPL modeling language can also be used for displaying data and results; a variety of options are available to format data for browsing, printing reports, or preparing input to other programs.

Through its emphasis on AMPL, this book differs considerably from the presentation of modeling in standard mathematical programming texts. The approach taken by a typical textbook is still strongly influenced by the circumstances of 30 years ago, when a student might be lucky to have the opportunity to solve a few small linear programs on any actual computer. As encountered in such textbooks, mathematical programming often appears to require only the conversion of a ''word problem'' into a small system of inequalities and an objective function, which are then presented to a simple optimization package that prints a short listing of answers. While this can be a good approach for introductory purposes, it is not workable for dealing with the hundreds or thousands of variables and constraints that are found in most real-world mathematical programs.

The availability of an algebraic modeling language makes it possible to emphasize the kinds of general models that can be used to describe large-scale optimization problems. Each AMPL model in this book describes a whole class of mathematical programming problems, whose members correspond to different choices of indexing sets and numerical data. Even though we use relatively small data sets for illustration, the resulting problems tend to be larger than those of the typical textbook. More important, the same approach, using still larger data sets, works just as well for mathematical programs of realistic size and practical value.

We have not attempted to cover the optimization theory and algorithmic details that comprise the greatest part of most mathematical programming texts. Thus, for readers who want to study the whole field in some depth, this book is a complement to existing textbooks, not a replacement. On the other hand, for those whose immediate concern is to apply mathematical programming to a particular problem, the book can provide a useful introduction on its own.

In addition, AMPL software is readily available for experiment: the AMPL web site, www.ampl.com, provides free downloadable ''student'' versions of AMPL and representative solvers that run on Windows, Unix/Linux, and Mac OS X. These can easily handle problems of a few hundred variables and constraints, including all of the examples in the book. Versions that support much larger problems and additional solvers are also available from a variety of vendors; again, details may be found on the web site.

Outline of the book

The second edition, like the first, is organized conceptually into four parts. Chapters 1 through 4 are a tutorial introduction to models for linear programming:

1. Production Models: Maximizing Profits
2. Diet and Other Input Models: Minimizing Costs
3. Transportation and Assignment Models
4. Building Larger Models

These chapters are intended to get you started using AMPL as quickly as possible. They include a brief review of linear programming and a discussion of a handful of simple modeling ideas that underlie most large-scale optimization problems. They also illustrate how to provide the data that convert a model into a specific problem instance, how to solve a problem, and how to display the answers.

The next four chapters describe the fundamental components of an AMPL linear programming model in detail, using more complex examples to examine major aspects of the language systematically:

5. Simple Sets and Indexing
6. Compound Sets and Indexing
7. Parameters and Expressions
8. Linear Programs: Variables, Objectives and Constraints

We have tried to cover the most important features, so that these chapters can serve as a general user's guide. Each feature is introduced by one or more examples, building on previous examples wherever possible.

The following six chapters describe how to use AMPL in more sophisticated ways:

9. Specifying Data
10. Database Access
11. Modeling Commands
12. Display Commands
13. Command Scripts
14. Interactions with Solvers

The first two of these chapters explain how to provide the data values that define a specific instance of a model; Chapter 9 describes AMPL's text file data format, while Chapter 10 presents features for access to information in relational database systems. Chapter 11 explains the commands that read models and data, and invoke solvers; Chapter 12 shows how to display and save results. AMPL provides facilities for creating scripts of commands, and for writing loops and conditional statements; these are covered in Chapter 13. Chapter 14 goes into more detail on how to interact with solvers so as to make the best use of their capabilities and the information they provide.

Finally, we turn to the rich variety of problems and applications beyond purely linear models. The remaining chapters deal with six important special cases and generalizations:

Chapters 15 and 16 describe additional language features that help AMPL represent particular kinds of linear programs more naturally, and that may help to speed translation and solution. The last four chapters cover generalizations that can help models to be more realistic than linear programs, although they can also make the resulting optimization problems harder to solve.

Appendix A is the AMPL reference manual; it describes all language features, including some not mentioned elsewhere in the text. Bibliography and exercises may be found in most of the chapters.

About the second edition

AMPL has evolved a lot in ten years, but its core remains essentially unchanged, and almost all of the models from the first edition work with the current program. Although we have made substantial revisions throughout the text, much of the brand new material is concentrated in the third part, where the original single chapter on the command environment has been expanded into five chapters. In particular, database access, scripts and programming constructs represent completely new material, and many additional AMPL commands for examining models and accessing solver information have been added.

The first edition was written in 1992, just before the explosion in Internet and web use, and while personal computers were still rather limited in their capabilities; the first student versions of AMPL ran on DOS on tiny, slow machines, and were distributed on floppy disks.

Today, the web site at www.ampl.com is the central source for all AMPL information and software. Pages at this site cover all that you need to learn about and experiment with optimization and the use of AMPL:

- Free versions of AMPL for a variety of operating systems.
- Free versions of several solvers for a variety of problem types.
- All of the model and data files used as examples in this book.

The free software is fully functional, save that it can only handle problems of a few hundred variables and constraints. Unrestricted commercial versions of AMPL and solvers are available as well; see the web site for a list of vendors.

You can also try AMPL without downloading any software, through browser interfaces at www.ampl.com/TRYAMPL and the NEOS Server (neos.mcs.anl.gov). The AMPL web site also provides information on graphical user interfaces and new AMPL language features, which are under continuing development.

Acknowledgements to the first edition

We are deeply grateful to Jon Bentley and Margaret Wright, who made extensive comments on several drafts of the manuscript. We also received many helpful suggestions on AMPL and the book from Collette Coullard, Gary Cramer, Arne Drud, Grace Emlin, Gus Gassmann, Eric Grosse, Paul Kelly, Mark Kernighan, Todd Lowe, Bob Richton, Michael Saunders, Robert Seip, Lakshman Sinha, Chris Van Wyk, Juliana Vignali, Thong Vukhac, and students in the mathematical programming classes at Northwestern University. Lorinda Cherry helped with indexing, and Jerome Shepheard with typesetting. Our sincere thanks to all of them.

Bibliography

E. M. L. Beale, "Matrix Generators and Output Analyzers." In Harold W. Kuhn (ed.), *Proceedings of the Princeton Symposium on Mathematical Programming*, Princeton University Press (Princeton, NJ, 1970) pp. 25–36. A history and explanation of matrix generator software for linear programming.

Johannes Bisschop and Alexander Meeraus, "On the Development of a General Algebraic Modeling System in a Strategic Planning Environment." Mathematical Programming Study **20** (1982) pp. 1–29. An introduction to GAMS, one of the first and most widely used algebraic modeling languages.

Robert E. Bixby, "Solving Real-World Linear Programs: A Decade and More of Progress." Operations Reearch **50** (2002) pp. 3)–15. A history of recent advances in solvers for linear programming. Also in this issue are accounts of the early days of mathematical programming by pioneers of the field.

George B. Dantzig, "Linear Programming: The Story About How It Began." In Jan Karel Lenstra, Alexander H. G. Rinnooy Kan and Alexander Schrijver, eds., *History of Mathematical Programming: A Collection of Personal Reminiscences.* North-Holland (Amsterdam, 1991) pp. 19–31. A source for our brief account of the history of linear programming. Dantzig was a pioneer of such key ideas as objective functions and the simplex algorithm.

Robert Fourer, "Modeling Languages versus Matrix Generators for Linear Programming." ACM Transactions on Mathematical Software **9** (1983) pp. 143–183. The case for modeling languages.

C. A. C. Kuip, "Algebraic Languages for Mathematical Programming." European Journal of Operational Research **67** (1993) 25–51. A survey.

1

Production Models:
Maximizing Profits

As we stated in the Introduction, mathematical programming is a technique for solving certain kinds of problems — notably maximizing profits and minimizing costs — subject to constraints on resources, capacities, supplies, demands, and the like. AMPL is a language for specifying such optimization problems. It provides an algebraic notation that is very close to the way that you would describe a problem mathematically, so that it is easy to convert from a familiar mathematical description to AMPL.

We will concentrate initially on linear programming, which is the best known and easiest case; other kinds of mathematical programming are taken up later in the book. This chapter addresses one of the most common applications of linear programming: maximizing the profit of some operation, subject to constraints that limit what can be produced. Chapters 2 and 3 are devoted to two other equally common kinds of linear programs, and Chapter 4 shows how linear programming models can be replicated and combined to produce truly large-scale problems. These chapters are written with the beginner in mind, but experienced practitioners of mathematical programming should find them useful as a quick introduction to AMPL.

We begin with a linear program (or LP for short) in only two decision variables, motivated by a mythical steelmaking operation. This will provide a quick review of linear programming to refresh your memory if you already have some experience, or to help you get started if you're just learning. We'll show how the same LP can be represented as a general algebraic model of production, together with specific data. Then we'll show how to express several linear programming problems in AMPL and how to run AMPL and a solver to produce a solution.

The separation of model and data is the key to describing more complex linear programs in a concise and understandable fashion. The final example of the chapter illustrates this by presenting several enhancements to the model.

1

1.1 A two-variable linear program

An (extremely simplified) steel company must decide how to allocate next week's time on a rolling mill. The mill takes unfinished slabs of steel as input, and can produce either of two semi-finished products, which we will call bands and coils. (The terminology is not entirely standard; see the bibliography at the end of the chapter for some accounts of realistic LP applications in steelmaking.) The mill's two products come off the rolling line at different rates:

> Tons per hour: Bands   200
>   Coils   140

and they also have different profitabilities:

> Profit per ton: Bands $25
>   Coils $30

To further complicate matters, the following weekly production amounts are the most that can be justified in light of the currently booked orders:

> Maximum tons: Bands   6,000
>   Coils   4,000

The question facing the company is as follows: If 40 hours of production time are available this week, how many tons of bands and how many tons of coils should be produced to bring in the greatest total profit?

While we are given numeric values for production rates and per-unit profits, the tons of bands and of coils to be produced are as yet unknown. These quantities are the decision *variables* whose values we must determine so as to maximize profits. The purpose of the linear program is to specify the profits and production limitations as explicit formulas involving the variables, so that the desired values of the variables can be determined systematically.

In an algebraic statement of a linear program, it is customary to use a mathematical shorthand for the variables. Thus we will write X_B for the number of tons of bands to be produced, and X_C for tons of coils. The total hours to produce all these tons is then given by

> (hours to make a ton of bands) $\times X_B$ + (hours to make a ton of coils) $\times X_C$

This number cannot exceed the 40 hours available. Since hours per ton is the reciprocal of the tons per hour given above, we have a *constraint* on the variables:

> $(1/200) X_B + (1/140) X_C \leq 40.$

There are also production limits:

> $0 \leq X_B \leq 6000$
> $0 \leq X_C \leq 4000$

In the statement of the problem above, the upper limits were specified, but the lower limits were assumed — it was obvious that a negative production of bands or coils would be meaningless. Dealing with a computer, however, it is necessary to be quite explicit.

By analogy with the formula for total hours, the total profit must be

$$\text{(profit per ton of bands)} \times X_B + \text{(profit per ton of coils)} \times X_C$$

That is, our objective is to maximize $25\ X_B + 30\ X_C$. Putting this all together, we have the following linear program:

Maximize $25\ X_B + 30\ X_C$

Subject to $(1/200)\ X_B + (1/140)\ X_C \leq 40$
$$0 \leq X_B \leq 6000$$
$$0 \leq X_C \leq 4000$$

This is a very simple linear program, so we'll solve it by hand in a couple of ways, and then check the answer with AMPL.

First, by multiplying profit per ton times tons per hour, we can determine the profit per hour of mill time for each product:

Profit per hour: Bands $5,000
                    Coils $4,200

Bands are clearly a more profitable use of mill time, so to maximize profit we should produce as many bands as the production limit will allow — 6,000 tons, which takes 30 hours. Then we should use the remaining 10 hours to make coils — 1,400 tons in all. The profit is $25 times 6,000 tons plus $30 times 1,400 tons, for a total of $192,000.

Alternatively, since there are only two variables, we can show the possibilities graphically. If X_B values are plotted along the horizontal axis, and X_C values along the vertical axis, each point represents a choice of values, or solution, for the decision variables:

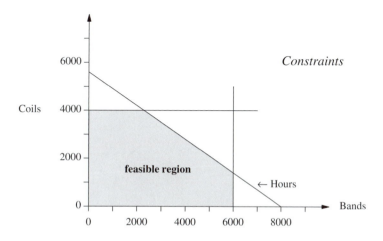

The horizontal line represents the production limit on coils, the vertical on bands. The diagonal line is the constraint on hours; each point on that line represents a combination of bands and coils that requires exactly 40 hours of production time, and any point downward and to the left requires less than 40 hours.

The shaded region bounded by the axes and these three lines corresponds exactly to the *feasible* solutions — those that satisfy all three constraints. Among all the feasible solutions represented in this region, we seek the one that maximizes the profit.

For this problem, a line of slope $-25/30$ represents combinations that produce the same profit; for example, in the figure below, the line from (0, 4500) to (5400, 0) represents combinations that yield \$135,000 profit. Different profits give different but parallel lines in the figure, with higher profits giving lines that are higher and further to the right.

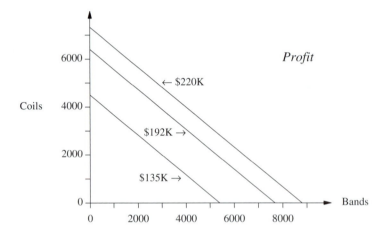

If we combine these two plots, we can see the profit-maximizing, or *optimal*, feasible solution:

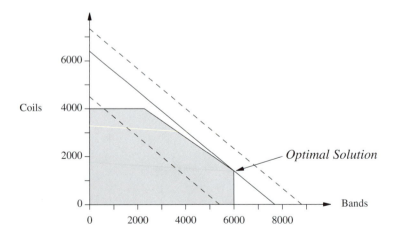

The line segment for profit equal to $135,000 is partly within the feasible region; any point on this line and within the region corresponds to a solution that achieves a profit of $135,000. On the other hand, the line for $220,000 does not intersect the feasible region at all; this tells us that there is no way to achieve a profit as high as $220,000. Viewed in this way, solving the linear program reduces to answering the following question: Among all profit lines that intersect the feasible region, which is highest and furthest to the right? The answer is the middle line, which just touches the region at one of the corners. This point corresponds to 6,000 tons of bands and 1,400 tons of coils, and a profit of $192,000 — the same as we found before.

1.2 The two-variable linear program in AMPL

Solving this linear program with AMPL can be as simple as typing AMPL's description of the linear program,

```
var XB;
var XC;
maximize Profit: 25 * XB + 30 * XC;
subject to Time: (1/200) * XB + (1/140) * XC <= 40;
subject to B_limit: 0 <= XB <= 6000;
subject to C_limit: 0 <= XC <= 4000;
```

into a file — call it prod0.mod — and then typing a few AMPL commands:

```
ampl: model prod0.mod;

ampl: solve;
MINOS 5.5: optimal solution found.
2 iterations, objective 192000

ampl: display XB, XC;
XB = 6000
XC = 1400

ampl: quit;
```

The invocation and appearance of an AMPL session will depend on your operating environment and interface, but you will always have the option of typing AMPL statements in response to the ampl: prompt, until you leave AMPL by typing *quit*. (Throughout the book, material you type is shown in *this slanted font*.)

The AMPL linear program that you type into the file parallels the algebraic form in every respect. It specifies the decision variables, defines the objective, and lists the constraints. It differs mainly in being somewhat more formal and regular, to facilitate computer processing. Each variable is named in a var statement, and each constraint by a statement that begins with subject to and a name like Time or B_limit for the constraint. Multiplication requires an explicit * operator, and the ≤ relation is written <=.

The first command of your AMPL session, model prod0.mod, reads the file into AMPL, just as if you had typed it line-by-line at ampl: prompts. You then need only type `solve` to have AMPL translate your linear program, send it to a linear program solver, and return the answer. A final command, `display`, is used to show the optimal values of the variables.

The message `MINOS 5.5` directly following the `solve` command indicates that AMPL used version 5.5 of a solver called MINOS. We have used MINOS and several other solvers for the examples in this book. You may have a different collection of solvers available on your computer, but any solver should give you the same optimal objective value for a linear program. Often there is more than one solution that achieves the optimal objective, however, in which case different solvers may report different optimal values for the variables. (Commands for choosing and controlling solvers will be explained in Section 11.2.)

Procedures for running AMPL can vary from one computer and operating system to another. Details are provided in supplementary instructions that come with your version of the AMPL software, rather than in this book. For subsequent examples, we will assume that AMPL has been started up, and that you have received the first `ampl:` prompt. If you are using a graphical interface for AMPL, like one of those mentioned briefly in Section 1.7, many of the AMPL commands may have equivalent menu or dialog entries. You will still have the option of typing the commands as shown in this book, but you may have to open a ''command window'' of some kind to see the prompts.

1.3 A linear programming model

The simple approach employed so far in this chapter is helpful for understanding the fundamentals of linear programming, but you can see that if our problem were only slightly more realistic — a few more products, a few more constraints — it would be a nuisance to write down and impossible to illustrate with pictures. And if the problem were subject to frequent change, either in form or merely in the data values, it would be hard to update as well.

If we are to progress beyond the very tiniest linear programs, we must adopt a more general and concise way of expressing them. This is where mathematical notation comes to the rescue. We can write a compact description of the general form of the problem, which we call a *model*, using algebraic notation for the objective and the constraints. Figure 1-1 shows the production problem in algebraic notation.

Figure 1-1 is a symbolic linear programming model. Its components are fundamental to all models:

- **sets**, like the products
- **parameters**, like the production and profit rates
- **variables**, whose values the solver is to determine
- an **objective**, to be maximized or minimized
- **constraints** that the solution must satisfy.

Given: P, a set of products
                 a_j = tons per hour of product j, for each $j \in P$
                 b = hours available at the mill
                 c_j = profit per ton of product j, for each $j \in P$
                 u_j = maximum tons of product j, for each $j \in P$

Define variables: X_j = tons of product j to be made, for each $j \in P$

Maximize: $$\sum_{j \in P} c_j X_j$$

Subject to: $$\sum_{j \in P} (1/a_j) X_j \leq b$$

                 $$0 \leq X_j \leq u_j, \text{ for each } j \in P$$

Figure 1-1: Basic production model in algebraic form.

The model describes an infinite number of related optimization problems. If we provide specific values for data, however, the model becomes a specific problem, or *instance* of the model, that can be solved. Each different collection of data values defines a different instance; the example in the previous section was one such instance.

It might seem that we have made things less rather than more concise, since our model is longer than the original statement of the linear program in Section 1.1. Consider what would happen, however, if the set P had 42 products rather than 2. The linear program would have 120 more data values (40 each for a_j, c_j, and u_j); there would be 40 more variables, with new lower and upper limits for each; and there would be 40 more terms in the objective and the hours constraint. Yet the abstract model, as shown above, would be no different. Without this ability of a short model to describe a long linear program, larger and more complex instances of linear programming would become impossible to deal with.

A mathematical model like this is thus usually the best compromise between brevity and comprehension; and fortunately, it is easy to convert into a language that a computer can process. From now on, we'll assume models are given in the algebraic form. As always, reality is rarely so simple, so most models will have more sets, parameters and variables, and more complicated objectives and constraints. In fact, in any real situation, formulating a correct model and providing accurate data are by far the hardest tasks; solving a specific problem requires only a solver and enough computing power.

1.4 The linear programming model in AMPL

Now we can talk about AMPL. The AMPL language is intentionally as close to the mathematical form as it can get while still being easy to type on an ordinary keyboard and

```
set P;

param a {j in P};
param b;
param c {j in P};
param u {j in P};

var X {j in P};

maximize Total_Profit: sum {j in P} c[j] * X[j];

subject to Time: sum {j in P} (1/a[j]) * X[j] <= b;

subject to Limit {j in P}: 0 <= X[j] <= u[j];
```

Figure 1-2: Basic production model in AMPL (file prod.mod).

to process by a program. There are AMPL constructions for each of the basic components listed above — sets, parameters, variables, objectives, and constraints — and ways to write arithmetic expressions, sums over sets, and so on.

We first give an AMPL model that resembles our algebraic model as much as possible, and then present an improved version that takes better advantage of the language.

The basic model

For the basic production model of Figure 1-1, a direct transcription into AMPL would look like Figure 1-2.

The keyword set declares a set name, as in

```
set P;
```

The members of set P will be provided in separate data statements, which we'll show in a moment.

The keyword param declares a parameter, which may be a single scalar value, as in

```
param b;
```

or a collection of values indexed by a set. Where algebraic notation says that "there is an a_j for each j in P", one writes in AMPL

```
param a {j in P};
```

which means that a is a collection of parameter values, one for each member of the set P. Subscripts in algebraic notation are written with square brackets in AMPL, so an individual value like a_j is written a[j].

The var declaration

```
var X {j in P};
```

names a collection of variables, one for each member of P, whose values the solver is to determine.

The objective is given by the declaration

```
maximize Total_Profit: sum {j in P} c[j] * X[j];
```

The name `Total_Profit` is arbitrary; a name is required by the syntax, but any name will do. The precedence of the `sum` operator is lower than that of *, so the expression is indeed a sum of products, as intended.

Finally, the constraints are given by

```
subject to Time: sum {j in P} (1/a[j]) * X[j] <= b;

subject to Limit {j in P}: 0 <= X[j] <= u[j];
```

The `Time` constraint says that a certain sum over the set P may not exceed the value of parameter b. The `Limit` constraint is actually a family of constraints, one for each member j of P: each X[j] is bounded by zero and the corresponding u[j].

The construct {j in P} is called an *indexing expression*. As you can see from our example, indexing expressions are used not only in declaring parameters and variables, but in any context where the algebraic model does something "for each j in P". Thus the `Limit` constraints are declared

```
subject to Limit {j in P}
```

because we want to impose a different restriction 0 <= X[j] <= u[j] for each different product *j* in the set *P*. In the same way, the summation in the objective is written

```
sum {j in P} c[j] * X[j]
```

to indicate that the different terms c[j] * X[j], for each j in the set P, are to be added together in computing the profit.

The layout of an AMPL model is quite free. Sets, parameters, and variables must be declared before they are used but can otherwise appear in any order. Statements end with semicolons and can be spaced and split across lines to enhance readability. Upper and lower case letters are different, so `time`, `Time`, and `TIME` are three different names.

You have undoubtedly noticed several places where traditional mathematical notation has been adapted in AMPL to the limitations of normal keyboards and character sets. AMPL uses the word `sum` instead of Σ to express a summation, and `in` rather than $\in$ for set membership. Set specifications are enclosed in braces, as in {j in P}. Where mathematical notation uses adjacency to signify multiplication in $c_j X_j$, AMPL uses the * operator of most programming languages, and subscripts are denoted by brackets, so $c_j X_j$ becomes c[j]*X[j].

You will find that the rest of AMPL is similar — a few more arithmetic operators, a few more key words like `sum` and `in`, and many more ways to specify indexing expressions. Like any other computer language, AMPL has a precise grammar, but we won't stress the rules too much here; most will become clear as we go along, and full details are given in the reference manual, Appendix A.

Our original two-variable linear program is one of the many LPs that are instances of the Figure 1-2 model. To specify it or any other such instance, we need to supply the

```
set P := bands coils;

param:       a      c      u   :=
   bands    200    25    6000
   coils    140    30    4000 ;

param b := 40;
```

Figure 1-3: Production model data (file `prod.dat`).

membership of P and the values of the various parameters. There is no standard way to describe these data values in algebraic notation; usually some kind of informal tables are used, such as the ones we showed earlier. In AMPL, there is a specific syntax for data tables, which is sufficiently regular and unambiguous to be translated by a computer. Figure 1-3 gives data for the basic production model in that form. A `set` statement supplies the members (`bands` and `coils`) of set P, and a `param` table gives the corresponding values for a, c, and u. A simple `param` statement gives the value for b. These data statements, which are described in detail in Chapter 9, have a variety of options that let you list or tabulate parameters in convenient ways.

An improved model

We could go on immediately to solve the linear program defined by Figures 1-2 and 1-3. Once we have written the model in AMPL, however, we need not feel constrained by all the conventions of algebra, and we can instead consider changes that might make the model easier to work with. Figures 1-4a and 1-4b show a possible "improved" version. The short "mathematical" names for the sets, parameters and variables have been replaced by longer, more meaningful ones. The indexing expressions have become {p in PROD}, or just {PROD} in those declarations that do not use the index p. The bounds on variables have been placed within their `var` declaration, rather than in a separate constraint; analogous bounds have been placed on the parameters, to indicate the ones that must be positive or nonnegative in any meaningful linear program derived from the model.

Finally, comments have been added to help explain the model to a reader. Comments begin with # and end at the end of the line. As in any programming language, judicious use of meaningful names, comments and formatting helps to make AMPL models more readable and understandable.

There are always many ways to describe a particular model in AMPL. It is left to the modeler to pick the way that seems clearest or most convenient. Our earlier, mathematical approach is often preferred for working quickly with a familiar model. On the other hand, the second version is more attractive for a model that will be maintained and modified by several people over months or years.

```
set PROD;   # products

param rate {PROD} > 0;      # tons produced per hour
param avail >= 0;           # hours available in week

param profit {PROD};        # profit per ton
param market {PROD} >= 0;   # limit on tons sold in week

var Make {p in PROD} >= 0, <= market[p]; # tons produced

maximize Total_Profit: sum {p in PROD} profit[p] * Make[p];
                # Objective: total profits from all products

subject to Time: sum {p in PROD} (1/rate[p]) * Make[p] <= avail;
                # Constraint: total of hours used by all
                # products may not exceed hours available
```

Figure 1-4a: Steel production model (steel.mod).

```
set PROD := bands coils;

param:      rate  profit  market :=
    bands    200    25     6000
    coils    140    30     4000 ;

param avail := 40;
```

Figure 1-4b: Data for steel production model (steel.dat).

If we put all of the model declarations into a file called steel.mod, and the data specification into a file steel.dat, then as before a solution can be found and displayed by typing just a few statements:

```
ampl: model steel.mod;
ampl: data steel.dat;
ampl: solve;
MINOS 5.5: optimal solution found.
2 iterations, objective 192000

ampl: display Make;
Make [*] :=
bands   6000
coils   1400
;
```

The model and data commands each specify a file to be read, in this case the model from steel.mod, and the data from steel.dat. The use of two file-reading commands encourages a clean separation of model from data.

Filenames can have any form recognized by your computer's operating system; AMPL doesn't check them for correctness. The filenames here and in the rest of the book refer to example files that are available from the AMPL web site and other AMPL distributions.

Once the model has been solved, we can show the optimal values of all of the variables `Make[p]`, by typing ***display Make***. The output from `display` uses the same formats as AMPL data input, so that there is only one set of formats to learn. (The `[*]` indicates a variable or parameter with a single subscript. It is not strictly necessary for input, since `Make` is one-dimensional, but `display` prints it as a reminder.)

Catching errors

You will inevitably make some mistakes as you develop a model. AMPL detects various kinds of incorrect statements, which are reported in error messages following the `model`, `data` or `solve` commands.

AMPL catches many errors as soon as the model is read. For example, if you use the wrong syntax for the bounds in the declaration of the variable `Make`, you will receive an error message like this, right after you enter the `model` command:

```
steel.mod, line 8 (offset 250):
        syntax error
context: var Make {p in PROD} >>> 0 <<< <= Make[p] <= market[p];
```

If you inadvertently use `make` instead of `Make` in an expression like `profit[p] * make[p]`, you will receive this message:

```
steel.mod, line 11 (offset 339):
        make is not defined
context: maximize Total_Profit:
               sum {p in PROD} profit[p] *  >>> make[p] <<< ;
```

In each case, the offending line is printed, with the approximate location of the error surrounded by `>>>` and `<<<`.

Other common sources of error messages include a model component used before it is declared, a missing semicolon at the end of a command, or a reserved word like `sum` or `in` used in the wrong context. (Section A.1 contains a list of reserved words.) Syntax errors in data statements are similarly reported right after you enter a `data` command.

Errors in the data values are caught after you type `solve`. If the number of hours were given as –40, for instance, you would see:

```
ampl: model steel.mod;
ampl: data steel.dat;
ampl: solve;
Error executing "solve" command:
error processing param avail:
        failed check: param avail = -40
                is not >= 0;
```

It is good practice to include as many validity checks as possible in the model, so that errors are caught at an early stage.

Despite your best efforts to formulate the model correctly and to include validity checks on the data, sometimes a model that generates no error messages and that elicits an "optimal solution" report from the solver will nonetheless produce a clearly wrong or meaningless solution. All of the production levels might be zero, for example, or the product with a lower profit per hour may be produced at a higher volume. In cases like these, you may have to spend some time reviewing your formulation before you discover what is wrong.

The `expand` command can be helpful in your search for errors, by showing you how AMPL instantiated your symbolic model. To see what AMPL generated for the objective `Total_Profit`, for example, you could type:

```
ampl: expand Total_Profit;
maximize Total_Profit:
        25*Make['bands'] + 30*Make['coils'];
```

This corresponds directly to our explicit formulation back in Section 1.1. Expanding the constraint works similarly:

```
ampl: expand Time;
subject to Time:
        0.005*Make['bands'] + 0.00714286*Make['coils'] <= 40;
```

Expressions in the symbolic model, such as the coefficients `1/rate[p]` in this example, are evaluated before the expansion is displayed. You can expand the objective and all of the constraints at once by typing `expand` by itself.

The expressions above show that the symbolic model's `Make[j]` expands to the explicit variables `Make['bands']` and `Make['coils']`. You can use expressions like these in AMPL commands, for example to expand a particular variable to see what coefficients it has in the objective and constraints:

```
ampl: expand Make['coils'];
Coefficients of Make['coils']:
        Time             0.00714286
        Total_Profit  30
```

Either single quotes (') or double quotes (") may surround the subscript.

1.5 Adding lower bounds to the model

Once the model and data have been set up, it is a simple matter to change them and then re-solve. Indeed, we would not expect to find an LP application in which the model and data are prepared and solved just once, or even a few times. Most commonly, numerous refinements are introduced as the model is developed, and changes to the data continue for as long as the model is used.

Let's conclude this chapter with a few examples of changes and refinements. These examples also highlight some additional features of AMPL.

Suppose first that we add another product, steel plate. The model stays the same, but in the data we have to add `plate` to the list of members for the set PROD, and we have to add a line of parameter values for `plate`:

```
set PROD := bands coils plate;

param:      rate  profit  market :=
   bands     200     25     6000
   coils     140     30     4000
   plate     160     29     3500 ;

param avail := 40;
```

We put this version of the data in a file called `steel2.dat`, and use AMPL as before to get the solution:

```
ampl: model steel.mod; data steel2.dat; solve;
MINOS 5.5: optimal solution found.
2 iterations, objective 196400

ampl: display Make;
Make [*] :=
bands  6000
coils     0
plate  1600
;
```

Profits have increased compared to the two-variable version, but now it is best to produce no coils at all! On closer examination, this result is not so surprising. Plate yields a profit of $4640 per hour, which is less than for bands but more than for coils. Thus plate is produced to absorb the capacity not taken by bands; coils would be produced only if both bands and plate reached their market limits before the available hours were exhausted.

In reality, a whole product line cannot be shut down solely to increase weekly profits. The simplest way to reflect this in the model is to add lower bounds on the production amounts, as shown in Figures 1-5a and 1-5b. We have declared a new collection of parameters named `commit`, to represent the lower bounds on production that are imposed by sales commitments, and we have changed `>= 0` to `>= commit[p]` in the declaration of the variables `Make[p]`.

After these changes are made, we can run AMPL again to get a more realistic solution:

```
ampl: model steel3.mod; data steel3.dat; solve;
MINOS 5.5: optimal solution found.
2 iterations, objective 194828.5714

ampl: display commit, Make, market;
:      commit    Make    market    :=
bands    1000    6000      6000
coils     500     500      4000
plate     750   1028.57    3500
;
```

For comparison, we have displayed `commit` and `market` on either side of the actual production, `Make`. As expected, after the commitments are met, it is most profitable to

```
set PROD;   # products

param rate {PROD} > 0;      # produced tons per hour
param avail >= 0;           # hours available in week
param profit {PROD};        # profit per ton

param commit {PROD} >= 0;   # lower limit on tons sold in week
param market {PROD} >= 0;   # upper limit on tons sold in week

var Make {p in PROD} >= commit[p], <= market[p]; # tons produced

maximize Total_Profit: sum {p in PROD} profit[p] * Make[p];
                  # Objective: total profits from all products

subject to Time: sum {p in PROD} (1/rate[p]) * Make[p] <= avail;
                  # Constraint: total of hours used by all
                  # products may not exceed hours available
```

Figure 1-5a: Lower bounds on production (`steel3.mod`).

```
set PROD := bands coils plate;

param:    rate  profit   commit   market :=
   bands   200    25      1000     6000
   coils   140    30       500     4000
   plate   160    29       750     3500 ;

param avail := 40;
```

Figure 1-5b: Data for lower bounds on production (`steel3.dat`).

produce bands up to the market limit, and then to produce plate with the remaining available time.

1.6 Adding resource constraints to the model

Processing of steel slabs is not a single operation, but a series of steps that may proceed at different rates. To motivate a more general model, imagine that we divide production into a reheat stage that can process the incoming slabs at 200 tons per hour, and a rolling stage that makes bands, coils or plate at the rates previously given. Further imagine that there are only 35 hours of reheat time, even though there are 40 hours of rolling time.

To cover this kind of situation, we can add a set STAGE of production stages to our model. The parameter and constraint declarations are modified accordingly, as shown in Figure 1-6a. Since there is a potentially different number of hours available in each stage, the parameter avail is now indexed over STAGE. Since there is a potentially different production rate for each product in each stage, the parameter rate is indexed over both PROD and STAGE. In the Time constraint, the production rate for product p in

```
set PROD;    # products
set STAGE;   # stages

param rate {PROD,STAGE} > 0; # tons per hour in each stage
param avail {STAGE} >= 0;     # hours available/week in each stage
param profit {PROD};          # profit per ton

param commit {PROD} >= 0;     # lower limit on tons sold in week
param market {PROD} >= 0;     # upper limit on tons sold in week

var Make {p in PROD} >= commit[p], <= market[p]; # tons produced

maximize Total_Profit: sum {p in PROD} profit[p] * Make[p];

                # Objective: total profits from all products

subject to Time {s in STAGE}:
   sum {p in PROD} (1/rate[p,s]) * Make[p] <= avail[s];

                # In each stage: total of hours used by all
                # products may not exceed hours available
```

Figure 1-6a: Additional resource constraints (steel4.mod).

stage s is referred to as rate[p,s]; this is AMPL's version of a doubly subscripted entity like a_{ps} in algebraic notation.

The only other change is to the constraint declaration, where we no longer have a single constraint, but a constraint for each stage, imposed by limited time available at that stage. In algebraic notation, this might have been written

$$\text{Subject to } \sum_{p \in P} (1/a_{ps}) X_p \leq b_s, \text{ for each } s \in S.$$

Compare the AMPL version:

```
subject to Time {s in STAGE}:
   sum {p in PROD} (1/rate[p,s]) * Make[p] <= avail[s];
```

As in the other examples, this is a straightforward analogue, adapted to the requirements of a computer language. In almost all models, most of the constraints are indexed collections like this one.

Since rate is now indexed over combinations of two indices, it requires a data table all to itself, as in Figure 1-6b. The data file must also include the membership for the new set STAGE, and values of avail for both reheat and roll.

After these changes are made, we use AMPL to get another revised solution:

```
ampl: reset;
ampl: model steel4.mod; data steel4.dat; solve;
MINOS 5.5: optimal solution found.
4 iterations, objective 190071.4286
```

```
set PROD := bands coils plate;
set STAGE := reheat roll;

param rate:   reheat  roll :=
   bands         200    200
   coils         200    140
   plate         200    160 ;

param:     profit  commit  market :=
   bands      25    1000    6000
   coils      30     500    4000
   plate      29     750    3500 ;

param avail :=  reheat 35   roll   40 ;
```

Figure 1-6b: Data for additional resource constraints (`steel4.dat`).

```
ampl: display Make.lb, Make, Make.ub, Make.rc;
:       Make.lb    Make    Make.ub     Make.rc         :=
bands    1000     3357.14   6000     5.32907e-15
coils     500      500      4000    -1.85714
plate     750     3142.86   3500     3.55271e-15
;

ampl: display Time;
Time [*] :=
reheat  1800
  roll  3200
;
```

The `reset` command erases the previous model so a new one can be read in.

At the end of the example above we have displayed the ''marginal values'' (also called ''dual values'' or ''shadow prices'') associated with the `Time` constraints. The marginal value of a constraint measures how much the value of the objective would improve if the constraint were relaxed by a small amount. For example, here we would expect that up to some point, additional reheat time would produce another $1800 of extra profit per hour, and additional rolling time would produce $3200 per hour; decreasing these times would decrease the profit correspondingly. In output commands like `display`, AMPL interprets a constraint's name alone as referring to the associated marginal values.

We also display several quantities associated with the variables `Make`. First there are lower bounds `Make.lb` and upper bounds `Make.ub`, which in this case are the same as `commit` and `market`. We also show the ''reduced cost'' `Make.rc`, which has the same meaning with respect to the bounds that the marginal values have with respect to the constraints. Thus we see that, again up to some point, each increase of a ton in the lower bound (or commitment) for coil production should reduce profits by about $1.86; each one-ton decrease in the lower bound should improve profits by about $1.86. The production levels for bands and plates are between their bounds, so their reduced costs are essentially zero (recall that `e-15` means $\times 10^{-15}$), and changing their levels will have no

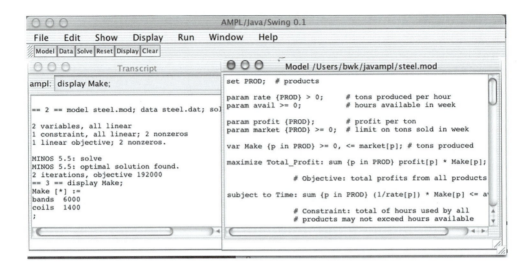

Figure 1-7a: A Java-based AMPL graphical user interface (Macintosh).

effect. Bounds, marginal (or dual) values, reduced costs and other quantities associated with variables and constraints are explored further in Section 12.5.

Comparing this session with our previous one, we see that the additional reheat time restriction reduces profits by about \$4750, and forces a substantial change in the optimal solution: much higher production of plate and lower production of bands. Moreover, the logic underlying the optimum is no longer so obvious. It is the difficulty of solving LPs by logical reasoning alone that necessitates computer-based systems such as AMPL.

1.7 AMPL interfaces

The examples that we have presented so far all use AMPL's command interface: the user types textual commands and the system responds with textual results. This is what we will use throughout the book to illustrate AMPL's capabilities. It permits access to all of AMPL's rich collection of features, and it will be the same in all environments. A text-based interface is most natural for creating scripts of frequently used commands and for writing programs that use AMPL's programming constructs (the topics of Chapter 13). And text commands are used in applications where AMPL is a hidden or behind-the-scenes part of some larger process.

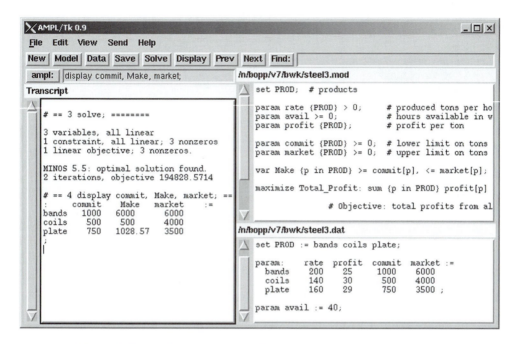

Figure 1-7b: A Tcl/Tk-based AMPL graphical user interface (Unix).

All that said, however, there are plenty of times where a graphical user interface can make a program easier to use, helping novices to get started and casual or infrequent users to recall details. AMPL is no exception. Thus there are a variety of graphical interfaces for AMPL, loosely analogous to the "integrated development environments" for conventional programming languages, though AMPL's environments are much less elaborate. An AMPL graphical interface typically provides a way to easily execute standard commands, set options, invoke solvers, and display the results, often by pushing buttons and selecting menu items instead of by typing commands.

Interfaces exist for standard operating system platforms. For example, Figure 1-7a shows a simple interface based on Java that runs on Unix and Linux, Windows, and Macintosh, presenting much the same appearance on each. (The Mac interface is shown.) Figure 1-7b shows a similar interface based on Tcl/Tk, shown running on Unix but also portable to Windows and Macintosh. Figure 1-7c shows another interface, created with Visual Basic and running on Windows.

There are also web-based interfaces that provide client-server access to AMPL or solvers over network connections, and a number of application program interfaces (API's) for calling AMPL from other programs. The AMPL web site, www.ampl.com, provides up to date information on all types of available interfaces.

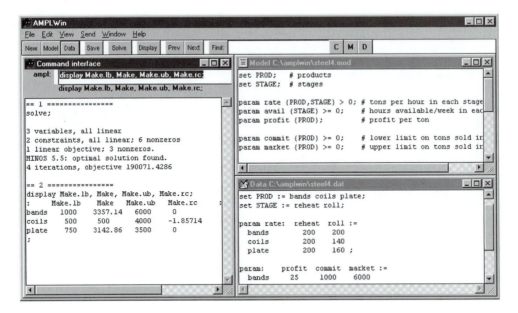

Figure 1-7c: A Visual Basic AMPL graphical user interface (Windows).

Bibliography

Julius S. Aronofsky, John M. Dutton and Michael T. Tayyabkhan, *Managerial Planning with Linear Programming: In Process Industry Operations.* John Wiley & Sons (New York, NY, 1978). A detailed account of a variety of profit-maximizing applications, with emphasis on the petroleum and petrochemical industries.

Vašek Chvátal, *Linear Programming*, W. H. Freeman (New York, NY, 1983). A concise and economical introduction to theoretical and algorithmic topics in linear programming.

Tibor Fabian, "A Linear Programming Model of Integrated Iron and Steel Production." Management Science **4** (1958) pp. 415–449. An application to all stages of steelmaking — from coal and ore through finished products — from the early days of linear programming.

Robert Fourer and Goutam Dutta, "A Survey of Mathematical Programming Applications in Integrated Steel Plants." Manufacturing & Service Operations Management **4** (2001) pp. 387–400.

David A. Kendrick, Alexander Meeraus and Jaime Alatorre, *The Planning of Investment Programs in the Steel Industry.* The Johns Hopkins University Press (Baltimore, MD, 1984). Several detailed mathematical programming models, using the Mexican steel industry as an example.

Robert J. Vanderbei, *Linear Programming: Foundations and Extensions* (2nd edition). Kluwer Academic Publishers (Dordrecht, The Netherlands, 2001). An updated survey of linear programming theory and methods.

Exercises

1-1. This exercise starts with a two-variable linear program similar in structure to the one of Sections 1.1 and 1.2, but with a quite different story behind it.

(a) You are in charge of an advertising campaign for a new product, with a budget of $1 million. You can advertise on TV or in magazines. One minute of TV time costs $20,000 and reaches 1.8 million potential customers; a magazine page costs $10,000 and reaches 1 million. You must sign up for at least 10 minutes of TV time. How should you spend your budget to maximize your audience? Formulate the problem in AMPL and solve it. Check the solution by hand using at least one of the approaches described in Section 1.1.

(b) It takes creative talent to create effective advertising; in your organization, it takes three person-weeks to create a magazine page, and one person-week to create a TV minute. You have only 100 person-weeks available. Add this constraint to the model and determine how you should now spend your budget.

(c) Radio advertising reaches a quarter million people per minute, costs $2,000 per minute, and requires only 1 person-day of time. How does this medium affect your solutions?

(d) How does the solution change if you have to sign up for at least two magazine pages? A maximum of 120 minutes of radio?

1-2. The steel model of this chapter can be further modified to reflect various changes in production requirements. For each part below, explain the modifications to Figures 1-6a and 1-6b that would be required to achieve the desired changes. (Make each change separately, rather than accumulating the changes from one part to the next.)

(a) How would you change the constraints so that total hours used by all products must *equal* the total hours available for each stage? Solve the linear program with this change, and verify that you get the same results. Explain why, in this case, there is no difference in the solution.

(b) How would you add to the model to restrict the total weight of all products to be less than a new parameter, `max_weight`? Solve the linear program for a weight limit of 6500 tons, and explain how this extra restriction changes the results.

(c) The incentive system for mill managers may tend to encourage them to produce as many tons as possible. How would you change the objective function to maximize total tons? For the data of our example, does this make a difference to the optimal solution?

(d) Suppose that instead of the lower bounds represented by `commit[p]` in our model, we want to require that each product represent a certain share of the total tons produced. In the algebraic notation of Figure 1-1, this new constraint might be represented as

$$X_j \geq s_j \sum_{k \in P} X_k, \text{ for each } j \in P$$

where s_j is the minimum share associated with project j. How would you change the AMPL model to use this constraint in place of the lower bounds `commit[p]`? If the minimum shares are 0.4 for bands and plate, and 0.1 for coils, what is the solution?

Verify that if you change the minimum shares to 0.5 for bands and plate, and 0.1 for coils, the linear program gives an optimal solution that produces nothing, at zero profit. Explain why this makes sense.

(e) Suppose there is an additional finishing stage for plates only, with a capacity of 20 hours and a rate of 150 tons per hour. Explain how you could modify the data, without changing the model, to incorporate this new stage.

1-3. This exercise deals with some issues of "sensitivity" in the steel models.

(a) For the linear program of Figures 1-5a and 1-5b, display `Time` and `Make.rc`. What do these values tell you about the solution? (You may wish to review the explanation of marginal values and reduced costs in Section 1.6.)

(b) Explain why the reheat time constraints added in Figure 1-6a result in a higher production of plate and a lower production of bands.

(c) Use AMPL to verify the following statements: If the available reheat time is increased from 35 to 36 in the data of Figure 1-6b, then the profit goes up by $1800 as predicted in Section 1.6. If the reheat time is further increased to 37, the profit goes up by another $1800. However, if the reheat time is increased to 38, there is a smaller increase in the profit, and further increases past 38 have no effect on the optimal profit at all. To change the reheat time to, say, 26 without changing and reading the data file over again, type the command

```
let avail["reheat"] := 36;
```

By trying some other values of the reheat time, confirm that the profit increases by $1800 per extra hour for any number of hours between 35 and 37 $9/14$, but that any increase in the reheat time beyond 37 $9/14$ hours doesn't give any further profit.

Draw a plot of the profit versus the number of reheat hours available, for hours ≥ 35.

(d) To find the slope of the plot from (c) — profit versus reheat time available — at any particular reheat time value, you need only look at the marginal value of `Time["reheat"]`. Using this observation as an aid, extend your plot from (c) down to 25 hours of reheat time. Verify that the slope of the plot remains at $6000 per hour from 25 hours down to less than 12 hours of reheat time. Explain what happens when the available reheat time drops to 11 hours.

1-4. Here is a similar profit-maximizing model, but in a different context. An automobile manufacturer produces several kinds of cars. Each kind requires a certain amount of factory time per car to produce, and yields a certain profit per car. A certain amount of factory time has been scheduled for the next week, and it is desired to use all this time; but at least a certain number of each kind of car must be manufactured to meet dealer requirements.

(a) What are the data values that define this problem? How would you declare the sets and parameter values for this problem in AMPL? What are the decision variables, and how would you declare them in AMPL?

(b) Assuming that the objective is to maximize total profit, how would you declare an objective in AMPL for this problem? How would you declare the constraints?

(c) For purposes of experiment, suppose that there are three kinds of cars, known at the factory as T, C and L, that 120 hours are available, and that the time per car, profit per car and dealer orders for each kind of car are as follows:

Car	time	profit	orders
T	1	200	10
C	2	500	20
L	3	700	15

How much of each car should be produced, and what is the maximum profit? You should find that your solution specifies a fractional amount of one of the cars. As a practical matter, how could you make use of this solution?

(d) If you maximize the total number of cars produced instead of the total profit, how many more cars do you make? How much less profit?

(e) Each kind of car achieves a certain fuel efficiency, and the manufacturer is required by law to maintain a certain ''fleet average'' efficiency. The fleet average is computed by multiplying the efficiency of each kind of car times the number of that kind produced, summing all of the resulting products, and dividing by the total of all cars produced. Extend your AMPL model to contain a minimum fleet average efficiency constraint. Rearrange the constraint as necessary to make it linear — no variables divided into other variables.

(f) Find the optimal solution for the case where cars T, C and L achieve fuel efficiencies of 50, 30 and 20 miles/gallon, and the fleet average efficiency must be at least 35 miles/gallon. Explain how this changes the production amounts and the total profit. Dealing with the fractional amounts in the solution is not so easy in this case. What might you do?

If you had 10 more hours of production time, you could make more profit. Does the addition of the fleet average efficiency constraint make the extra 10 hours more or less valuable?

(g) Explain how you could further refine this model to account for different production stages that have different numbers of hours available per stage, much as in the steel model of Section 1.6.

1-5. A group of young entrepreneurs earns a (temporarily) steady living by acquiring inadequately supervised items from electronics stores and re-selling them. Each item has a street value, a weight, and a volume; there are limits on the numbers of available items, and on the total weight and volume that can be managed at one time.

(a) Formulate an AMPL model that will help to determine how much of each item to pick up, to maximize one day's profit.

(b) Find a solution for the case given by the following table,

	Value	Weight	Volume	Available
TV	50	35	8	20
radio	15	5	1	50
camera	85	4	2	20
CD player	40	3	1	30
VCR	50	15	5	30
camcorder	120	20	4	15

and by limits of 500 pounds and 300 cubic feet.

(c) Suppose that it is desirable to acquire some of each item, so as to always have stock available for re-sale. Suppose in addition that there are upper bounds on how many of each item you can reasonably expect to sell. How would you add these conditions to the model?

(d) How could the group use the dual variables on the maximum-weight and maximum-volume constraints to evaluate potential new partners for their activities?

(e) Through adverse circumstances the group has been reduced to only one member, who can carry a mere 75 pounds and five cubic feet. What is the optimum strategy now? Given that this requires a non-integral number of acquisitions, what is the best all-integer solution? (The integrality constraint converts this from a standard linear programming problem into a much harder problem called a Knapsack Problem. See Chapter 20.)

1-6. Profit-maximizing models of oil refining were one of the first applications of linear programming. This exercise asks you to model a simplified version of the final stage of the refining process.

A refinery breaks crude oil into some collection of intermediate materials, then blends these materials back together into finished products. Given the volumes of intermediates that will be available, we want to determine how to blend the intermediates so that the resulting products are most profitable. The decision is made more complicated, however, by the existence of upper limits on certain attributes of the products, which must be respected in any feasible solution.

To formulate an algebraic linear programming model for this problem, we can start by defining sets I of intermediates, J of final products, and K of attributes. The relevant technological data may be represented by

> a_i barrels of intermediate i available, for each $i \in I$
>
> r_{ik} units of attribute k contributed per barrel of intermediate i, for each $i \in I$ and $k \in K$
>
> u_{jk} maximum allowed units of attribute k per barrel of final product j,
>   for each $j \in J$ and $k \in K$
>
> δ_{ij} 1 if intermediate i is allowed in the blend for product j, or 0 otherwise,
>   for each $i \in I$ and $j \in J$

and the economic data can be given by

> c_j revenue per barrel of product j, for each $j \in J$

There are two collections of decision variables:

> X_{ij} barrels of intermediate i used to make product j, for each $i \in I$ and $j \in J$
>
> Y_j barrels of product j made, for each $j \in J$

The objective is to

$$\text{maximize} \quad \sum_{j \in J} c_j \, Y_j,$$

which is the sum of the revenues from the various products.

It remains to specify the constraints. The amount of each intermediate used to make products must equal the amount available:

$$\sum_{j \in J} X_{ij} = a_i, \text{ for each } i \in I.$$

The amount of a product made must equal the sum of amounts of the components blended into it:

$$\sum_{i \in I} X_{ij} = Y_j, \text{ for each } j \in J.$$

For each product, the total attributes contributed by all intermediates must not exceed the total allowed:

$$\sum_{i \in I} r_{ik} X_{ij} \leq u_{jk} Y_j, \text{ for each } j \in J \text{ and } k \in K.$$

Finally, we bound the variables as follows:

$$0 \leq X_{ij} \leq \delta_{ij} a_i, \text{ for each } i \in I, j \in J,$$
$$0 \leq Y_j, \text{ for each } j \in J.$$

The upper bound on X_{ij} assures that only the appropriate intermediates will be used in blending. If intermediate i is not allowed in the blend for product j, as indicated by δ_{ij} being zero, then the upper bound on X_{ij} is zero; this ensures that X_{ij} cannot be positive in any solution. Otherwise, the upper bound on X_{ij} is just a_i, which has no effect since there are only a_i barrels of intermediate i available for blending in any case.

(a) Transcribe this model to AMPL, using the same names as in the algebraic form for the sets, parameters and variables as much as possible.

(b) Re-write the AMPL model using meaningful names and comments, in the style of Figure 1-4a.

(c) In a representative small-scale instance of this model, the intermediates are SRG (straight run gasoline), N (naphtha), RF (reformate), CG (cracked gasoline), B (butane), DI (distillate intermediate), GO (gas oil), and RS (residuum). The final products are PG (premium gasoline), RG (regular gasoline), D (distillate), and HF (heavy fuel oil). Finally, the attributes are vap (vapor pressure), oct (research octane), den (density), and sul (sulfur).

The following amounts of the intermediates are scheduled to be available:

SRG	N	RF	CG	B	DI	GO	RS
21170	500	16140	4610	370	250	11600	25210

The intermediates that can be blended into each product, and the amounts of the attributes that they possess, are as follows (with blank entries representing zeros):

	Premium & regular gasoline		Distillate		Heavy fuel oil	
	vap	oct	den	sul	den	sul
SRG	18.4	−78.5				
N	6.54	−65.0	272	.283		
RF	2.57	−104.0				
CG	6.90	−93.7				
B	199.2	−91.8				
DI			292	.526		
GO			295	.353	295	.353
RS					343	4.70

The attribute limits and revenues/barrel for the products are:

	vap	oct	den	sul	revenue
PG	12.2	−90			10.50
RG	12.7	−86			9.10
D			306	0.5	7.70
HF			352	3.5	6.65

Limits left blank, such as density for gasoline, are irrelevant and may be set to some relatively large number.

Create a data file for your AMPL model and determine the optimal blend and production amounts.

(d) It looks a little strange that the attribute amounts for research octane are negative. What is the limit constraint for this attribute really saying?

2

Diet and Other Input Models: Minimizing Costs

To complement the profit-maximizing models of Chapter 1, we now consider linear programming models in which the objective is to minimize costs. Where the constraints of maximization models tend to be upper limits on the availability of resources, the constraints in minimization models are more likely to be lower limits on the amounts of certain ''qualities'' in the solution.

As an intuitive example of a cost-minimizing model, this chapter uses the well-known ''diet problem'', which finds a mix of foods that satisfies requirements on the amounts of various vitamins. We will again construct a small, explicit linear program, and then show how a general model can be formulated for all linear programs of that kind. Since you are now more familiar with AMPL, however, we will spend more time on AMPL and less with algebraic notation.

After formulating the diet model, we will discuss a few changes that might make it more realistic. The full power of this model, however, derives from its applicability to many situations that have nothing to do with diets. Thus we conclude this chapter by rewriting the model in a more general way, and discussing its application to blending, economics, and scheduling.

2.1 A linear program for the diet problem

Consider the problem of choosing prepared foods to meet certain nutritional requirements. Suppose that precooked dinners of the following kinds are available for the following prices per package:

BEEF	beef	$3.19
CHK	chicken	2.59
FISH	fish	2.29
HAM	ham	2.89
MCH	macaroni & cheese	1.89
MTL	meat loaf	1.99
SPG	spaghetti	1.99
TUR	turkey	2.49

These dinners provide the following percentages, per package, of the minimum daily requirements for vitamins A, C, B1 and B2:

	A	C	B1	B2
BEEF	60%	20%	10%	15%
CHK	8	0	20	20
FISH	8	10	15	10
HAM	40	40	35	10
MCH	15	35	15	15
MTL	70	30	15	15
SPG	25	50	25	15
TUR	60	20	15	10

The problem is to find the cheapest combination of packages that will meet a week's requirements — that is, at least 700% of the daily requirement for each nutrient.

Let us write X_{BEEF} for the number of packages of beef dinner to be purchased, X_{CHK} for the number of packages of chicken dinner, and so forth. Then the total cost of the diet will be:

$$\text{total cost} =$$
$$3.19\,X_{BEEF} + 2.59\,X_{CHK} + 2.29\,X_{FISH} + 2.89\,X_{HAM} +$$
$$1.89\,X_{MCH} + 1.99\,X_{MTL} + 1.99\,X_{SPG} + 2.49\,X_{TUR}$$

The total percentage of the vitamin A requirement is given by a similar formula, except that X_{BEEF}, X_{CHK}, and so forth are multiplied by the percentage per package instead of the cost per package:

$$\text{total percentage of vitamin A daily requirement met} =$$
$$60\,X_{BEEF} + 8\,X_{CHK} + 8\,X_{FISH} + 40\,X_{HAM} +$$
$$15\,X_{MCH} + 70\,X_{MTL} + 25\,X_{SPG} + 60\,X_{TUR}$$

This amount needs to be greater than or equal to 700 percent. There is a similar formula for each of the other vitamins, and each of these also needs to be ≥ 700.

Putting these all together, we have the following linear program:

Minimize
$$3.19\,X_{BEEF} + 2.59\,X_{CHK} + 2.29\,X_{FISH} + 2.89\,X_{HAM} + $$
$$1.89\,X_{MCH} + 1.99\,X_{MTL} + 1.99\,X_{SPG} + 2.49\,X_{TUR}$$

Subject to
$$60\,X_{BEEF} + 8\,X_{CHK} + 8\,X_{FISH} + 40\,X_{HAM} + $$
$$15\,X_{MCH} + 70\,X_{MTL} + 25\,X_{SPG} + 60\,X_{TUR} \geq 700$$

$$20\,X_{BEEF} + 0\,X_{CHK} + 10\,X_{FISH} + 40\,X_{HAM} + $$
$$35\,X_{MCH} + 30\,X_{MTL} + 50\,X_{SPG} + 20\,X_{TUR} \geq 700$$

$$10\,X_{BEEF} + 20\,X_{CHK} + 15\,X_{FISH} + 35\,X_{HAM} + $$
$$15\,X_{MCH} + 15\,X_{MTL} + 25\,X_{SPG} + 15\,X_{TUR} \geq 700$$

$$15\,X_{BEEF} + 20\,X_{CHK} + 10\,X_{FISH} + 10\,X_{HAM} + $$
$$15\,X_{MCH} + 15\,X_{MTL} + 15\,X_{SPG} + 10\,X_{TUR} \geq 700$$

$$X_{BEEF} \geq 0,\, X_{CHK} \geq 0,\, X_{FISH} \geq 0,\, X_{HAM} \geq 0,$$
$$X_{MCH} \geq 0,\, X_{MTL} \geq 0,\, X_{SPG} \geq 0,\, X_{TUR} \geq 0$$

At the end we have added the common-sense requirement that no fewer than zero packages of a food can be purchased.

As we first did with the production LP of Chapter 1, we can transcribe to a file, say `diet0.mod`, an AMPL statement of the explicit diet LP:

```
var Xbeef >= 0; var Xchk >= 0; var Xfish >= 0;
var Xham >= 0;  var Xmch >= 0; var Xmtl >= 0;
var Xspg >= 0;  var Xtur >= 0;

minimize cost:
   3.19*Xbeef + 2.59*Xchk + 2.29*Xfish + 2.89*Xham +
   1.89*Xmch  + 1.99*Xmtl + 1.99*Xspg  + 2.49*Xtur;

subject to A:
   60*Xbeef +  8*Xchk +  8*Xfish + 40*Xham +
   15*Xmch  + 70*Xmtl + 25*Xspg  + 60*Xtur >= 700;

subject to C:
   20*Xbeef +  0*Xchk + 10*Xfish + 40*Xham +
   35*Xmch  + 30*Xmtl + 50*Xspg  + 20*Xtur >= 700;

subject to B1:
   10*Xbeef + 20*Xchk + 15*Xfish + 35*Xham +
   15*Xmch  + 15*Xmtl + 25*Xspg  + 15*Xtur >= 700;

subject to B2:
   15*Xbeef + 20*Xchk + 10*Xfish + 10*Xham +
   15*Xmch  + 15*Xmtl + 15*Xspg  + 10*Xtur >= 700;
```

Again a few AMPL commands then suffice to read the file, send the LP to a solver, and retrieve the results:

```
ampl: model diet0.mod;
ampl: solve;
MINOS 5.5: optimal solution found.
6 iterations, objective 88.2
ampl: display Xbeef,Xchk,Xfish,Xham,Xmch,Xmtl,Xspg,Xtur;
Xbeef = 0
Xchk = 0
Xfish = 0
Xham = 0
Xmch = 46.6667
Xmtl = -3.69159e-18
Xspg = -4.05347e-16
Xtur = 0
```

The optimal solution is found quickly, but it is hardly what we might have hoped for. The cost is minimized by a monotonous diet of $46\frac{2}{3}$ packages of macaroni and cheese! You can check that this neatly provides $15\% \times 46\frac{2}{3} = 700\%$ of the requirement for vitamins A, B1 and B2, and a lot more vitamin C than necessary; the cost is only $\$1.89 \times 46\frac{2}{3} = \$88.20$. (The tiny negative values for meat loaf and spaghetti can be regarded as zeros, like the tiny positive values we saw in Section 1.6.)

You might guess that a better solution would be generated by requiring the amount of each vitamin to equal 700% exactly. Such a requirement can easily be imposed by changing each >= to = in the AMPL constraints. If you go ahead and solve the changed LP, you will find that the diet does indeed become more varied: approximately 19.5 packages of chicken, 16.3 of macaroni and cheese, and 4.3 of meat loaf. But since equalities are more restrictive than inequalities, the cost goes up to $89.99.

2.2 An AMPL model for the diet problem

Clearly we will have to consider more extensive modifications to our linear program in order to produce a diet that is even remotely acceptable. We will probably want to change the sets of food and nutrients, as well as the nature of the constraints and bounds. As in the production example of the previous chapter, this will be much easier to do if we rely on a general model that can be coupled with a variety of specific data files.

This model deals with two things: nutrients and foods. Thus we begin an AMPL model by declaring sets of each:

```
set NUTR;
set FOOD;
```

Next we need to specify the numbers required by the model. Certainly a positive cost should be given for each food:

```
param cost {FOOD} > 0;
```

We also specify that for each food there are lower and upper limits on the number of packages in the diet:

```
param f_min {FOOD} >= 0;
param f_max {j in FOOD} >= f_min[j];
```

Notice that we need a dummy index j to run over FOOD in the declaration of f_max, in order to say that the maximum for each food must be greater than or equal to the corresponding minimum.

To make this model somewhat more general than our examples so far, we also specify similar lower and upper limits on the amount of each nutrient in the diet:

```
param n_min {NUTR} >= 0;
param n_max {i in NUTR} >= n_min[i];
```

Finally, for each combination of a nutrient and a food, we need a number that represents the amount of the nutrient in one package of the food. You may recall from Chapter 1 that such a ''product'' of two sets is written by listing them both:

```
param amt {NUTR,FOOD} >= 0;
```

References to this parameter require two indices. For example, amt[i,j] is the amount of nutrient i in a package of food j.

The decision variables for this model are the numbers of packages to buy of the different foods:

```
var Buy {j in FOOD} >= f_min[j], <= f_max[j];
```

The number of packages of some food j to be bought will be called Buy[j]; in any acceptable solution it will have to lie between f_min[j] and f_max[j].

The total cost of buying a food j is the cost per package, cost[j], times the number of packages, Buy[j]. The objective to be minimized is the sum of this product over all foods j:

```
minimize Total_Cost:  sum {j in FOOD} cost[j] * Buy[j];
```

This minimize declaration works the same as maximize did in Chapter 1.

Similarly, the amount of a nutrient i supplied by a food j is the nutrient per package, amt[i,j], times the number of packages Buy[j]. The total amount of nutrient i supplied is the sum of this product over all foods j:

```
sum {j in FOOD} amt[i,j] * Buy[j]
```

To complete the model, we need only specify that each such sum must lie between the appropriate bounds. Our constraint declaration begins

```
subject to Diet {i in NUTR}:
```

to say that a constraint named Diet[i] must be imposed for each member i of NUTR. The rest of the declaration gives the algebraic statement of the constraint for nutrient i: the variables must satisfy

```
n_min[i] <= sum {j in FOOD} amt[i,j] * Buy[j] <= n_max[i]
```

```
set NUTR;
set FOOD;

param cost {FOOD} > 0;
param f_min {FOOD} >= 0;
param f_max {j in FOOD} >= f_min[j];

param n_min {NUTR} >= 0;
param n_max {i in NUTR} >= n_min[i];

param amt {NUTR,FOOD} >= 0;

var Buy {j in FOOD} >= f_min[j], <= f_max[j];

minimize Total_Cost:  sum {j in FOOD} cost[j] * Buy[j];

subject to Diet {i in NUTR}:
   n_min[i] <= sum {j in FOOD} amt[i,j] * Buy[j] <= n_max[i];
```

Figure 2-1: Diet model in AMPL (`diet.mod`).

A "double inequality" like this is interpreted in the obvious way: the value of the sum in the middle must lie between `n_min[i]` and `n_max[i]`. The complete model is shown in Figure 2-1.

2.3 Using the AMPL diet model

By specifying appropriate data, we can solve any of the linear programs that correspond to the above model. Let's begin by using the data from the beginning of this chapter, which is shown in AMPL format in Figure 2-2.

The values of `f_min` and `n_min` are as given originally, while `f_max` and `n_max` are set, for the time being, to large values that won't affect the optimal solution. In the table for `amt`, the notation `(tr)` indicates that we have "transposed" the table so the columns correspond to the first index (nutrients), and the rows to the second (foods). Alternatively, we could have changed the model to say

```
param amt {FOOD,NUTR}
```

in which case we would have had to write `amt[j,i]` in the constraint.

Suppose that model and data are stored in the files `diet.mod` and `diet.dat`, respectively. Then AMPL is used as follows to read these files and to solve the resulting linear program:

```
ampl: model diet.mod;
ampl: data diet.dat;

ampl: solve;
MINOS 5.5: optimal solution found.
6 iterations, objective 88.2
```

```
set NUTR := A B1 B2 C ;
set FOOD := BEEF CHK FISH HAM MCH MTL SPG TUR ;

param:    cost   f_min   f_max :=
   BEEF   3.19      0     100
   CHK    2.59      0     100
   FISH   2.29      0     100
   HAM    2.89      0     100
   MCH    1.89      0     100
   MTL    1.99      0     100
   SPG    1.99      0     100
   TUR    2.49      0     100 ;

param:    n_min   n_max :=
   A       700    10000
   C       700    10000
   B1      700    10000
   B2      700    10000 ;

param amt (tr):
          A    C    B1   B2 :=
   BEEF  60   20   10   15
   CHK    8    0   20   20
   FISH   8   10   15   10
   HAM   40   40   35   10
   MCH   15   35   15   15
   MTL   70   30   15   15
   SPG   25   50   25   15
   TUR   60   20   15   10 ;
```

Figure 2-2: Data for diet model (`diet.dat`).

```
ampl: display Buy;
Buy [*] :=
BEEF    0
 CHK    0
FISH    0
 HAM    0
 MCH   46.6667
 MTL   -1.07823e-16
 SPG   -1.32893e-16
 TUR    0
;
```

Naturally, the result is the same as before.

Now suppose that we want to make the following enhancements. To promote variety, the weekly diet must contain between 2 and 10 packages of each food. The amount of sodium and calories in each package is also given; total sodium must not exceed 40,000 mg, and total calories must be between 16,000 and 24,000. All of these changes can be made through a few modifications to the data, as shown in Figure 2-3. Putting this new data in file `diet2.dat`, we can run AMPL again:

```
set NUTR := A B1 B2 C NA CAL ;
set FOOD := BEEF CHK FISH HAM MCH MTL SPG TUR ;

param:    cost   f_min   f_max :=
   BEEF   3.19    2       10
   CHK    2.59    2       10
   FISH   2.29    2       10
   HAM    2.89    2       10
   MCH    1.89    2       10
   MTL    1.99    2       10
   SPG    1.99    2       10
   TUR    2.49    2       10   ;

param:    n_min   n_max :=
   A        700   20000
   C        700   20000
   B1       700   20000
   B2       700   20000
   NA         0   40000
   CAL    16000   24000 ;

param amt (tr):
           A    C    B1   B2    NA    CAL :=
   BEEF   60   20    10   15    938   295
   CHK     8    0    20   20   2180   770
   FISH    8   10    15   10    945   440
   HAM    40   40    35   10    278   430
   MCH    15   35    15   15   1182   315
   MTL    70   30    15   15    896   400
   SPG    25   50    25   15   1329   370
   TUR    60   20    15   10   1397   450 ;
```

Figure 2-3: Data for enhanced diet model (`diet2.dat`).

```
ampl: model diet.mod;
ampl: data diet2.dat;

ampl: solve;
MINOS 5.5: infeasible problem.
9 iterations
```

The message `infeasible problem` tells us that we have constrained the diet too tightly; there is no way that all of the restrictions can be satisfied.

AMPL lets us examine a variety of values produced by a solver as it attempts to find a solution. In Chapter 1, we used marginal (or dual) values to investigate the sensitivity of an optimum solution to changes in the constraints. Here there is no optimum, but the solver does return the last solution that it found while attempting to satisfy the constraints. We can look for the source of the infeasibility by displaying some values associated with this solution:

```
ampl: display Diet.lb, Diet.body, Diet.ub;
:    Diet.lb   Diet.body Diet.ub    :=
A        700    1993.09    20000
B1       700     841.091   20000
B2       700     601.091   20000
C        700    1272.55    20000
CAL    16000   17222.9     24000
NA         0   40000       40000
;
```

For each nutrient, Diet.body is the sum of the terms amt[i,j] * Buy[j] in the constraint Diet[i]. The Diet.lb and Diet.ub values are the "lower bounds" and "upper bounds" on the sum in Diet[i] — in this case, just the values n_min[i] and n_max[i]. We can see that the diet returned by the solver does not supply enough vitamin B2, while the amount of sodium (NA) has reached its upper bound.

At this point, there are two obvious choices: we could require less B2 or we could allow more sodium. If we try the latter, and relax the sodium limit to 50,000 mg, a feasible solution becomes possible:

```
ampl: let n_max["NA"] := 50000;

ampl: solve;
MINOS 5.5: optimal solution found.
5 iterations, objective 118.0594032

ampl: display Buy;
Buy [*] :=
BEEF    5.36061
 CHK    2
FISH    2
 HAM   10
 MCH   10
 MTL   10
 SPG    9.30605
 TUR    2
;
```

This is at least a start toward a palatable diet, although we have to spend $118.06, compared to $88.20 for the original, less restricted case. Clearly it would be easy, now that the model is set up, to try many other possibilities. (Section 11.3 describes ways to quickly change the data and re-solve.)

One still disappointing aspect of the solution is the need to buy 5.36061 packages of beef, and 9.30605 of spaghetti. How can we find the best possible solution in terms of whole packages? You might think that we could simply round the optimal values to whole numbers — or integers, as they're often called in the context of optimization — but it is not so easy to do so in a feasible way. Using AMPL to modify the reported solution, we can observe that rounding up to 6 packages of beef and 10 of spaghetti, for example, will violate the sodium limit:

```
ampl: let Buy["BEEF"] := 6;
ampl: let Buy["SPG"] := 10;

ampl: display Diet.lb, Diet.body, Diet.ub;
:    Diet.lb Diet.body Diet.ub      :=
A        700     2012    20000
B1       700     1060    20000
B2       700      720    20000
C        700     1730    20000
CAL    16000    20240    24000
NA         0    51522    50000
;
```

(The `let` statement, which permits modifications of data, is described in Section 11.3.)
You can similarly check that rounding the solution down to 5 of beef and 9 of spaghetti
will provide insufficient vitamin B2. Rounding one up and the other down doesn't work
either. With enough experimenting you can find a nearby all-integer solution that does
satisfy the constraints, but still you will have no guarantee that it is the least-cost all-
integer solution.

AMPL does provide for putting the integrality restriction directly into the declaration
of the variables:

```
var Buy {j in FOOD} integer >= f_min[j], <= f_max[j];
```

This will only help, however, if you use a solver that can deal with problems whose variables must be integers. For this, we turn to CPLEX, a solver that can handle these so-
called integer programs. If we add `integer` to the declaration of variable `Buy` as
above, save the resulting model in the file `dieti.mod`, and add the higher sodium limit
to `diet2a.dat`, then we can re-solve as follows:

```
ampl: reset;
ampl: model dieti.mod;
ampl: data diet2a.dat;
ampl: option solver cplex;
ampl: solve;
CPLEX 8.0.0: optimal integer solution; objective 119.3
11 MIP simplex iterations
1 branch-and-bound nodes

ampl: display Buy;
Buy [*] :=
BEEF    9
 CHK    2
FISH    2
 HAM    8
 MCH   10
 MTL   10
 SPG    7
 TUR    2
;
```

Since integrality is an added constraint, it is no surprise that the best integer solution costs about $1.24 more than the best "continuous" one. But the difference between the diets is unexpected; the amounts of 3 foods change, each by two or more packages. In general, integrality and other "discrete" restrictions make solutions for a model much harder to find. We discuss this at length in Chapter 20.

2.4 Generalizations to blending, economics and scheduling

Your personal experience probably suggests that diet models are not widely used by people to choose their dinners. These models would be much better suited to situations in which packaging and personal preferences don't play such a prominent role — for example, the blending of animal feed or perhaps food for college dining halls.

The diet model is a convenient, intuitive example of a linear programming formulation that appears in many contexts. Suppose that we rewrite the model in a more general way, as shown in Figure 2-4. The objects that were called foods and nutrients in the diet model are now referred to more generically as "inputs" and "outputs". For each input `j`, we must decide to use a quantity `X[j]` that lies between `in_min[j]` and `in_max[j]`; as a result we incur a cost equal to `cost[j] * X[j]`, and we create `io[i,j] * X[j]` units of each output `i`. Our goal is to find the least-cost combination of inputs that yields, for each output `i`, an amount between `out_min[i]` and `out_max[i]`.

In one common class of applications for this model, the inputs are raw materials to be mixed together. The outputs are qualities of the resulting blend. The raw materials could be the components of an animal feed, but they could equally well be the crude oil derivatives that are blended to make gasoline, or the different kinds of coal that are mixed as input to a coke oven. The qualities can be amounts of something (sodium or calories for animal feed), or more complex measures (vapor pressure or octane rating for gasoline), or even physical properties such as weight and volume.

In another well-known application, the inputs are production activities of some sector of an economy, and the outputs are various products. The `in_min` and `in_max` parameters are limits on the levels of the activities, while `out_min` and `out_max` are regulated by demands. Thus the goal is to find levels of the activities that meet demand at the lowest cost. This interpretation is related to the concept of an economic equilibrium, as we will explain in Chapter 19.

In still another, quite different application, the inputs are work schedules, and the outputs correspond to hours worked on certain days of a month. For a particular work schedule `j`, `io[i,j]` is the number of hours that a person following schedule `j` will work on day `i` (zero if none), `cost[j]` is the monthly salary for a person following schedule `j`, and `X[j]` is the number of workers assigned that schedule. Under this interpretation, the objective becomes the total cost of the monthly payroll, while the constraints say that for each day `i`, the total number of workers assigned to work that day must lie between the limits `out_min[i]` and `out_max[i]`. The same approach can

```
set INPUT;    # inputs
set OUTPUT;   # outputs

param cost {INPUT} > 0;
param in_min {INPUT} >= 0;
param in_max {j in INPUT} >= in_min[j];

param out_min {OUTPUT} >= 0;
param out_max {i in OUTPUT} >= out_min[i];

param io {OUTPUT,INPUT} >= 0;

var X {j in INPUT} >= in_min[j], <= in_max[j];

minimize Total_Cost:  sum {j in INPUT} cost[j] * X[j];

subject to Outputs {i in OUTPUT}:
    out_min[i] <= sum {j in INPUT} io[i,j] * X[j] <= out_max[i];
```

Figure 2-4: Least-cost input model (`blend.mod`).

be used in a variety of other scheduling contexts, where the hours, days or months are replaced by other periods of time.

Although linear programming can be very useful in applications like these, we need to keep in mind the assumptions that underlie the LP model. We have already mentioned the "continuity" assumption whereby `X[j]` is allowed to take on any value between `in_min[j]` and `in_max[j]`. This may be a lot more reasonable for blending than for scheduling.

As another example, in writing the objective as

```
sum {j in INPUT} cost[j] * X[j]
```

we are assuming "linearity of costs", that is, that the cost of an input is proportional to the amount of the input used, and that the total cost is the sum of the inputs' individual costs.

In writing the constraints as

```
out_min[i] <= sum {j in INPUT} io[i,j] * X[j] <= out_max[i]
```

we are also assuming that the yield of an output i from a particular input is proportional to the amount of the input used, and that the total yield of an output i is the sum of the yields from the individual inputs. This "linearity of yield" assumption poses no problem when the inputs are schedules, and the outputs are hours worked. But in the blending example, linearity is a physical assumption about the nature of the raw materials and the qualities, which may or may not hold. In early applications to refineries, for example, it was recognized that the addition of lead as an input had a nonlinear effect on the quality known as octane rating in the resulting blend.

AMPL makes it easy to express discrete or nonlinear models, but any departure from continuity or linearity is likely to make an optimal solution much harder to obtain. At the least, it takes a more powerful solver to optimize the resulting mathematical programs. Chapters 17 through 20 discuss these issues in more detail.

Bibliography

George B. Dantzig, ''The Diet Problem.'' Interfaces **20**, 4 (1990) pp. 43–47. An entertaining account of the origins of the diet problem.

Susan Garner Garille and Saul I. Gass, ''Stigler's Diet Problem Revisited.'' Operations Research **49**, 1 (2001) pp. 1–13. A review of the diet problem's origins and its influence over the years on linear programming and on nutritionists.

Said S. Hilal and Warren Erikson, ''Matching Supplies to Save Lives: Linear Programming the Production of Heart Valves.'' Interfaces **11**, 6 (1981) pp. 48–56. A less appetizing equivalent of the diet problem, involving the choice of pig heart suppliers.

Exercises

2-1. Suppose the foods listed below have calories, protein, calcium, vitamin A, and costs per pound as shown. In what amounts should these food be purchased to meet at least the daily requirements listed while minimizing the total cost? (This problem comes from George B. Dantzig's classic book, *Linear Programming and Extensions*, page 118. We will take his word on nutritional values, and for nostalgic reasons have left the prices as they were when the book was published in 1963.)

	bread	meat	potatoes	cabbage	milk	gelatin	required
calories	1254	1457	318	46	309	1725	3000
protein	39	73	8	4	16	43	70 g.
calcium	418	41	42	141	536	0	800 mg.
vitamin A	0	0	70	860	720	0	500 I.U.
cost/pound	$0.30	$1.00	$0.05	$0.08	$0.23	$0.48	

2-2. (a) You have been advised by your doctor to get more exercise, specifically, to burn off at least 2000 extra calories per week by some combination of walking, jogging, swimming, exercise-machine, collaborative indoor recreation, and pushing yourself away from the table at mealtimes. You have a limited tolerance for each activity in hours/week; each expends a certain number of calories per hour, as shown below:

	walking	jogging	swimming	machine	indoor	pushback
Calories	100	200	300	150	300	500
Tolerance	5	2	3	3.5	3	0.5

How should you divide your exercising among these activities to minimize the amount of time you spend?

(b) Suppose that you should also have some variety in your exercise — you must do at least one hour of each of the first four exercises, but no more than four hours total of walking, jogging, and exercise-machine. Solve the problem in this form.

2-3. (a) A manufacturer of soft drinks wishes to blend three sugars in approximately equal quantities to ensure uniformity of taste in a product. Suppliers only provide combinations of the sugars, at varying costs/ton:

				SUPPLIER			
Sugar	A	B	C	D	E	F	G
Cane	10%	10	20	30	40	20	60
Corn	30%	40	40	20	60	70	10
Beet	60%	50	40	50	0	10	30
Cost/ton	$10	11	12	13	14	12	15

Formulate an AMPL model that minimizes the cost of supply while producing a blend that contains 52 tons of cane sugar, 56 tons of corn sugar, and 59 tons of beet sugar.

(b) The manufacturer feels that to ensure good relations with suppliers it is necessary to buy at least 10 tons from each. How does this change the model and the minimum-cost solution?

(c) Formulate an alternative to the model in (a) that finds the lowest-cost way to blend one ton of supplies so that the amount of each sugar is between 30 and 37 percent of the total.

2-4. At the end of Chapter 1, we indicated how to interpret the marginal (or dual) values of constraints and the reduced costs of variables in a production model. The same ideas can be applied to this chapter's diet model.

(a) Going back to the diet problem that was successfully solved in Section 2.3, we can display the marginal values as follows:

```
ampl: display Diet.lb,Diet.body,Diet.ub,Diet;
:    Diet.lb  Diet.body Diet.ub     Diet          :=
A       700     1956.29   20000    0
B1      700     1036.26   20000    0
B2      700        700    20000    0.404585
C       700     1682.51   20000    0
CAL   16000     19794.6   24000    0
NA        0       50000   50000   -0.00306905
;
```

How can you interpret the two that are nonzero?

(b) For the same problem, this listing gives the reduced costs:

```
ampl: display Buy.lb,Buy,Buy.ub,Buy.rc;
:     Buy.lb      Buy   Buy.ub       Buy.rc        :=
BEEF      2    5.36061      10    8.88178e-16
CHK       2          2      10    1.18884
FISH      2          2      10    1.14441
HAM       2         10      10   -0.302651
MCH       2         10      10   -0.551151
MTL       2         10      10   -1.3289
SPG       2    9.30605      10    0
TUR       2          2      10    2.73162
;
```

Based on this information, if you want to save money by eating more than 10 packages of some food, which one is likely to be your best choice?

2-5. A chain of fast-food restaurants operates 7 days a week, and requires the following minimum number of kitchen employees from Monday through Sunday: 45, 45, 40, 50, 65, 35, 35. Each employee is scheduled to work one weekend day (Saturday or Sunday) and four other days in a week. The management wants to know the minimum total number of employees needed to satisfy the requirements on every day.

(a) Set up and solve this problem as a linear program.

(b) In light of the discussion in Section 2.4, explain how this problem can be viewed as a special case of the blending model in Figure 2-4.

2-6. The output of a paper mill consists of standard rolls 110 inches (110") wide, which are cut into smaller rolls to meet orders. This week there are orders for rolls of the following widths:

Width	Orders
20"	48
45"	35
50"	24
55"	10
75"	8

The owner of the mill wants to know what cutting patterns to apply so as to fill the orders using the smallest number of 110" rolls.

(a) A cutting pattern consists of a certain number of rolls of each width, such as two of 45" and one of 20", or one of 50" and one of 55" (and 5" of waste). Suppose, to start with, that we consider only the following six patterns:

Width	1	2	3	4	5	6
20"	3	1	0	2	1	3
45"	0	2	0	0	0	1
50"	1	0	1	0	0	0
55"	0	0	1	1	0	0
75"	0	0	0	0	1	0

How many rolls should be cut according to each pattern, to minimize the number of 110" rolls used? Formulate and solve this problem as a linear program, assuming that the number of smaller rolls produced need only be greater than or equal to the number ordered.

(b) Re-solve the problem, with the restriction that the number of rolls produced in each size must be between 10% under and 40% over the number ordered.

(c) Find another pattern that, when added to those above, improves the optimal solution.

(d) All of the solutions above use fractional numbers of rolls. Can you find solutions that also satisfy the constraints, but that cut a whole number of rolls in each pattern? How much does your whole-number solution cause the objective function value to go up in each case? (See Chapter 20 for a discussion of how to find optimal whole-number, or integer, solutions.)

2-7. In the refinery model of Exercise 1-6, the amount of premium gasoline to be produced is a decision variable. Suppose instead that orders dictate a production of 42,000 barrels. The octane rating of the product is permitted to be in the range of 89 to 91, and the vapor pressure in a range of 11.7 to 12.7. The five feedstocks that are blended to make premium gasoline have the following production and/or purchase costs:

SRG	9.57
N	8.87
RF	11.69
CG	10.88
B	6.75

Other data are as in Exercise 1-6. Construct a blending model and data file to represent this problem. Run them through AMPL to determine the optimal composition of the blend.

2-8. Recall that Figure 2-4 generalizes the diet model as a minimum-cost input selection model, with constraints on the outputs.

(a) In the same way, generalize the production model of Figure 1-6a as a maximum-revenue output selection model, with constraints on the inputs.

(b) The concept of an ''input-output'' model was one of the first applications of linear programming in economic analysis. Such a model can be described in terms of a set A of activities and a set M of materials. The decision variables are the levels $X_j \geq 0$ at which the activities are run; they have lower limits u_j^- and upper limits u_j^+.

Each activity j has either a revenue per unit $c_j > 0$, or a cost per unit represented by $c_j < 0$. Thus total profit from all activities is $\sum_{j \in A} c_j X_j$, which is to be maximized.

Each unit of activity j produces an amount of material i given by $a_{ij} \geq 0$, or consumes an amount of material i represented by $a_{ij} < 0$. Thus if $\sum_{j \in A} a_{ij} X_j$ is > 0 it is the total production of material i by all activities; if < 0, it is the total consumption of material i by all activities.

For each material i, there is either an upper limit on the total production given by $b_i^+ > 0$, or a lower limit on the total consumption given by $b_i^+ < 0$. Similarly, there is either a lower limit on the total production given by $b_i^- > 0$, or an upper limit on the total consumption given by $b_i^- < 0$.

Write out a formulation of this model in AMPL.

(c) Explain how the minimum-cost input selection model and maximum-revenue output-selection model can be viewed as special cases of the input-output model.

3

Transportation and Assignment Models

The linear programs in Chapters 1 and 2 are all examples of classical "activity" models. In such models the variables and constraints deal with distinctly different kinds of activities — tons of steel produced versus hours of mill time used, or packages of food bought versus percentages of nutrients supplied. To use these models you must supply coefficients like tons per hour or percentages per package that convert a unit of activity in the variables to the corresponding amount of activity in the constraints.

This chapter addresses a significantly different but equally common kind of model, in which something is shipped or assigned, but not converted. The resulting constraints, which reflect both limitations on availability and requirements for delivery, have an especially simple form.

We begin by describing the so-called transportation problem, in which a single good is to be shipped from several origins to several destinations at minimum overall cost. This problem gives rise to the simplest kind of linear program for minimum-cost flows. We then generalize to a transportation model, an essential step if we are to manage all the data, variables and constraints effectively.

As with the diet model, the power of the transportation model lies in its adaptability. We continue by considering some other interpretations of the "flow" from origins to destinations, and work through one particular interpretation in which the variables represent assignments rather than shipments.

The transportation model is only the most elementary kind of minimum-cost flow model. More general models are often best expressed as networks, in which nodes — some of which may be origins or destinations — are connected by arcs that carry flows of some kind. AMPL offers convenient features for describing network flow models, including `node` and `arc` declarations that specify network structure directly. Network models and the relevant AMPL features are the topic of Chapter 15.

3.1 A linear program for the transportation problem

Suppose that we have decided (perhaps by the methods described in Chapter 1) to produce steel coils at three mill locations, in the following amounts:

GARY	Gary, Indiana	1400
CLEV	Cleveland, Ohio	2600
PITT	Pittsburgh, Pennsylvania	2900

The total of 6,900 tons must be shipped in various amounts to meet orders at seven locations of automobile factories:

FRA	Framingham, Massachusetts	900
DET	Detroit, Michigan	1200
LAN	Lansing, Michigan	600
WIN	Windsor, Ontario	400
STL	St. Louis, Missouri	1700
FRE	Fremont, California	1100
LAF	Lafayette, Indiana	1000

We now have an optimization problem: What is the least expensive plan for shipping the coils from mills to plants?

To answer the question, we need to compile a table of shipping costs per ton:

	GARY	CLEV	PITT
FRA	39	27	24
DET	14	9	14
LAN	11	12	17
WIN	14	9	13
STL	16	26	28
FRE	82	95	99
LAF	8	17	20

Let GARY:FRA be the number of tons to be shipped from GARY to FRA, and similarly for the other city pairs. Then the objective can be written as follows:

Minimize
$$39 \text{ GARY:FRA} + 27 \text{ CLEV:FRA} + 24 \text{ PITT:FRA} +$$
$$14 \text{ GARY:DET} + 9 \text{ CLEV:DET} + 14 \text{ PITT:DET} +$$
$$11 \text{ GARY:LAN} + 12 \text{ CLEV:LAN} + 17 \text{ PITT:LAN} +$$
$$14 \text{ GARY:WIN} + 9 \text{ CLEV:WIN} + 13 \text{ PITT:WIN} +$$
$$16 \text{ GARY:STL} + 26 \text{ CLEV:STL} + 28 \text{ PITT:STL} +$$
$$82 \text{ GARY:FRE} + 95 \text{ CLEV:FRE} + 99 \text{ PITT:FRE} +$$
$$8 \text{ GARY:LAF} + 17 \text{ CLEV:LAF} + 20 \text{ PITT:LAF}$$

There are 21 decision variables in all. Even a small transportation problem like this one has a lot of variables, because there is one for each combination of mill and factory.

By supplying each factory from the mill that can ship most cheaply to it, we could achieve the lowest conceivable shipping cost. But we would then be shipping 900 tons from PITT, 1600 from CLEV, and all the rest from GARY — amounts quite inconsistent with the production levels previously decided upon. We need to add a constraint that the sum of the shipments from GARY to the seven factories is equal to the production level of 1400:

> GARY:FRA + GARY:DET + GARY:LAN + GARY:WIN +
>   GARY:STL + GARY:FRE + GARY:LAF = 1400

There are analogous constraints for the other two mills:

> CLEV:FRA + CLEV:DET + CLEV:LAN + CLEV:WIN +
>   CLEV:STL + CLEV:FRE + CLEV:LAF = 2600
>
> PITT:FRA + PITT:DET + PITT:LAN + PITT:WIN +
>   PITT:STL + PITT:FRE + PITT:LAF = 2900

There also have to be constraints like these at the factories, to ensure that the amounts shipped equal the amounts ordered. At FRA, the sum of the shipments received from the three mills must equal the 900 tons ordered:

> GARY:FRA + CLEV:FRA + PITT:FRA = 900

And similarly for the other six factories:

> GARY:DET + CLEV:DET + PITT:DET = 1200
> GARY:LAN + CLEV:LAN + PITT:LAN = 600
> GARY:WIN + CLEV:WIN + PITT:WIN = 400
> GARY:STL + CLEV:STL + PITT:STL = 1700
> GARY:FRE + CLEV:FRE + PITT:FRE = 1100
> GARY:LAF + CLEV:LAF + PITT:LAF = 1000

We have ten constraints in all, one for each mill and one for each factory. If we add the requirement that all variables be nonnegative, we have a complete linear program for the transportation problem.

We won't even try showing what it would be like to type all of these constraints into an AMPL model file. Clearly we want to set up a general model to deal with this problem.

3.2 An AMPL model for the transportation problem

Two fundamental sets of objects underlie the transportation problem: the sources or origins (mills, in our example) and the destinations (factories). Thus we begin the AMPL model with a declaration of these two sets:

```
set ORIG;
set DEST;
```

There is a supply of something at each origin (tons of steel coils produced, in our case), and a demand for the same thing at each destination (tons of coils ordered). AMPL defines nonnegative quantities like these with `param` statements indexed over a set; in this case we add one extra refinement, a `check` statement to test the data for validity:

```
param supply {ORIG} >= 0;
param demand {DEST} >= 0;

check: sum {i in ORIG} supply[i] = sum {j in DEST} demand[j];
```

The `check` statement says that the sum of the supplies has to equal the sum of the demands. The way that our model is to be set up, there can't possibly be any solutions unless this condition is satisfied. By putting it in a `check` statement, we tell AMPL to test this condition after reading the data, and to issue an error message if it is violated.

For each combination of an origin and a destination, there is a transportation cost and a variable representing the amount transported. Again, the ideas from previous chapters are easily adapted to produce the appropriate AMPL statements:

```
param cost {ORIG,DEST} >= 0;
var Trans {ORIG,DEST} >= 0;
```

For a particular origin i and destination j, we ship `Trans[i,j]` units from i to j, at a cost of `cost[i,j]` per unit; the total cost for this pair is

```
cost[i,j] * Trans[i,j]
```

Adding over all pairs, we have the objective function:

```
minimize Total_Cost:
    sum {i in ORIG, j in DEST} cost[i,j] * Trans[i,j];
```

which could also be written as

```
    sum {j in DEST, i in ORIG} cost[i,j] * Trans[i,j];
```

or as

```
    sum {i in ORIG} sum {j in DEST} cost[i,j] * Trans[i,j];
```

As long as you express the objective in some mathematically correct way, AMPL will sort out the terms.

It remains to specify the two collections of constraints, those at the origins and those at the destinations. If we name these collections `Supply` and `Demand`, their declarations will start as follows:

```
subject to Supply {i in ORIG}:   ...
subject to Demand {j in DEST}:   ...
```

To complete the `Supply` constraint for origin i, we need to say that the sum of all shipments out of i is equal to the supply available. Since the amount shipped out of i to a particular destination j is `Trans[i,j]`, the amount shipped to all destinations must be

```
set ORIG;    # origins
set DEST;    # destinations

param supply {ORIG} >= 0;    # amounts available at origins
param demand {DEST} >= 0;    # amounts required at destinations

    check: sum {i in ORIG} supply[i] = sum {j in DEST} demand[j];

param cost {ORIG,DEST} >= 0;    # shipment costs per unit
var Trans {ORIG,DEST} >= 0;     # units to be shipped

minimize Total_Cost:
    sum {i in ORIG, j in DEST} cost[i,j] * Trans[i,j];

subject to Supply {i in ORIG}:
    sum {j in DEST} Trans[i,j] = supply[i];

subject to Demand {j in DEST}:
    sum {i in ORIG} Trans[i,j] = demand[j];
```

Figure 3-1a: Transportation model (`transp.mod`).

```
sum {j in DEST} Trans[i,j]
```

Since we have already defined a parameter `supply` indexed over origins, the amount available at `i` is `supply[i]`. Thus the constraint is

```
subject to Supply {i in ORIG}:
    sum {j in DEST} Trans[i,j] = supply[i];
```

(Note that the names `supply` and `Supply` are unrelated; AMPL distinguishes upper and lower case.) The other collection of constraints is much the same, except that the roles of `i` in `ORIG`, and `j` in `DEST`, are exchanged, and the sum equals `demand[j]`.

We can now present the complete transportation model, Figure 3-1a. As you might have noticed, we have been consistent in using the index `i` to run over the set `ORIG`, and the index `j` to run over `DEST`. This is not an AMPL requirement, but such a convention makes it easier to read a model. You may name your own indices whatever you like, but keep in mind that the scope of an index — the part of the model where it has the same meaning — is to the end of the expression that defines it. Thus in the Demand constraint

```
subject to Demand {j in DEST}:
    sum {i in ORIG} Trans[i,j] = demand[j];
```

the scope of `j` runs to the semicolon at the end of the declaration, while the scope of `i` extends only through the summand `Trans[i,j]`. Since `i`'s scope is inside `j`'s scope, these two indices must have different names. Also an index may not have the same name as a set or other model component. Index scopes are discussed more fully, with further examples, in Section 5.5.

Data values for the transportation model are shown in Figure 3-1b. To define `DEST` and `demand`, we have used an input format that permits a set and one or more parameters indexed over it to be specified together. The set name is surrounded by colons. (We also show some comments, which can appear among data statements just as in a model.)

```
param: ORIG:    supply :=   # defines set "ORIG" and param "supply"
        GARY    1400
        CLEV    2600
        PITT    2900 ;

param: DEST:    demand :=   # defines "DEST" and "demand"
        FRA      900
        DET     1200
        LAN      600
        WIN      400
        STL     1700
        FRE     1100
        LAF     1000 ;

param cost:
        FRA   DET   LAN   WIN   STL   FRE   LAF :=
  GARY   39    14    11    14    16    82     8
  CLEV   27     9    12     9    26    95    17
  PITT   24    14    17    13    28    99    20 ;
```

Figure 3-1b: Data for transportation model (`transp.dat`).

If the model is stored in a file `transp.mod` and the data in `transp.dat`, we can solve the linear program and examine the output:

```
ampl: model transp.mod;
ampl: data transp.dat;
ampl: solve;
CPLEX 8.0.0: optimal solution; objective 196200
12 dual simplex iterations (0 in phase I)

ampl: display Trans;
Trans [*,*] (tr)
:     CLEV   GARY   PITT        :=
DET   1200      0      0
FRA      0      0    900
FRE      0   1100      0
LAF    400    300    300
LAN    600      0      0
STL      0      0   1700
WIN    400      0      0
;
```

By displaying the variable `Trans`, we see that most destinations are supplied from a single mill, but CLEV, GARY and PITT all ship to LAF.

It is instructive to compare this solution to one given by another solver, SNOPT:

```
ampl: option solver snopt;
ampl: solve;
SNOPT 6.1-1: Optimal solution found.
15 iterations, objective 196200
```

```
ampl: display Trans;
Trans [*,*] (tr)
:      CLEV   GARY   PITT      :=
DET    1200      0      0
FRA       0      0    900
FRE       0   1100      0
LAF     400      0    600
LAN     600      0      0
STL       0    300   1400
WIN     400      0      0
;
```

The minimum cost is still 196200, but it is achieved in a different way. Alternative optimal solutions such as these are often exhibited by transportation problems, particularly when the coefficients in the objective function are round numbers.

Unfortunately, there is no easy way to characterize all the optimal solutions. You may be able to get a better choice of optimal solution by working with several objectives, however, as we will illustrate in Section 8.3.

3.3 Other interpretations of the transportation model

As the name suggests, a transportation model is applicable whenever some material is being shipped from a set of origins to a set of destinations. Given certain amounts available at the origins, and required at the destinations, the problem is to meet the requirements at a minimum shipping cost.

Viewed more broadly, transportation models do not have to be concerned with the shipping of "materials". They can be applied to the transportation of anything, provided that the quantities available and required can be measured in some units, and that the transportation cost per unit can be determined. They might be used to model the shipments of automobiles to dealers, for example, or the movement of military personnel to new assignments.

In an even broader view, transportation models need not deal with "shipping" at all. The quantities at the origins may be merely associated with various destinations, while the objective measures some value of the association that has nothing to do with actually moving anything. Often the result is referred to as an "assignment" model.

As one particularly well-known example, consider a department that needs to assign some number of people to an equal number of offices. The origins now represent individual people, and the destinations represent individual offices. Since each person is assigned one office, and each office is occupied by one person, all of the parameter values supply[i] and demand[j] are 1. We interpret Trans[i,j] as the "amount" of person i that is assigned to office j; that is, if Trans[i,j] is 1 then person i will occupy office j, while if Trans[i,j] is 0 then person i will not occupy office j.

What of the objective? One possibility is to ask people to rank the offices, giving their first choice, second choice, and so forth. Then we can let cost[i,j] be the rank

```
set ORIG := Coullard Daskin Hazen Hopp Iravani Linetsky
            Mehrotra Nelson Smilowitz Tamhane White ;

set DEST := C118 C138 C140 C246 C250 C251 D237 D239 D241 M233 M239;

param supply default 1 ;

param demand default 1 ;

param cost:
           C118 C138 C140 C246 C250 C251 D237 D239 D241 M233 M239 :=
Coullard     6    9    8    7   11   10    4    5    3    2    1
Daskin      11    8    7    6    9   10    1    5    4    2    3
Hazen        9   10   11    1    5    6    2    7    8    3    4
Hopp        11    9    8   10    6    5    1    7    4    2    3
Iravani      3    2    8    9   10   11    1    5    4    6    7
Linetsky    11    9   10    5    3    4    6    7    8    1    2
Mehrotra     6   11   10    9    8    7    1    2    5    4    3
Nelson      11    5    4    6    7    8    1    9   10    2    3
Smilowitz   11    9   10    8    6    5    7    3    4    1    2
Tamhane      5    6    9    8    4    3    7   10   11    2    1
White       11    9    8    4    6    5    3   10    7    2    1 ;
```

Figure 3-2: Data for assignment problem (`assign.dat`).

that person i gives to office j. This convention lets each objective function term `cost[i,j] * Trans[i,j]` represent the preference of person i for office j, if person i is assigned to office j (`Trans[i,j]` equals 1), or zero if person i is not assigned to office j (`Trans[i,j]` equals 0). Since the objective is the sum of all these terms, it must equal the sum of all the nonzero terms, which is the sum of everyone's rankings for the offices to which they were assigned. By minimizing this sum, we can hope to find an assignment that will please a lot of people.

To use the transportation model for this purpose, we need only supply the appropriate data. Figure 3-2 is one example, with 11 people to be assigned to 11 offices. The `default` option has been used to set all the `supply` and `demand` values to 1 without typing all the 1's. If we store this data set in `assign.dat`, we can use it with the transportation model that we already have:

```
ampl: model transp.mod;
ampl: data assign.dat;
ampl: solve;
CPLEX 8.0.0: optimal solution; objective 28
24 dual simplex iterations (0 in phase I)
```

By setting the option `omit_zero_rows` to 1, we can print just the nonzero terms in the objective. (Options for displaying results are presented in Chapter 12.) This listing tells us each person's assigned room and his or her preference for it:

```
ampl: option omit_zero_rows 1;
ampl: display {i in ORIG, j in DEST} cost[i,j] * Trans[i,j];
cost[i,j]*Trans[i,j] :=
Coullard   C118    6
Daskin     D241    4
Hazen      C246    1
Hopp       D237    1
Iravani    C138    2
Linetsky   C250    3
Mehrotra   D239    2
Nelson     C140    4
Smilowitz  M233    1
Tamhane    C251    3
White      M239    1
;
```

The solution is reasonably successful, although it does assign two fourth choices and one sixth choice.

It is not hard to see that when all the `supply[i]` and `demand[j]` values are 1, any `Trans[i,j]` satisfying all the constraints must be between 0 and 1. But how did we know that every `Trans[i,j]` would equal either 0 or 1 in the optimal solution, rather than, say, ½? We were able to rely on a special property of transportation models, which guarantees that as long as all supply and demand values are integers, and all lower and upper bounds on the variables are integers, there will be an optimal solution that is entirely integral. Moreover, we used a solver that always finds one of these integral solutions. But don't let this favorable result mislead you into assuming that integrality can be assured in all other circumstances; even in examples that seem to be much like the transportation model, finding integral solutions can require a special solver, and a lot more work. Chapter 20 discusses issues of integrality at length.

A problem of assigning 100 people to 100 rooms has ten thousand variables; assigning 1000 people to 1000 rooms yields a million variables. In applications on this scale, however, most of the assignments can be ruled out in advance, so that the number of actual decision variables is not too large. After looking at an initial solution, you may want to rule out some more assignments — in our example, perhaps no assignment to lower than fifth choice should be allowed — or you may want to force some assignments to be made a certain way, in order to see how the rest could be done optimally. These situations require models that can deal with subsets of pairs (of people and offices, or origins and destinations) in a direct way. AMPL's features for describing pairs and other ''compound'' objects are the subject of Chapter 6.

Exercises

3-1. This transportation model, which deals with finding a least cost shipping schedule, comes from Dantzig's *Linear Programming and Extensions*. A company has plants in Seattle and San Diego, with capacities 350 and 600 cases per week respectively. It has customers in New York, Chicago, and Topeka, which order 325, 300, and 275 cases per week. The distances involved are:

	New York	Chicago	Topeka
Seattle	2500	1700	1800
San Diego	2500	1800	1400

The shipping cost is $90 per case per thousand miles. Formulate this model in AMPL and solve it to determine the minimum cost and the amounts to be shipped.

3-2. A small manufacturing operation produces six kinds of parts, using three machines. For the coming month, a certain number of each part is needed, and a certain number of parts can be accommodated on each machine; to complicate matters, it does not cost the same amount to make the same part on different machines. Specifically, the costs and related values are as follows:

	Part						
Machine	1	2	3	4	5	6	Capacity
1	3	3	2	5	2	1	80
2	4	1	1	2	2	1	30
3	2	2	5	1	1	2	160
Required	10	40	60	20	20	30	

(a) Using the model in Figure 3-1a, create a file of data statements for this problem; treat the machines as the origins, and the parts as the destinations. How many of each part should be produced on each machine, so as to minimize total cost?

(b) If the capacity of machine 2 is increased to 50, the manufacturer may be able to reduce the total cost of production somewhat. What small change to the model is necessary to analyze this situation? How much is the total cost reduced, and in what respects does the production plan change?

(c) Now suppose that the capacities are given in hours, rather than in numbers of parts, and that it takes a somewhat different number of hours to make the same part on different machines:

	Part						
Machine	1	2	3	4	5	6	Capacity
1	1.3	1.3	1.2	1.5	1.2	1.1	50
2	1.4	1.1	1.1	1.2	1.2	1.1	90
3	1.2	1.2	1.5	1.1	1.1	1.2	175

Modify the supply constraint so that it limits total time of production at each ''origin'' rather than the total quantity of production. How is the new optimal solution different? On which machines is all available time used?

(d) Solve the preceding problem again, but with the objective function changed to minimize total machine-hours rather than total cost.

3-3. This exercise deals with generalizations of the transportation model and data of Figure 3-1.

(a) Add two parameters, `supply_pct` and `demand_pct`, to represent the maximum fraction of a mill's supply that may be sent to any one factory, and the maximum fraction of a factory's demand that may be satisfied by any one mill. Incorporate these parameters into the model of Figure 3-1a.

Solve for the case in which no more than 50% of a mill's supply may be sent to any one factory, and no more than 85% of a factory's demand may be satisfied by any one mill. How does this change the minimum cost and the optimal amounts shipped?

(b) Suppose that the rolling mills do not produce their own slabs, but instead obtain slabs from two other plants, where the following numbers of tons are to be made available:

```
MIDTWN   2700
HAMLTN   4200
```

The cost per ton of shipping a slab from a plant to a mill is as follows:

```
          GARY     CLEV     PITT
MIDTWN    12        8       17
HAMLTN    10        5       13
```

All other data values are the same as before, but with `supply_pct` reinterpreted as the maximum fraction of a plant's supply that may be sent to any one mill.

Formulate this situation as an AMPL model. You will need two indexed collections of variables, one for the shipments from plants to mills, and one for the shipments from mills to factories. Shipments from each mill will have to equal supply, and shipments to each factory will have to equal demand as before; also, shipments out of each mill will have to equal shipments in.

Solve the resulting linear program. What are the shipment amounts in the minimum-cost solution?

(c) In addition to the differences in shipping costs, there may be different costs of production at the plants and mills. Explain how production costs could be incorporated into the model.

(d) When slabs are rolled, some fraction of the steel is lost as scrap. Assuming that this fraction may be different at each mill, revise the model to take scrap loss into account.

(e) In reality, scrap is not really lost, but is sold for recycling. Make a further change to the model to account for the value of the scrap produced at each mill.

3-4. This exercise considers variations on the assignment problem introduced in Section 3.3.

(a) Try reordering the list of members of DEST in the data (Figure 3-2), and solving again. Find a reordering that causes your solver to report a different optimal assignment.

(b) An assignment that gives even one person a very low-ranked office may be unacceptable, even if the total of the rankings is optimized. In particular, our solution gives one individual her sixth choice; to rule this out, change all preferences of six or larger in the cost data to 99, so that they will become very unattractive. (You'll learn more convenient features for doing the same thing in later chapters, but this crude approach will work for now.) Solve the assignment problem again, and verify that the result is an equally good assignment in which no one gets worse than fifth choice.

Now apply the same approach to try to give everyone no worse than fourth choice. What do you find?

(c) Suppose now that offices C118, C250 and C251 become unavailable, and you have to put two people each into C138, C140 and C246. Add 20 to each ranking for these three offices, to reflect the fact that anyone would prefer a private office to a shared one. What other modifications to the model and data would be necessary to handle this situation? What optimal assignment do you get?

(d) Some people may have seniority that entitles them to greater consideration in their choice of office. Explain how you could enhance the model to use seniority level data for each person.

4

Building Larger Models

The linear programs that we have presented so far have been quite small, so their data and solutions could fit onto a page. Most of the LPs found in practical applications, however, have hundreds or thousands of variables and constraints, and some are even larger.

How do linear programs get to be so large? They might be like the ones we have shown, but with larger indexing sets and more data. A steel mill could be considered to make hundreds of different products, for example, if every variation of width, thickness, and finish is treated separately. Or a large organization could have thousands of people involved in one assignment problem. Nevertheless, these kinds of applications are not as common as one might expect. As a model is refined to greater levels of detail, its data values become harder to maintain and its solutions harder to understand; past a certain point, extra detail offers no benefit. Thus to plan production for a few lines, considerable detail may be justifiable; but to plan for an entire company, it may be better to have a small aggregated, plant-level model that be run many times with different scenarios.

A more common source of large linear programs is the linking together of smaller ones. It is not unusual for an application to give rise to many simple LPs of the kinds we have discussed before; here are three possibilities:

- Many products are to be shipped, and there is a transportation problem (as in Chapter 3) for each product.
- Manufacturing is to be planned over many weeks, and there is a production problem (as in Chapter 1) for each week.
- Several products are made at several mills, and shipped to several factories; there is a production problem for each mill, and a transportation problem for each product.

When variables or constraints are added to tie these LPs together, the result can be one very large LP. No individual part need be particularly detailed; the size is more due to the large number of combinations of origins, destinations, products and weeks.

This chapter shows how AMPL models might be formulated for the three situations outlined above. The resulting models are necessarily more complicated than our previous ones, and require the use of a few more features from the AMPL language. Since they build on the terminology and logic of smaller models that have already been introduced, however, these larger models are still manageable.

4.1 A multicommodity transportation model

The transportation model of the previous chapter was concerned with shipping a single commodity from origins to destinations. Suppose now that we are shipping several different products. We can define a new set, PROD, whose members represent the different products, and we can add PROD to the indexing of every component in the model; the result can be seen in Figure 4-1. Because supply, demand, cost, and Trans are indexed over one more set in this version, they take one more subscript: supply[i,p] for the amount of product p shipped from origin i, Trans[i,j,p] for the amount of p shipped from i to j, and so forth. Even the check statement is now indexed over PROD, so that it verifies that supply equals demand for each separate product.

If we look at Supply, Demand and Trans, there are (origins + destinations) × (products) constraints in (origins) × (destinations) × (products) variables. The result could be quite a large linear program, even if the individual sets do not have many members. For example, 5 origins, 20 destinations and 10 products give 250 constraints in 1000 variables. The size of this LP is misleading, however, because the shipments of the products are independent. That is, the amounts we ship of one product do not affect the amounts we can ship of any other product, or the costs of shipping any other product. We would do better in this case to solve a smaller transportation problem for each individual product. In AMPL terms, we would use the simple transportation model from the previous chapter, together with a different data file for each product.

The situation would be different if some additional circumstances had the effect of tying together the different products. As an example, imagine that there are restrictions on the total shipments of products from an origin to a destination, perhaps because of limited shipping capacity. To accommodate such restrictions in our model, we declare a new parameter limit indexed over the combinations of origins and destinations:

```
param limit {ORIG,DEST} >= 0;
```

Then we have a new collection of (origins) × (destinations) constraints, one for each origin i and destination j, which say that the sum of shipments from i to j of all products p may not exceed limit[i,j]:

```
subject to Multi {i in ORIG, j in DEST}:
    sum {p in PROD} Trans[i,j,p] <= limit[i,j];
```

Subject to these constraints (also shown in Figure 4-1), we can no longer set the amount of one product shipped from i to j without considering the amounts of other products also shipped from i to j, since it is the sum of all products that is limited. Thus we have no choice but to solve the one large linear program.

For the steel mill in Chapter 1, the products were bands, coils, and plate. Thus the data for the multicommodity model could look like Figure 4-2. We invoke AMPL in the usual way to get the following solution:

```
ampl: model multi.mod; data multi.dat; solve;
CPLEX 8.0.0: optimal solution; objective 199500
41 dual simplex iterations (0 in phase I)
```

```
set ORIG;   # origins
set DEST;   # destinations
set PROD;   # products

param supply {ORIG,PROD} >= 0;  # amounts available at origins
param demand {DEST,PROD} >= 0;  # amounts required at destinations

   check {p in PROD}:
       sum {i in ORIG} supply[i,p] = sum {j in DEST} demand[j,p];

param limit {ORIG,DEST} >= 0;

param cost {ORIG,DEST,PROD} >= 0;  # shipment costs per unit
var Trans {ORIG,DEST,PROD} >= 0;    # units to be shipped

minimize Total_Cost:
   sum {i in ORIG, j in DEST, p in PROD}
      cost[i,j,p] * Trans[i,j,p];

subject to Supply {i in ORIG, p in PROD}:
   sum {j in DEST} Trans[i,j,p] = supply[i,p];

subject to Demand {j in DEST, p in PROD}:
   sum {i in ORIG} Trans[i,j,p] = demand[j,p];

subject to Multi {i in ORIG, j in DEST}:
   sum {p in PROD} Trans[i,j,p] <= limit[i,j];
```

Figure 4-1: Multicommodity transportation model (`multi.mod`).

```
ampl: display {p in PROD}: {i in ORIG, j in DEST} Trans[i,j,p];

Trans[i,j,'bands'] [*,*] (tr)
:    CLEV  GARY  PITT    :=
DET     0     0   300
FRA   225     0    75
FRE     0     0   225
LAF   225     0    25
LAN     0     0   100
STL   250   400     0
WIN     0     0    75
;

Trans[i,j,'coils'] [*,*] (tr)
:    CLEV  GARY  PITT    :=
DET   525     0   225
FRA     0     0   500
FRE   225   625     0
LAF     0   150   350
LAN   400     0     0
STL   300    25   625
WIN   150     0   100
;
```

```
set ORIG := GARY CLEV PITT ;
set DEST := FRA DET LAN WIN STL FRE LAF ;
set PROD := bands coils plate ;

param supply (tr):   GARY    CLEV    PITT :=
              bands   400     700     800
              coils   800    1600    1800
              plate   200     300     300 ;

param demand (tr):
            FRA   DET   LAN   WIN   STL   FRE   LAF :=
    bands   300   300   100    75   650   225   250
    coils   500   750   400   250   950   850   500
    plate   100   100     0    50   200   100   250 ;

param limit default 625 ;

param cost :=

   [*,*,bands]:  FRA   DET   LAN   WIN   STL   FRE   LAF :=
          GARY    30    10     8    10    11    71     6
          CLEV    22     7    10     7    21    82    13
          PITT    19    11    12    10    25    83    15

   [*,*,coils]:  FRA   DET   LAN   WIN   STL   FRE   LAF :=
          GARY    39    14    11    14    16    82     8
          CLEV    27     9    12     9    26    95    17
          PITT    24    14    17    13    28    99    20

   [*,*,plate]:  FRA   DET   LAN   WIN   STL   FRE   LAF :=
          GARY    41    15    12    16    17    86     8
          CLEV    29     9    13     9    28    99    18
          PITT    26    14    17    13    31   104    20 ;
```

Figure 4-2: Multicommodity transportation problem data (`multi.dat`).

```
Trans[i,j,'plate'] [*,*] (tr)
:    CLEV  GARY  PITT       :=
DET   100     0     0
FRA    50     0    50
FRE   100     0     0
LAF     0     0   250
LAN     0     0     0
STL     0   200     0
WIN    50     0     0
;
```

In both our specification of the shipping costs and AMPL's display of the solution, a three-dimensional collection of data (that is, indexed over three sets) must be represented on a two-dimensional screen or page. We accomplish this by "slicing" the data along one index, so that it appears as a collection of two-dimensional tables. The `display` command will make a guess as to the best index on which to slice, but by use of an

```
set PROD;       # products
param T > 0;    # number of weeks

param rate {PROD} > 0;             # tons per hour produced
param avail {1..T} >= 0;           # hours available in week
param profit {PROD,1..T};          # profit per ton
param market {PROD,1..T} >= 0;     # limit on tons sold in week

var Make {p in PROD, t in 1..T} >= 0, <= market[p,t];
                                   # tons produced

maximize Total_Profit:
    sum {p in PROD, t in 1..T} profit[p,t] * Make[p,t];
          # total profits from all products in all weeks

subject to Time {t in 1..T}:
    sum {p in PROD} (1/rate[p]) * Make[p,t] <= avail[t];
          # total of hours used by all products
          # may not exceed hours available, in each week
```

Figure 4-3: Production model replicated over periods (steelT0.mod).

explicit indexing expression as shown above, we can tell it to display a table for each product.

The optimal solution above ships only 25 tons of coils from GARY to STL and 25 tons of bands from PITT to LAF. It might be reasonable to require that, if any amount at all is shipped, it must be at least, say, 50 tons. In terms of our model, either Trans[i,j,p] = 0 or Trans[i,j,p] >= 50. Unfortunately, although it is possible to write such an ''either/or'' constraint in AMPL, it is not a linear constraint, and so there is no way that an LP solver can handle it. Chapter 20 explains how more powerful (but costlier) integer programming techniques can deal with this and related kinds of discrete restrictions.

4.2 A multiperiod production model

Another common way in which models are expanded is by replicating them over time. To illustrate, we consider how the model of Figure 1-4a might be used to plan production for the next T weeks, rather than for a single week.

We begin by adding another index set to most of the quantities of interest. The added set represents weeks numbered 1 through T, as shown in Figure 4-3. The expression 1..T is AMPL's shorthand for the set of integers from 1 through T. We have replicated all the parameters and variables over this set, except for rate, which is regarded as fixed over time. As a result there is a constraint for each week, and the profit terms are summed over weeks as well as products.

So far this is merely a separate LP for each week, unless something is added to tie the weeks together. Just as we were able to find constraints that involved all the products, we could look for constraints that involve production in all of the weeks. Most multiperiod models take a different approach, however, in which constraints relate each week's production to that of the following week only.

Suppose that we allow some of a week's production to be placed in inventory, for sale in any later week. We thus add new decision variables to represent the amounts inventoried and sold in each week. The variables `Make[j,t]` are retained, but they represent only the amounts produced, which are now not necessarily the same as the amounts sold. Our new variable declarations look like this:

```
var Make {PROD,1..T} >= 0;
var Inv {PROD,0..T} >= 0;

var Sell {p in PROD, t in 1..T} >= 0, <= market[p,t];
```

The bounds `market[p,t]`, which represent the maximum amounts that can be sold in a week, are naturally transferred to `Sell[p,t]`.

The variable `Inv[p,t]` will represent the inventory of product p at the end of period t. Thus the quantities `Inv[p,0]` will be the inventories at the end of week zero, or equivalently at the beginning of the first week — in other words, now. Our model assumes that these initial inventories are provided as part of the data:

```
param inv0 {PROD} >= 0;
```

A simple constraint guarantees that the variables `Inv[p,0]` take these values:

```
subject to Init_Inv {p in PROD}:  Inv[p,0] = inv0[p];
```

It may seem ''inefficient'' to devote a constraint like this to saying that a variable equals a constant, but when it comes time to send the linear program to a solver, AMPL will automatically substitute the value of `inv0[p]` for any occurrence of `Inv[p,0]`. In most cases, we can concentrate on writing the model in the clearest or easiest way, and leave matters of efficiency to the computer.

Now that we are distinguishing sales, production, and inventory, we can explicitly model the contribution of each to the profit, by defining three parameters:

```
param revenue {PROD,1..T} >= 0;
param prodcost {PROD} >= 0;
param invcost {PROD} >= 0;
```

These are incorporated into the objective as follows:

```
maximize Total_Profit:
    sum {p in PROD, t in 1..T} (revenue[p,t]*Sell[p,t] -
        prodcost[p]*Make[p,t] - invcost[p]*Inv[p,t]);
```

As you can see, `revenue[p,t]` is the amount received per ton of product p sold in week t; `prodcost[p]` and `invcost[p]` are the production and inventory carrying cost per ton of product p in any week.

Finally, with the sales and inventories fully incorporated into our model, we can add the key constraints that tie the weeks together: the amount of a product made available in a week, through production or from inventory, must equal the amount disposed of in that week, through sale or to inventory:

```
subject to Balance {p in PROD, t in 1..T}:
   Make[p,t] + Inv[p,t-1] = Sell[p,t] + Inv[p,t];
```

Because the index `t` is from a set of numbers, the period previous to `t` can be written as `t-1`. In fact, `t` can be used in any arithmetic expression; conversely, an AMPL expression such as `t-1` may be used in any context where it makes sense. Notice also that for a first-period constraint (`t` equal to 1), the inventory term on the left is `Inv[p,0]`, the initial inventory.

We now have a complete model, as shown in Figure 4-4. To illustrate a solution, we use the small sample data file shown in Figure 4-5; it represents a four-week expansion of the data from Figure 1-4b.

If we put the model and data into files `steelT.mod` and `steelT.dat`, then AMPL can be invoked to find a solution:

```
ampl: model steelT.mod;
ampl: data steelT.dat;
ampl: solve;
MINOS 5.5: optimal solution found.
20 iterations, objective 515033

ampl: option display_1col 0;

ampl: display Make;
Make [*,*] (tr)
:   bands   coils    :=
1   5990    1407
2   6000    1400
3   1400    3500
4   2000    4200
;

ampl: display Inv;
Inv [*,*] (tr)
: bands   coils    :=
0   10      0
1   0       1100
2   0       0
3   0       0
4   0       0
;

ampl: display Sell;
Sell [*,*] (tr)
:   bands   coils    :=
1   6000    307
2   6000    2500
3   1400    3500
4   2000    4200
;
```

```
set PROD;       # products
param T > 0;    # number of weeks

param rate {PROD} > 0;                 # tons per hour produced
param inv0 {PROD} >= 0;                # initial inventory
param avail {1..T} >= 0;               # hours available in week
param market {PROD,1..T} >= 0;   # limit on tons sold in week

param prodcost {PROD} >= 0;            # cost per ton produced
param invcost {PROD} >= 0;             # carrying cost/ton of inventory
param revenue {PROD,1..T} >= 0; # revenue per ton sold

var Make {PROD,1..T} >= 0;             # tons produced
var Inv {PROD,0..T} >= 0;              # tons inventoried
var Sell {p in PROD, t in 1..T} >= 0, <= market[p,t]; # tons sold

maximize Total_Profit:
   sum {p in PROD, t in 1..T} (revenue[p,t]*Sell[p,t] -
      prodcost[p]*Make[p,t] - invcost[p]*Inv[p,t]);

                   # Total revenue less costs in all weeks

subject to Time {t in 1..T}:
   sum {p in PROD} (1/rate[p]) * Make[p,t] <= avail[t];

                   # Total of hours used by all products
                   # may not exceed hours available, in each week

subject to Init_Inv {p in PROD}:  Inv[p,0] = inv0[p];

                   # Initial inventory must equal given value

subject to Balance {p in PROD, t in 1..T}:
   Make[p,t] + Inv[p,t-1] = Sell[p,t] + Inv[p,t];

                   # Tons produced and taken from inventory
                   # must equal tons sold and put into inventory
```

Figure 4-4: Multiperiod production model (steelT.mod).

```
param T := 4;
set PROD := bands coils;

param avail :=  1 40  2 40  3 32  4 40 ;

param rate :=  bands 200   coils 140 ;
param inv0 :=  bands  10   coils   0 ;

param prodcost :=  bands 10   coils  11 ;
param invcost  :=  bands 2.5  coils   3 ;

param revenue:   1     2     3     4 :=
        bands    25    26    27    27
        coils    30    35    37    39 ;

param market:    1     2     3     4 :=
        bands  6000  6000  4000  6500
        coils  4000  2500  3500  4200 ;
```

Figure 4-5: Data for multiperiod production model (steelT.dat).

Production of coils in the first week is held over to be sold at a higher price in the second week. In the second through fourth weeks, coils are more profitable than bands, and so coils are sold up to the limit, with bands filling out the capacity. (Setting option `display_1col` to zero permits this output to appear in a nicer format, as explained in Section 12.2.)

4.3 A model of production and transportation

Large linear programs can be created not only by tying together small models of one kind, as in the two examples above, but by linking different kinds of models. We conclude this chapter with an example that combines features of both production and transportation models.

Suppose that the steel products are made at several mills, from which they are shipped to customers at the various factories. For each mill we can define a separate production model to optimize the amounts of each product to make. For each product we can define a separate transportation model, with mills as origins and factories as destinations, to optimize the amounts of the product to be shipped. We would like to link all these separate models into a single integrated model of production and transportation.

To begin, we replicate the production model of Figure 1-4a over mills — that is, origins — rather than over weeks as in the previous example:

```
set PROD;   # products
set ORIG;   # origins (steel mills)

param rate {ORIG,PROD} > 0;   # tons per hour at origins
param avail {ORIG} >= 0;      # hours available at origins

var Make {ORIG,PROD} >= 0;    # tons produced at origins

subject to Time {i in ORIG}:
   sum {p in PROD} (1/rate[i,p]) * Make[i,p] <= avail[i];
```

We have temporarily dropped the components pertaining to the objective, to which we will return later. We have also dropped the market demand parameters, since the demands are now properly associated with the destinations in the transportation models.

The next step is to replicate the transportation model, Figure 3-1a, over products, as we did in the multicommodity example at the beginning of this chapter:

```
set ORIG;   # origins (steel mills)
set DEST;   # destinations (factories)
set PROD;   # products

param supply {ORIG,PROD} >= 0; # tons available at origins
param demand {DEST,PROD} >= 0; # tons required at destinations

var Trans {ORIG,DEST,PROD} >= 0; # tons shipped
```

```
set ORIG;    # origins (steel mills)
set DEST;    # destinations (factories)
set PROD;    # products

param rate {ORIG,PROD} > 0;        # tons per hour at origins
param avail {ORIG} >= 0;           # hours available at origins
param demand {DEST,PROD} >= 0;    # tons required at destinations

param make_cost {ORIG,PROD} >= 0;            # manufacturing cost/ton
param trans_cost {ORIG,DEST,PROD} >= 0;   # shipping cost/ton

var Make {ORIG,PROD} >= 0;          # tons produced at origins
var Trans {ORIG,DEST,PROD} >= 0;  # tons shipped

minimize Total_Cost:
    sum {i in ORIG, p in PROD} make_cost[i,p] * Make[i,p] +
    sum {i in ORIG, j in DEST, p in PROD}
                        trans_cost[i,j,p] * Trans[i,j,p];

subject to Time {i in ORIG}:
    sum {p in PROD} (1/rate[i,p]) * Make[i,p] <= avail[i];

subject to Supply {i in ORIG, p in PROD}:
    sum {j in DEST} Trans[i,j,p] = Make[i,p];

subject to Demand {j in DEST, p in PROD}:
    sum {i in ORIG} Trans[i,j,p] = demand[j,p];
```

Figure 4-6: Production/transportation model, 3rd version (steelP.mod).

```
    subject to Supply {i in ORIG, p in PROD}:
        sum {j in DEST} Trans[i,j,p] = supply[i,p];

    subject to Demand {j in DEST, p in PROD}:
        sum {i in ORIG} Trans[i,j,p] = demand[j,p];
```

Comparing the resulting production and transportation models, we see that the sets of origins (ORIG) and products (PROD) are the same in both models. Moreover, the "tons available at origins" (supply) in the transportation model are really the same thing as the "tons produced at origins" (Make) in the production model, since the steel available for shipping will be whatever is made at the mill.

We can thus merge the two models, dropping the definition of supply and substituting Make[i,p] for the occurrence of supply[i,p]:

```
    subject to Supply {i in ORIG, p in PROD}:
        sum {j in DEST} Trans[i,j,p] = Make[i,p];
```

There are several ways in which we might add an objective to complete the model. Perhaps the simplest is to define a cost per ton corresponding to each variable. We define a parameter make_cost so that there is a term make_cost[i,p] * Make[i,p] in the objective for each origin i and product p; and we define trans_cost so that there is a term trans_cost[i,j,p] * Trans[i,j,p] in the objective for each origin i, destination j and product p. The full model is shown in Figure 4-6.

```
set ORIG := GARY CLEV PITT ;
set DEST := FRA DET LAN WIN STL FRE LAF ;
set PROD := bands coils plate ;

param avail :=  GARY 20  CLEV 15  PITT 20 ;

param demand (tr):
              FRA    DET    LAN    WIN    STL    FRE    LAF :=
     bands    300    300    100     75    650    225    250
     coils    500    750    400    250    950    850    500
     plate    100    100      0     50    200    100    250 ;

param rate (tr):   GARY   CLEV   PITT :=
           bands    200    190    230
           coils    140    130    160
           plate    160    160    170 ;

param make_cost (tr):
                  GARY   CLEV   PITT :=
           bands   180    190    190
           coils   170    170    180
           plate   180    185    185 ;

param trans_cost :=

  [*,*,bands]:  FRA   DET   LAN   WIN   STL   FRE   LAF :=
        GARY    30    10     8    10    11    71     6
        CLEV    22     7    10     7    21    82    13
        PITT    19    11    12    10    25    83    15

  [*,*,coils]:  FRA   DET   LAN   WIN   STL   FRE   LAF :=
        GARY    39    14    11    14    16    82     8
        CLEV    27     9    12     9    26    95    17
        PITT    24    14    17    13    28    99    20

  [*,*,plate]:  FRA   DET   LAN   WIN   STL   FRE   LAF :=
        GARY    41    15    12    16    17    86     8
        CLEV    29     9    13     9    28    99    18
        PITT    26    14    17    13    31   104    20 ;
```

Figure 4-7: Data for production/transportation model (steelP.dat).

Reviewing this formulation, we might observe that, according to the Supply declaration, the nonnegative expression

```
sum {j in DEST} Trans[i,j,p]
```

can be substituted for Make[i,p]. If we make this substitution for all occurrences of Make[i,p] in the objective and in the Time constraints, we no longer need to include the Make variables or the Supply constraints in our model, and our linear programs will be smaller as a result. Nevertheless, in most cases we will be better off leaving the model as it is shown above. By ''substituting out'' the Make variables we render the model harder to read, and not a great deal easier to solve.

As an instance of solving a linear program based on this model, we can adapt the data from Figure 4-2, as shown in Figure 4-7. Here are some representative result values:

```
ampl: model steelP.mod; data steelP.dat; solve;
CPLEX 8.0.0: optimal solution; objective 1392175
27 dual simplex iterations (0 in phase I)

ampl: option display_1col 5;
ampl: option omit_zero_rows 1, omit_zero_cols 1;

ampl: display Make;
Make [*,*]
:        bands   coils plate      :=
CLEV         0    1950     0
GARY      1125    1750   300
PITT       775     500   500
;

ampl: display Trans;
Trans [CLEV,*,*]
:   coils      :=
DET   750
LAF   500
LAN   400
STL    50
WIN   250

  [GARY,*,*]
:    bands coils plate      :=
FRE   225    850    100
LAF   250      0      0
STL   650    900    200

  [PITT,*,*]
:    bands coils plate      :=
DET   300      0    100
FRA   300    500    100
LAF     0      0    250
LAN   100      0      0
WIN    75      0     50
;

ampl: display Time;
Time [*] :=
CLEV  -1300
GARY  -2800
;
```

As one might expect, the optimal solution does not ship all products from all mills to all factories. We have used the options omit_zero_rows and omit_zero_cols to suppress the printing of table rows and columns that are all zeros. The dual values for Time show that additional capacity is likely to have the greatest impact on total cost if it is placed at GARY, and no impact if it is placed at PITT.

We can also investigate the relative costs of production and shipping, which are the two components of the objective:

```
ampl: display sum {i in ORIG, p in PROD}
           make_cost[i,p] * Make[i,p];
sum{i in ORIG, p in PROD} make_cost[i,p]*Make[i,p] = 1215250

ampl: display sum {i in ORIG, j in DEST, p in PROD}
           trans_cost[i,j,p] * Trans[i,j,p];
sum{i in ORIG, j in DEST, p in PROD}
   trans_cost[i,j,p]*Trans[i,j,p] = 176925
```

Clearly the production costs dominate in this case. These examples point up the ability of AMPL to evaluate and display any valid expression.

Bibliography

H. P. Williams, *Model Building in Mathematical Programming* (4th edition). John Wiley & Sons (New York, 1999). An extended compilation of many kinds of models and combinations of them.

Exercises

4-1. Formulate a multi-period version of the transportation model, in which inventories are kept at the origins.

4-2. Formulate a combination of a transportation model for each of several foods, and a diet model at each destination.

4-3. The following questions pertain to the multiperiod production model and data of Section 4.2.

(a) Display the marginal values associated with the constraints Time[t]. In which periods does it appear that additional production capacity would be most valuable?

(b) By soliciting additional sales, you might be able to raise the upper bounds market[p,t]. Display the reduced costs Sell[p,t].rc, and use them to suggest whether you would prefer to go after more orders of bands or of coils in each week.

(c) If the inventory costs are all positive, any optimal solution will have zero inventories after the last week. Why is this so?

This phenomenon is an example of an "end effect". Because the model comes to an end after period T, the solution tends to behave as if production is to be shut down after that point. One way of dealing with end effects is to increase the number of weeks modeled; then the end effects should have little influence on the solution for the earlier weeks. Another approach is to modify the model to better reflect the realities of inventories. Describe some modifications you might make to the constraints, and to the objective.

4-4. A producer of packaged cookies and crackers runs several shifts each month at its large bakery. This exercise is concerned with a multiperiod planning model for deciding how many crews to employ each month. In the algebraic description of the model, there are sets S of shifts and P of products, and the planning horizon is T four-week periods. The relevant operational data are as follows:

l number of production lines: maximum number of crews that can work in any shift
r_p production rate for product p, in crew-hours per 1000 boxes
h_t number of hours that a crew works in planning period t

The following data are determined by market or managerial considerations:

w_s total wages for a crew on shift s in one period
d_{pt} demand for product p that must be met in period t
M maximum change in number of crews employed from one period to the next

The decision variables of the model are:

$X_{pt} \geq d_{pt}$ total boxes (in 1000s) of product p baked in period t
$0 \leq Y_{st} \leq l$ number of crews employed on shift s in period t

The objective is to minimize the total cost of all crews employed,

$$\sum_{s \in S} \sum_{t=1}^{T} w_s Y_{st} .$$

Total hours required for production in each period may not exceed total hours available from all shifts,

$$\sum_{p \in P} r_p X_{pt} \leq h_t \sum_{s \in S} Y_{st}, \text{ for each } t = 1, \ldots, T.$$

The change in number of crews is restricted by

$$-M \leq \sum_{s \in S} (Y_{s,t+1} - Y_{st}) \leq M, \text{ for each } t = 1, \ldots, T-1.$$

As required by the definition of M, this constraint restricts any change to lie between a reduction of M crews and an increase of M crews.

(a) Formulate this model in AMPL, and solve the following instance. There are $T = 13$ periods, $l = 8$ production lines, and a maximum change of $M = 3$ crews per period. The products are 18REG, 24REG, and 24PRO, with production rates r_p of 1.194, 1.509 and 1.509 respectively. Crews work either a day shift with wages w_s of \$44,900, or a night shift with wages \$123,100. The demands and working hours are given as follows by period:

Period t	$d_{18REG,t}$	$d_{24REG,t}$	$d_{24PRO,t}$	h_t
1	63.8	1212.0	0.0	156
2	76.0	306.2	0.0	152
3	88.4	319.0	0.0	160
4	913.8	208.4	0.0	152
5	115.0	298.0	0.0	156
6	133.8	328.2	0.0	152
7	79.6	959.6	0.0	152
8	111.0	257.6	0.0	160
9	121.6	335.6	0.0	152
10	470.0	118.0	1102.0	160
11	78.4	284.8	0.0	160
12	99.4	970.0	0.0	144
13	140.4	343.8	0.0	144

Display the numbers of crews required on each shift. You will find many fractional numbers of crews; how would you convert this solution to an optimal one in whole numbers?

(b) To be consistent, you should also require at most a change of M between the known initial number of crews (already employed in the period before the first) and the number of crews to be employed in the first planning period. Add a provision for this restriction to the model.

Re-solve with 11 initial crews. You should get the same solution.

(c) Because of the limit on the change in crews from period to period, more crews than necessary are employed in some periods. One way to deal with this is to carry inventories from one period to the next, so as to smooth out the amount of production required in each period. Add a variable for the amount of inventory of each product after each period, as in the model of Figure 4-4, and add constraints that relate inventory to production and demand. (Because inventories can be carried forward, production X_{pt} need not be $\geq$ demand d_{pt} in every period as required by the previous versions.) Also make a provision for setting initial inventories to zero. Finally, add an inventory cost per period per 1000 boxes to the objective.

Let the inventory costs be $34.56 for product 18REG, and $43.80 for 24REG and 24PRO. Solve the resulting linear program; display the crew sizes and inventory levels. How different is this solution? How much of a saving is achieved in labor cost, at how much expense in inventory cost?

(d) The demands in the given data peak at certain periods, when special discount promotions are in effect. Big inventories are built up in advance of these peaks, particularly before period 4. Baked goods are perishable, however, so that building up inventories past a certain number of periods is unrealistic.

Modify the model so that the inventory variables are indexed by product, period and age, where age runs from 1 to a specified limit A. Add constraints that the inventories of age 1 after any period cannot exceed the amounts just produced, and that inventories of age $a > 1$ after period t cannot exceed the inventories of age $a - 1$ after period $t - 1$.

Verify that, with a maximum inventory age of 2 periods, you can use essentially the same solution as in (c), but that with a maximum inventory age of 1 there are some periods that require more crews.

(e) Suppose now that instead of adding a third index on inventory variables as in (d), you impose the following inventory constraint: The amount of product p in inventory after period t may not exceed the total production of product p in periods $t - A + 1$ through t.

Explain why this constraint is sufficient to prevent any inventory from being more than A periods old, provided that inventories are managed on a *first-in, first-out* basis. Support your conclusion by showing that you get the same results as in (d) when solving with a maximum inventory age of 2 or of 1.

(f) Explain how you would modify the models in (c), (d), and (e) to account for initial inventories that are *not* zero.

4-5. Multiperiod linear programs can be especially difficult to develop, because they require data pertaining to the future. To hedge against the uncertainty of the future, a user of these LPs typically develops various scenarios, containing different forecasts of certain key parameters. This exercise asks you to develop what is known as a stochastic program, which finds a solution that can be considered robust over all scenarios.

(a) The revenues per ton might be particularly hard to predict, because they depend on fluctuating market conditions. Let the revenue data in Figure 4-5 be scenario 1, and also consider scenario 2:

```
param revenue:    1     2     3     4  :=
         bands    23    24    25    25
         coils    30    33  · 35    36 ;
```

and scenario 3:

```
param revenue:    1     2     3     4  :=
         bands    21    27    33    35
         coils    30    32    33    33 ;
```

By solving the three associated linear programs, verify that each of these scenarios leads to a different optimal production and sales strategy, even for the first week. You need one strategy, however, not three. The purpose of the stochastic programming approach is to determine a single solution that produces a good profit ''on average'' in a certain sense.

(b) As a first step toward formulating a stochastic program, consider how the three scenarios could be brought together into one linear program. Define a parameter S as the number of scenarios, and replicate the revenue data over the set 1..S:

```
param S > 0;
param revenue {PROD,1..T,1..S} >= 0;
```

Replicate all the variables and constraints in a similar way. (The idea is the same as earlier in this chapter, where we replicated model components over products or weeks.)

Define a new collection of parameters prob[s], to represent your estimate of the probability that a scenario s takes place:

```
param prob {1..S} >= 0, <= 1;
    check: 0.99999 < sum {s in 1..S} prob[s] < 1.00001;
```

The objective function is the expected profit, which equals the sum over all scenarios of the probability of each scenario times the optimum profit under that scenario:

```
maximize Expected_Profit:
    sum {s in 1..S} prob[s] *
        sum {p in PROD, t in 1..T} (revenue[p,t,s]*Sell[p,t,s] -
            prodcost[p]*Make[p,t,s] - invcost[p]*Inv[p,t,s]);
```

Complete the formulation of this multiscenario linear program, and put together the data for it. Let the probabilities of scenarios 1, 2 and 3 be 0.45, 0.35 and 0.20, respectively. Show that the solution consists of a production strategy for each scenario that is the same as the strategy in (a).

(c) The formulation in (b) is no improvement because it makes no connection between the scenarios. One way to make the model usable is to add ''nonanticipativity'' constraints that require each week-1 variable to be given the same value across all scenarios. Then the result will give you the best single strategy for the first week, in the sense of maximizing expected profit for all weeks. The strategies will still diverge after the first week — but a week from now you can update your data and run the stochastic program again to generate a second week's strategy.

A nonanticipativity constraint for the Make variables can be written

```
subject to Make_na {p in PROD, s in 1..S-1}:
    Make[p,1,s] = Make[p,1,s+1];
```

Add the analogous constraints for the Inv and Sell variables. Solve the stochastic program, and verify that the solution consists of a single period-1 strategy for all three scenarios.

(d) After getting your solution in (c), use the following command to look at the profits that the recommended strategy will achieve under the three scenarios:

```
display {s in 1..S}
    sum {p in PROD, t in 1..T} (revenue[p,t,s]*Sell[p,t,s] -
        prodcost[p]*Make[p,t,s] - invcost[p]*Inv[p,t,s]);
```

Which scenario will be most profitable, and which will be least profitable?

Repeat the analysis with probabilities of 0.0001, 0.0001 and 0.9998 for scenarios 1, 2 and 3. You should find that profit from strategy 3 goes up, but profits from the other two go down. Explain what these profits represent, and why the results are what you would expect.

5

Simple Sets and Indexing

The next four chapters of this book are a comprehensive presentation of AMPL's facilities for linear programming. The organization is by language features, rather than by model types as in the four preceding tutorial chapters. Since the basic features of AMPL tend to be closely interrelated, we do not attempt to explain any one feature in isolation. Rather, we assume at the outset a basic knowledge of AMPL such as Chapters 1 through 4 provide.

We begin with sets, the most fundamental components of an AMPL model. Almost all of the parameters, variables, and constraints in a typical model are indexed over sets, and many expressions contain operations (usually summations) over sets. Set indexing is the feature that permits a concise model to describe a large mathematical program.

Because sets are so fundamental, AMPL offers a broad variety of set types and operations. A set's members may be strings or numbers, ordered or unordered; they may occur singly, or as ordered pairs, triples or longer ''tuples''. Sets may be defined by listing or computing their members explicitly, by applying operations like union and intersection to other sets, or by specifying arbitrary arithmetic or logical conditions for membership.

Any model component or iterated operation can be indexed over any set, using a standard form of indexing expression. Even sets themselves may be declared in collections indexed over other sets.

This chapter introduces the simpler kinds of sets, as well as set operations and indexing expressions; it concludes with a discussion of ordered sets. Chapter 6 shows how these ideas are extended to compound sets, including sets of pairs and triples, and indexed collections of sets. Chapter 7 is devoted to parameters and expressions, and Chapter 8 to the variables, objectives and constraints that make up a linear program.

5.1 Unordered sets

The most elementary kind of AMPL set is an unordered collection of character strings. Usually all of the strings in a set are intended to represent instances of the same kind of entity — such as raw materials, products, factories or cities. Often the strings are chosen to have recognizable meanings (`coils`, `FISH`, `New_York`), but they could just as well be codes known only to the modeler (`23RPFG`, `486/33C`). A literal string that appears in an AMPL model must be delimited by quotes, either single (`'A&P'`) or double (`"Bell+Howell"`). In all contexts, upper case and lower case letters are distinct, so that for example `"fish"`, `"Fish"`, and `"FISH"` represent different set members.

The declaration of a set need only contain the keyword `set` and a name. For example, a model may declare

```
set PROD;
```

to indicate that a certain set will be referred to by the name `PROD` in the rest of the model. A name may be any sequence of letters, numerals, and underscore (_) characters that is not a legal number. A few names have special meanings in AMPL, and may only be used for specific purposes, while a larger number of names have predefined meanings that can be changed if they are used in some other way. For example, `sum` is reserved for the iterated addition operator; but `prod` is merely pre-defined as the iterated multiplication operator, so you can redefine `prod` as a set of products:

```
set prod;
```

A list of reserved words is given in Section A.1.

A declared set's membership is normally specified as part of the data for the model, in the manner to be described in Chapter 9; this separation of model and data is recommended for most mathematical programming applications. Occasionally, however, it is desirable to refer to a particular set of strings within a model. A literal set of this kind is specified by listing its members within braces:

```
{"bands", "coils", "plate"}
```

This expression may be used anywhere that a set is valid, for example in a model statement that gives the set `PROD` a fixed membership:

```
set PROD = {"bands", "coils", "plate"};
```

This sort of declaration is best limited to cases where a set's membership is small, is a fundamental aspect of the model, or is not expected to change often. Nevertheless we will see that the = phrase is often useful in set declarations, for the purpose of defining a set in terms of other sets and parameters. The operator = may be replaced by `default` to initialize the set while allowing its value to be overridden by a data statement or changed by subsequent assignments. These options are more important for parameters, however, so we discuss them more fully in Section 7.5.

Notice that AMPL makes a distinction between a string such as `"bands"` and a set like {`"bands"`} that has a membership of one string. The set that has no members (the empty set) is denoted { }.

5.2 Sets of numbers

Set members may also be numbers. In fact a set's members may be a mixture of numbers and strings, though this is seldom the case. In an AMPL model, a literal number is written in the customary way as a sequence of digits, optionally preceded by a sign, containing an optional decimal point, and optionally followed by an exponent; the exponent consists of a d, D, e, or E, optionally a sign, and a sequence of digits. A number (1) and the corresponding string ("1") are distinct; by contrast, different representations of the same number, such as 100 and 1E+2, stand for the same set member.

A set of numbers is often a sequence that corresponds to some progression in the situation being modeled, such as a series of weeks or years. Just as for strings, the numbers in a set can be specified as part of the data, or can be specified within a model as a list between braces, such as {1,2,3,4,5,6}. This sort of set can be described more concisely by the notation 1..6. An additional by clause can be used to specify an interval other than 1 between the numbers; for instance,

```
1990 .. 2020 by 5
```

represents the set

```
{1990, 1995, 2000, 2005, 2010, 2015, 2020}
```

This kind of expression can be used anywhere that a set is appropriate, and in particular within the assignment phrase of a set declaration:

```
set YEARS = 1990 .. 2020 by 5;
```

By giving the set a short and meaningful name, this declaration may help to make the rest of the model more readable.

It is not good practice to specify all the numbers within a .. expression by literals like 2020 and 5, unless the values of these numbers are fundamental to the model or will rarely change. A better arrangement is seen in the multiperiod production example of Figures 4-4 and 4-5, where a parameter T is declared to represent the number of periods, and the expressions 1..T and 0..T are used to represent sets of periods over which parameters, variables, constraints and sums are indexed. The value of T is specified in the data, and is thus easily changed from one run to the next. As a more elaborate example, we could write

```
param start integer;
param end > start integer;
param interval > 0 integer;

set YEARS = start .. end by interval;
```

If subsequently we were to give the data as

```
param start := 1990;
param end := 2020;
param interval := 5;
```

then YEARS would be the same set as in the previous example (as it would also be if end were 2023.) You may use any arithmetic expression to represent any of the values in a .. expression.

The members of a set of numbers have the same properties as any other numbers, and hence can be used in arithmetic expressions. A simple example is seen in Figure 4-4, where the material balance constraint is declared as

```
subject to Balance {p in PROD, t in 1..T}:
    Make[p,t] + Inv[p,t-1] = Sell[p,t] + Inv[p,t];
```

Because t runs over the set 1..T, we can write Inv[p,t-1] to represent the inventory at the end of the previous week. If t instead ran over a set of strings, the expression t-1 would be rejected as an error.

Set members need not be integers. AMPL attempts to store each numerical set member as the nearest representable floating-point number. You can see how this works out on your computer by trying an experiment like the following:

```
ampl: option display_width 50;
ampl: display -5/3 .. 5/3 by 1/3;
set -5/3 .. 5/3 by 1/3 :=
-1.6666666666666667        0.33333333333333326
-1.3333333333333335        0.6666666666666663
-1                         0.9999999999999998
-0.6666666666666667        1.3333333333333333
-0.3333333333333335        1.6666666666666663
-2.220446049250313e-16;
```

You might expect 0 and 1 to be members of this set, but things do not work out that way due to rounding error in the floating-point computations. It is unwise to use fractional numbers in sets, if your model relies on set members having precise values. There should be no comparable problem with integer members of reasonable size; integers are represented exactly for magnitudes up to 2^{53} (approximately 10^{16}) for IEEE standard arithmetic, and up to 2^{47} (approximately 10^{14}) for almost any computer in current use.

5.3 Set operations

AMPL has four operators that construct new sets from existing ones:

A union B	union: in either A or B
A inter B	intersection: in both A and B
A diff B	difference: in A but not B
A symdiff B	symmetric difference: in A or B but not both

The following excerpt from an AMPL session shows how these work:

```
ampl: set Y1 = 1990 .. 2020 by 5;
ampl: set Y2 = 2000 .. 2025 by 5;
ampl: display Y1 union Y2, Y1 inter Y2;
set Y1 union Y2 := 1990 1995 2000 2005 2010 2015 2020 2025;
set Y1 inter Y2 := 2000 2005 2010 2015 2020;

ampl: display Y1 diff Y2, Y1 symdiff Y2;
set Y1 diff Y2 := 1990 1995;
set Y1 symdiff Y2 := 1990 1995 2025;
```

The operands of set operators may be other set expressions, allowing more complex expressions to be built up:

```
ampl: display Y1 symdiff (Y1 symdiff Y2);
set Y1 symdiff (Y1 symdiff Y2) :=
2000    2005    2010    2015    2020    2025;

ampl: display (Y1 union {2025,2035,2045}) diff Y2;
set Y1 union   {2025, 2035, 2045} diff Y2 :=
1990    1995    2035    2045;

ampl: display 2000..2040 by 5 symdiff (Y1 union Y2);
set 2000 .. 2040 by 5 symdiff (Y1 union Y2) :=
2030    2035    2040    1990    1995;
```

The operands must always represent sets, however, so that for example you must write Y1 union {2025}, not Y1 union 2025.

Set operators group to the left unless parentheses are used to indicate otherwise. The union, diff, and symdiff operators have the same precedence, just below that of inter. Thus, for example,

```
A union B inter C diff D
```

is parsed as

```
(A union (B inter C)) diff D
```

A precedence hierarchy of all AMPL operators is given in Table A-1 of Section A.4.

Set operations are often used in the assignment phrase of a set declaration, to define a new set in terms of already declared sets. A simple example is provided by a variation on the diet model of Figure 2-1. Rather than specifying a lower limit and an upper limit on the amount of every nutrient, suppose that you want to specify a set of nutrients that have a lower limit, and a set of nutrients that have an upper limit. (Every nutrient is in one set or the other; some nutrients might be in both.) You could declare:

```
set MINREQ;    # nutrients with minimum requirements
set MAXREQ;    # nutrients with maximum requirements
set NUTR;      # all nutrients (DUBIOUS)
```

But then you would be relying on the user of the model to make sure that NUTR contains exactly all the members of MINREQ and MAXREQ. At best this is unnecessary work, and at worst it will be done incorrectly. Instead you can define NUTR as the union:

```
set NUTR = MINREQ union MAXREQ;
```

```
set MINREQ;    # nutrients with minimum requirements
set MAXREQ;    # nutrients with maximum requirements

set NUTR = MINREQ union MAXREQ;     # nutrients
set FOOD;                           # foods

param cost {FOOD} > 0;
param f_min {FOOD} >= 0;
param f_max {j in FOOD} >= f_min[j];

param n_min {MINREQ} >= 0;
param n_max {MAXREQ} >= 0;

param amt {NUTR,FOOD} >= 0;

var Buy {j in FOOD} >= f_min[j], <= f_max[j];

minimize Total_Cost:  sum {j in FOOD} cost[j] * Buy[j];

subject to Diet_Min {i in MINREQ}:
   sum {j in FOOD} amt[i,j] * Buy[j] >= n_min[i];

subject to Diet_Max {i in MAXREQ}:
   sum {j in FOOD} amt[i,j] * Buy[j] <= n_max[i];
```

Figure 5-1: Diet model using `union` operator (`dietu.mod`).

All three of these sets are needed, since the nutrient minima and maxima are indexed over MINREQ and MAXREQ,

```
param n_min {MINREQ} >= 0;
param n_max {MAXREQ} >= 0;
```

while the amounts of nutrients in the foods are indexed over NUTR:

```
param amt {NUTR,FOOD} >= 0;
```

The modification of the rest of the model is straightforward; the result is shown in Figure 5-1.

As a general principle, it is a bad idea to set up a model so that redundant information has to be provided. Instead a minimal necessary collection of sets should be chosen to be supplied in the data, while other relevant sets are defined by expressions in the model.

5.4 Set membership operations and functions

Two other AMPL operators, `in` and `within`, test the membership of sets. As an example, the expression

```
"B2" in NUTR
```

is true if and only if the string `"B2"` is a member of the set NUTR. The expression

```
MINREQ within NUTR
```

is true if all members of the set MINREQ are also members of NUTR — that is, if MINREQ is a subset of (or is the same as) NUTR. The in and within operators are the AMPL counterparts of $\in$ and $\subseteq$ in traditional algebraic notation. The distinction between members and sets is especially important here; the left operand of in must be an expression that evaluates to a string or number, whereas the left operand of within must be an expression that evaluates to a set.

AMPL also provides operators not in and not within, which reverse the truth value of their result.

You may apply within directly to a set you are declaring, to say that it must be a subset of some other set. Returning to the diet example, if all nutrients have a minimum requirement, but only some subset of nutrients has a maximum requirement, it would make sense to declare the sets as:

```
set NUTR;
set MAXREQ within NUTR;
```

AMPL will reject the data for this model if any member specified for MAXREQ is not also a member of NUTR.

The built-in function card computes the number of members in (or cardinality of) a set; for example, card(NUTR) is the number of members in NUTR. The argument of the card function may be any expression that evaluates to a set.

5.5 Indexing expressions

In algebraic notation, the use of sets is indicated informally by phrases such as "for all $i \in P$" or "for $t = 1, \ldots, T$" or "for all $j \in R$ such that $c_j > 0$." The AMPL counterpart is the *indexing expression* that appears within braces { ... } in nearly all of our examples. An indexing expression is used whenever we specify the set over which a model component is indexed, or the set over which a summation runs. Since an indexing expression defines a set, it can be used in any place where a set is appropriate.

The simplest form of indexing expression is just a set name or expression within braces. We have seen this in parameter declarations such as these from the multiperiod production model of Figure 4-4:

```
param rate {PROD} > 0;
param avail {1..T} >= 0;
```

Later in the model, references to these parameters are subscripted with a single set member, in expressions such as avail[t] and rate[p]. Variables can be declared and used in exactly the same way, except that the keyword var takes the place of param.

The names such as t and i that appear in subscripts and other expressions in our models are examples of *dummy indices* that have been defined by indexing expressions. In fact, any indexing expression may optionally define a dummy index that runs over the specified set. Dummy indices are convenient in specifying bounds on parameters:

```
param f_min {FOOD} >= 0;
param f_max {j in FOOD} >= f_min[j];
```

and on variables:

```
var Buy {j in FOOD} >= f_min[j], <= f_max[j];
```

They are also essential in specifying the sets over which constraints are defined, and the sets over which summations are done. We have often seen these uses together, in declarations such as

```
subject to Time {t in 1..T}:
    sum {p in PROD} (1/rate[p]) * Make[p,t] <= avail[t];
```

and

```
subject to Diet_Min {i in MINREQ}:
    sum {j in FOOD} amt[i,j] * Buy[j] >= n_min[i];
```

An indexing expression consists of an index name, the keyword `in`, and a set expression as before. We have been using single letters for our index names, but this is not a requirement; an index name can be any sequence of letters, digits, and underscores that is not a valid number, just like the name for a model component.

Although a name defined by a model component's declaration is known throughout all subsequent statements in the model, the definition of a dummy index name is effective only within the *scope* of the defining indexing expression. Normally the scope is evident from the context. For instance, in the `Diet_Min` declaration above, the scope of `{i in MINREQ}` runs to the end of the statement, so that `i` can be used anywhere in the description of the constraint. On the other hand, the scope of `{j in FOOD}` covers only the summand `amt[i,j] * Buy[j]`. The scope of indexing expressions for sums and other iterated operators is discussed further in Chapter 7.

Once an indexing expression's scope has ended, its dummy index becomes undefined. Thus the same index name can be defined again and again in a model, and in fact it is good practice to use relatively few different index names. A common convention is to associate certain index names with certain sets, so that for example `i` always runs over NUTR and `j` always runs over FOOD. This is merely a convention, however, not a restriction imposed by AMPL. Indeed, when we modified the diet model so that there was a subset MINREQ of NUTR, we used `i` to run over MINREQ as well as NUTR. The opposite situation occurs, for example, if we want to specify a constraint that the amount of each food `j` in the diet is at least some fraction `min_frac[j]` of the total food in the diet:

```
subject to Food_Ratio {j in FOOD}:
    Buy[j] >= min_frac[j] * sum {jj in FOOD} Buy[jj];
```

Since the scope of `j in FOOD` extends to the end of the declaration, a different index `jj` is defined to run over the set FOOD in the summation within the constraint.

As a final option, the set in an indexing expression may be followed by a colon (`:`) and a logical condition. The indexing expression then represents only the subset of members that satisfy the condition. For example,

```
{j in FOOD: f_max[j] - f_min[j] < 1}
```

describes the set of all foods whose minimum and maximum amounts are nearly the same, and

```
{i in NUTR: i in MAXREQ or n_min[i] > 0}
```

describes the set of nutrients that are either in MAXREQ or for which n_min is positive. The use of operators such as or and < to form logical conditions will be fully explained in Chapter 7.

By specifying a condition, an indexing expression defines a new set. You can use the indexing expression to represent this set not only in indexed declarations and summations, but anywhere else that a set expression may appear. For example, you could say either of

```
set NUTREQ = {i in NUTR: i in MAXREQ or n_min[i] > 0};
set NUTREQ = MAXREQ union {i in MINREQ: n_min[i] > 0};
```

to define NUTREQ to represent our preceding example of a set expression, and you could use either of

```
set BOTHREQ = {i in MINREQ: i in MAXREQ};
set BOTHREQ = MINREQ inter MAXREQ;
```

to define BOTHREQ to be the set of all nutrients that have both minimum and maximum requirements. It's not unusual to find that there are several ways of describing some complicated set, depending on how you combine set operations and indexing expression conditions. Of course, some possibilities are easier to read than others, so it's worth taking some trouble to find the most readable. In Chapter 6 we also discuss efficiency considerations that sometimes make one alternative preferable to another in specifying compound sets.

In addition to being valuable within the model, indexing expressions are useful in display statements to summarize characteristics of the data or solution. The following example is based on the model of Figure 5-1 and the data of Figure 5-2:

```
ampl: model dietu.mod;
ampl: data dietu.dat;

ampl: display MAXREQ union {i in MINREQ: n_min[i] > 0};
set MAXREQ union {i in MINREQ: n_min[i] > 0}   := A NA CAL C;

ampl: solve;
CPLEX 8.0.0: optimal solution; objective 74.27382022
2 dual simplex iterations (0 in phase I)

ampl: display {j in FOOD: Buy[j] > f_min[j]};
set {j in FOOD: Buy[j] > f_min[j]}   := CHK MTL SPG;

ampl: display {i in MINREQ: Diet_Min[i].slack = 0};
set {i in MINREQ: (Diet_Min[i].slack) == 0}   := C CAL;
```

AMPL interactive commands are allowed to refer to variables and constraints in the condition phrase of an indexing expression, as illustrated by the last two display state-

```
set MINREQ := A B1 B2 C CAL ;
set MAXREQ := A NA CAL ;
set FOOD := BEEF CHK FISH HAM MCH MTL SPG TUR ;

param:    cost   f_min   f_max :=
  BEEF    3.19     2      10
  CHK     2.59     2      10
  FISH    2.29     2      10
  HAM     2.89     2      10
  MCH     1.89     2      10
  MTL     1.99     2      10
  SPG     1.99     2      10
  TUR     2.49     2      10   ;

param:    n_min   n_max :=
  A        700    20000
  C        700       .
  B1         0       .
  B2         0       .
  NA         .     50000
  CAL    16000     24000 ;

param amt (tr):    A     C    B1    B2     NA    CAL :=
          BEEF    60    20    10    15    938    295
          CHK      8     0    20    20   2180    770
          FISH     8    10    15    10    945    440
          HAM     40    40    35    10    278    430
          MCH     15    35    15    15   1182    315
          MTL     70    30    15    15    896    400
          SPG     25    50    25    15   1329    370
          TUR     60    20    15    10   1397    450 ;
```

Figure 5-2: Data for diet model (`dietu.dat`).

ments above. Within a model, however, only sets, parameters and dummy indices may be mentioned in any indexing expression.

The set BOTHREQ above might well be empty, in the case where every nutrient has either a minimum or a maximum requirement in the data, but not both. Indexing over an empty set is not an error. When a model component is declared to be indexed over a set that turns out to be empty, AMPL simply skips generating that component. A sum over an empty set is zero, and other iterated operators over empty sets have the obvious interpretations (see A.4).

5.6 Ordered sets

Any set of numbers has a natural ordering, so numbers are often used to represent entities, like time periods, whose ordering is essential to the specification of a model. To describe the difference between this week's inventory and the previous week's inventory, for example, we need the weeks to be ordered so that the "previous" week is always well defined.

An AMPL model can also define its own ordering for any set of numbers or strings, by adding the keyword `ordered` or `circular` to the set's declaration. The order in which you give the set's members, in either the model or the data, is then the order in which AMPL works with them. In a set declared `circular`, the first member is considered to follow the last one, and the last to precede the first; in an `ordered` set, the first member has no predecessor and the last member has no successor.

Ordered sets of strings often provide better documentation for a model's data than sets of numbers. Returning to the multiperiod production model of Figure 4-4, we observe that there is no way to tell from the data which weeks the numbers 1 through T refer to, or even that they are weeks instead of days or months. Suppose that instead we let the weeks be represented by an ordered set that contains, say, 27sep, 04oct, 11oct and 18oct. The declaration of T is replaced by

```
set WEEKS ordered;
```

and all subsequent occurrences of `1..T` are replaced by `WEEKS`. In the `Balance` constraint, the expression `t-1` is replaced by `prev(t)`, which selects the member before `t` in the set's ordering:

```
subject to Balance {p in PROD, t in WEEKS}:
   Make[p,t] + Inv[p,prev(t)] = Sell[p,t] + Inv[p,t]; # WRONG
```

This is not quite right, however, because when `t` is the first week in `WEEKS`, the member `prev(t)` is not defined. When you try to solve the problem, you will get an error message like this:

```
error processing constraint Balance['bands','27sep']:
        can't compute prev('27sep', WEEKS) --
          '27sep' is the first member
```

One way to fix this is to give a separate balance constraint for the first period, in which `Inv[p,prev(t)]` is replaced by the initial inventory, `inv0[p]`:

```
subject to Balance0 {p in PROD}:
   Make[p,first(WEEKS)] + inv0[p]
       = Sell[p,first(WEEKS)] + Inv[p,first(WEEKS)];
```

The regular balance constraint is limited to the remaining weeks:

```
subject to Balance {p in PROD, t in WEEKS: ord(t) > 1}:
   Make[p,t] + Inv[p,prev(t)] = Sell[p,t] + Inv[p,t];
```

The complete model and data are shown in Figures 5-3 and 5-4. As a tradeoff for more meaningful week names, we have to write a slightly more complicated model.

As our example demonstrates, AMPL provides a variety of functions that apply specifically to ordered sets. These functions are of three basic types.

```
set PROD;            # products
set WEEKS ordered;   # number of weeks

param rate {PROD} > 0;              # tons per hour produced
param inv0 {PROD} >= 0;             # initial inventory
param avail {WEEKS} >= 0;           # hours available in week
param market {PROD,WEEKS} >= 0;     # limit on tons sold in week

param prodcost {PROD} >= 0;         # cost per ton produced
param invcost {PROD} >= 0;          # carrying cost/ton of inventory
param revenue {PROD,WEEKS} >= 0;    # revenue/ton sold

var Make {PROD,WEEKS} >= 0;         # tons produced
var Inv {PROD,WEEKS} >= 0;          # tons inventoried
var Sell {p in PROD, t in WEEKS} >= 0, <= market[p,t]; # tons sold

maximize Total_Profit:
    sum {p in PROD, t in WEEKS} (revenue[p,t]*Sell[p,t] -
        prodcost[p]*Make[p,t] - invcost[p]*Inv[p,t]);

            # Objective: total revenue less costs in all weeks

subject to Time {t in WEEKS}:
    sum {p in PROD} (1/rate[p]) * Make[p,t] <= avail[t];

            # Total of hours used by all products
            # may not exceed hours available, in each week

subject to Balance0 {p in PROD}:
    Make[p,first(WEEKS)] + inv0[p]
        = Sell[p,first(WEEKS)] + Inv[p,first(WEEKS)];

subject to Balance {p in PROD, t in WEEKS: ord(t) > 1}:
    Make[p,t] + Inv[p,prev(t)] = Sell[p,t] + Inv[p,t];

            # Tons produced and taken from inventory
            # must equal tons sold and put into inventory
```

Figure 5-3: Production model with ordered sets (steelT2.mod).

First, there are functions that return a member from some absolute position in a set. You can write first(WEEKS) and last(WEEKS) for the first and last members of the ordered set WEEKS. To pick out other members, you can use member(5,WEEKS), say, for the 5th member of WEEKS. The arguments of these functions must evaluate to an ordered set, except for the first argument of member, which can be any expression that evaluates to a positive integer.

A second kind of function returns a member from a position relative to another member. Thus you can write prev(t,WEEKS) for the member immediately before t in WEEKS, and next(t,WEEKS) for the member immediately after. More generally, expressions such as prev(t,WEEKS,5) and next(t,WEEKS,3) refer to the 5th member before and the 3rd member after t in WEEKS. There are also "wraparound" versions prevw and nextw that work the same except that they treat the end of the set as wrapping around to the beginning; in effect, they treat all ordered sets as if their decla-

```
set PROD := bands coils ;
set WEEKS := 27sep 04oct 11oct 18oct ;

param avail :=  27sep 40   04oct 40   11oct 32   18oct 40 ;

param rate :=  bands 200   coils 140 ;
param inv0 :=  bands  10   coils   0 ;

param prodcost :=  bands 10     coils 11 ;
param invcost  :=  bands  2.5   coils  3 ;

param revenue: 27sep   04oct   11oct   18oct :=
        bands     25      26      27      27
        coils     30      35      37      39 ;

param market: 27sep   04oct   11oct   18oct :=
        bands    6000    6000    4000    6500
        coils    4000    2500    3500    4200 ;
```

Figure 5-4: Data for production model (steelT2.dat).

rations were circular. In all of these functions, the first argument must evaluate to a number or string, the second argument to an ordered set, and the third to an integer. Normally the integer is positive, but zero and negative values are interpreted in a consistent way; for instance, next(t,WEEKS,0) is the same as t, and next(t,WEEKS,-5) is the same as prev(t,WEEKS,5).

Finally, there are functions that return the position of a member within a set. The expression ord(t,WEEKS) returns the numerical position of t within the set WEEKS, or gives you an error message if t is not a member of WEEKS. The alternative ord0(t,WEEKS) is the same except that it returns 0 if t is not a member of WEEKS. For these functions the first argument must evaluate to a positive integer, and the second to an ordered set.

If the first argument of next, nextw, prev, prevw, or ord is a dummy index that runs over an ordered set, its associated indexing set is assumed if a set is not given as the second argument. Thus in the constraint

```
    subject to Balance {p in PROD, t in WEEKS: ord(t) > 1}:
        Make[p,t] + Inv[p,prev(t)] = Sell[p,t] + Inv[p,t];
```

the functions ord(t) and prev(t) are interpreted as if they had been written ord(t,WEEKS) and prev(t,WEEKS).

Ordered sets can also be used with any of the AMPL operators and functions that apply to sets generally. The result of a diff operation preserves the ordering of the left operand, so the material balance constraint in our example could be written:

```
    subject to Balance {p in PROD, t in WEEKS diff {first(WEEKS)}}:
        Make[p,t] + Inv[p,prev(t)] = Sell[p,t] + Inv[p,t];
```

For union, inter and symdiff, however, the ordering of the result is not well defined; AMPL treats the result as an unordered set.

For a set that is contained in an ordered set, AMPL provides a way to say that the ordering should be inherited. Suppose for example that you want to try running the multiperiod production model with horizons of different lengths. In the following declarations, the ordered set `ALL_WEEKS` and the parameter `T` are given in the data, while the subset `WEEKS` is defined by an indexing expression to include only the first `T` weeks:

```
set ALL_WEEKS ordered;
param T > 0 integer;

set WEEKS = {t in ALL_WEEKS: ord(t) <= T} ordered by ALL_WEEKS;
```

We specify `ordered by ALL_WEEKS` so that `WEEKS` becomes an ordered set, with its members having the same ordering as they do in `ALL_WEEKS`. The `ordered by` and `circular by` phrases have the same effect as the `within` phrase of Section 5.4 together with `ordered` or `circular`, except that they also cause the declared set to inherit the ordering from the containing set. There are also `ordered by reversed` and `circular by reversed` phrases, which cause the declared set's ordering to be the opposite of the containing set's ordering. All of these phrases may be used either with a subset supplied in the data, or with a subset defined by an expression as in the example above.

Predefined sets and interval expressions

AMPL provides special names and expressions for certain common intervals and other sets that are either infinite or potentially very large. Indexing expressions may not iterate over these sets, but they can be convenient for specifying the conditional phrases in `set` and `param` declarations.

AMPL intervals are sets containing all numbers between two bounds. There are intervals of real (floating-point) numbers and of integers, introduced by the keywords `interval` and `integer` respectively. They may be specified as closed, open, or half-open, following standard mathematical notation,

$$
\begin{aligned}
\texttt{interval [a, b]} &\equiv \{x: a \le x \le b\}, \\
\texttt{interval (a, b]} &\equiv \{x: a < x \le b\}, \\
\texttt{interval [a, b)} &\equiv \{x: a \le x < b\}, \\
\texttt{interval (a, b)} &\equiv \{x: a < x < b\}, \\
\texttt{integer [a, b]} &\equiv \{x \in I : a \le x \le b\}, \\
\texttt{integer (a, b]} &\equiv \{x \in I : a < x \le b\}, \\
\texttt{integer [a, b)} &\equiv \{x \in I : a \le x < b\}, \\
\texttt{integer (a, b)} &\equiv \{x \in I : a < x < b\}
\end{aligned}
$$

where `a` and `b` are any arithmetic expressions, and I denotes the set of integers. In the declaration phrases

```
in interval
within interval
ordered by [ reversed ] interval
circular by [ reversed ] interval
```

the keyword `interval` may be omitted.

As an example, in declaring Chapter 1's parameter `rate`, you can declare

```
param rate {PROD} in interval (0,maxrate];
```

to say that the production rates have to be greater than zero and not more than some pre-viously defined parameter `maxrate`; you could write the same thing more concisely as

```
param rate {PROD} in (0,maxrate];
```

or equivalently as

```
param rate {PROD} > 0, <= maxrate;
```

An open-ended interval can be specified by using the predefined AMPL parameter `Infinity` as the right-hand bound, or `-Infinity` as the left-hand bound, so that

```
param rate {PROD} in (0,Infinity];
```

means exactly the same thing as

```
param rate {PROD} > 0;
```

in Figure 1-4a. In general, intervals do not let you say anything new in set or parameter declarations; they just give you alternative ways to say things. (They have a more essen-tial role in defining imported functions, as discussed in Section A.22.)

The predefined infinite sets `Reals` and `Integers` are the sets of all floating-point numbers and integers, respectively, in numeric order. The predefined infinite sets `ASCII`, `EBCDIC`, and `Display` all represent the universal set of strings and numbers from which members of any one-dimensional set are drawn. `ASCII` and `EBCDIC` are ordered by the ASCII and EBCDIC collating sequences, respectively. `Display` has the ordering used in AMPL's `display` command (Section A.16): numbers precede literals and are ordered numerically; literals are sorted by the ASCII collating sequence.

As an example, you can declare

```
set PROD ordered by ASCII;
```

to make AMPL's ordering of the members of `PROD` alphabetical, regardless of their order-ing in the data. This reordering of the members of `PROD` has no effect on the solutions of the model in Figure 1-4a, but it causes AMPL listings of most entities indexed over `PROD` to appear in the same order (see A.6.2).

Exercises

5-1. (a) Display the sets

```
-5/3 .. 5/3 by 1/3
0 .. 1 by .1
```

Explain any evidence of rounding error in your computer's arithmetic.

(b) Try the following commands from Sections 5.2 and 5.4 on your computer:

```
ampl: set HUGE = 1..1e7;
ampl: display card(HUGE);
```

When AMPL runs out of memory, how many bytes does it say were available? (If your computer really does have enough memory, try `1..1e8`.) Experiment to see how big a set HUGE your computer can hold without running out of memory.

5-2. Revise the model of Exercise 1-6 so that it makes use of two different attribute sets: a set of attributes that have lower limits, and a set of attributes that have upper limits. Use the same approach as in Figure 5-1.

5-3. Use the `display` command, together with indexing expressions as demonstrated in Section 5.5, to determine the following sets relating to the diet model of Figures 5-1 and 5-2:

– Foods that have a unit cost greater than \$2.00.

– Foods that have a sodium (NA) content of more than 1000.

– Foods that contribute more than \$10 to the total cost in the optimal solution.

– Foods that are purchased at more than the minimum level but less than the maximum level in the optimal solution.

– Nutrients that the optimal diet supplies in exactly the minimum allowable amount.

– Nutrients that the optimal diet supplies in exactly the maximum allowable amount.

– Nutrients that the optimal diet supplies in more than the minimum allowable amount but less than the maximum allowable amount.

5-4. This exercise refers to the multiperiod production model of Figure 4-4.

(a) Suppose that we define two additional scalar parameters,

```
param Tbegin integer >= 1;
param Tend integer > Tbegin, <= T;
```

We want to solve the linear program that covers only the weeks from Tbegin through Tend. We still want the parameters to use the indexing `1..T`, however, so that we don't need to change the data tables every time we try a different value for Tbegin or Tend.

To start with, we can change every occurrence of `1..T` in the variable, objective and constraint declarations to `Tbegin..Tend`. By making these and other necessary changes, create a model that correctly covers only the desired weeks.

(b) Now suppose that we define a different scalar parameter,

```
param Tagg integer >= 1;
```

We want to "aggregate" the model, so that one "period" in our LP becomes Tagg weeks long, rather than one week. This would be appropriate if we have, say, a year of weekly data, which would yield an LP too large to be convenient for analysis.

To aggregate properly, we must define the availability of hours in each period to be the sum of the availabilities in all weeks of the period:

```
param avail_agg {t in 1..T by Tagg}
    = sum {u in t..t+Tagg-1} avail[u];
```

The parameters `market` and `revenue` must be similarly summed. Make all the necessary changes to the model of Figure 4-4 so that the resulting LP is properly aggregated.

(c) Re-do the models of (a) and (b) to use an ordered set of strings for the periods, as in Figure 5-3.

5-5. Extend the transportation model of Figure 3-1a to a multiperiod version, in which the periods are months represented by an ordered set of character strings such as "Jan", "Feb" and so forth. Use inventories at the origins to link the periods.

5-6. Modify the model of Figure 5-3 to merge the Balance0 and Balance constraints, as in Figure 4-4. Hint: 0..T and 1..T are analogous to

```
set WEEKS0 ordered;
set WEEKS = {i in WEEKS0: ord(i) > 1} ordered by WEEKS0;
```

6

Compound Sets and Indexing

Most linear programming models involve indexing over combinations of members from several different sets. Indeed, the interaction of indexing sets is often the most complicated aspect of a model; once you have the arrangement of sets worked out, the rest of the model can be written clearly and concisely.

All but the simplest models employ compound sets whose members are pairs, triples, quadruples, or even longer "tuples" of objects. This chapter begins with the declaration and use of sets of ordered pairs. We concentrate first on the set of all pairs from two sets, then move on to subsets of all pairs and to "slices" through sets of pairs. Subsequent sections explore sets of longer tuples, and extensions of AMPL's set operators and indexing expressions to sets of tuples.

The final section of this chapter introduces sets that are declared in collections indexed over other sets. An indexed collection of sets often plays much the same role as a set of tuples, but it represents a somewhat different way of thinking about the formulation. Each kind of set is appropriate in certain situations, and we offer some guidelines for choosing between them.

6.1 Sets of ordered pairs

An ordered pair of objects, whether numbers or strings, is written with the objects separated by a comma and surrounded by parentheses:

```
("PITT","STL")
("bands",5)
(3,101)
```

As the term "ordered" suggests, it makes a difference which object comes first; ("STL","PITT") is not the same as ("PITT","STL"). The same object may appear both first and second, as in ("PITT","PITT").

Pairs can be collected into sets, just like single objects. A comma-separated list of pairs may be enclosed in braces to denote a literal set of ordered pairs:

```
{("PITT","STL"),("PITT","FRE"),("PITT","DET"),("CLEV","FRE")}
{(1,1),(1,2),(1,3),(2,1),(2,2),(2,3),(3,1),(3,2),(3,3)}
```

Because sets of ordered pairs are often large and subject to change, however, they seldom appear explicitly in AMPL models. Instead they are described symbolically in a variety of ways.

The set of all ordered pairs from two given sets appears frequently in our examples. In the transportation model of Figure 3-1a, for instance, the set of all origin-destination pairs is written as either of

```
{ORIG, DEST}
{i in ORIG, j in DEST}
```

depending on whether the context requires dummy indices i and j. The multiperiod production model of Figure 4-4 uses a set of all pairs from a set of strings (representing products) and a set of numbers (representing weeks):

```
{PROD, 1..T}
{p in PROD, t in 1..T}
```

Various collections of model components, such as the parameter revenue and the variable Sell, are indexed over this set. When individual components are referenced in the model, they must have two subscripts, as in revenue[p,t] or Sell[p,t]. The order of the subscripts is always the same as the order of the objects in the pairs; in this case the first subscript must refer to a string in PROD, and the second to a number in 1..T.

An indexing expression like {p in PROD, t in 1..T} is the AMPL transcription of a phrase like "for all p in P, $t = 1, \ldots, T$" from algebraic notation. There is no compelling reason to think in terms of ordered pairs in this case, and indeed we did not mention ordered pairs when introducing the multiperiod production model in Chapter 4. On the other hand, we can modify the transportation model of Figure 3-1a to emphasize the role of origin-destination pairs as "links" between cities, by defining this set of pairs explicitly:

```
set LINKS = {ORIG,DEST};
```

The shipment costs and amounts can then be indexed over links:

```
param cost {LINKS} >= 0;
var Trans {LINKS} >= 0;
```

In the objective, the sum of costs over all shipments can be written like this:

```
minimize Total_Cost:
    sum {(i,j) in LINKS} cost[i,j] * Trans[i,j];
```

Notice that when dummy indices run over a set of pairs like LINKS, they must be defined in a pair like (i,j). It would be an error to sum over {k in LINKS}. The complete model is shown in Figure 6-1, and should be compared with Figure 3-1a. The specification of the data could be the same as in Figure 3-1b.

```
set ORIG;    # origins
set DEST;    # destinations

set LINKS = {ORIG,DEST};

param supply {ORIG} >= 0;   # amounts available at origins
param demand {DEST} >= 0;   # amounts required at destinations

    check: sum {i in ORIG} supply[i] = sum {j in DEST} demand[j];

param cost {LINKS} >= 0;    # shipment costs per unit
var Trans {LINKS} >= 0;     # units to be shipped

minimize Total_Cost:
    sum {(i,j) in LINKS} cost[i,j] * Trans[i,j];

subject to Supply {i in ORIG}:
    sum {j in DEST} Trans[i,j] = supply[i];

subject to Demand {j in DEST}:
    sum {i in ORIG} Trans[i,j] = demand[j];
```

Figure 6-1: Transportation model with all pairs (`transp2.mod`).

6.2 Subsets and slices of ordered pairs

In many applications, we are concerned only with a subset of all ordered pairs from two sets. For example, in the transportation model, shipments may not be possible from every origin to every destination. The shipping costs per unit may be provided only for the usable origin-destination pairs, so that it is desirable to index the costs and the variables only over these pairs. In AMPL terms, we want the set `LINKS` defined above to contain just a subset of pairs that are given in the data, rather than all pairs from `ORIG` and `DEST`.

It is not sufficient to declare `set LINKS`, because that declares only a set of single members. At a minimum, we need to say

```
set LINKS dimen 2;
```

to indicate that the data must consist of members of "dimension" two — that is, pairs. Better yet, we can say that `LINKS` is a subset of the set of all pairs from `ORIG` and `DEST`:

```
set LINKS within {ORIG,DEST};
```

This has the advantage of making the model's intent clearer; it also helps catch errors in the data. The subsequent declarations of parameter `cost`, variable `Trans`, and the objective function remain the same as they are in Figure 6-1. But the components `cost[i,j]` and `Trans[i,j]` will now be defined only for those pairs given in the data as members of `LINKS`, and the expression

```
sum {(i,j) in LINKS} cost[i,j] * Trans[i,j]
```

will represent a sum over the specified pairs only.

How are the constraints written? In the original transportation model, the supply limit constraint was:

```
subject to Supply {i in ORIG}:
    sum {j in DEST} Trans[i,j] = supply[i];
```

This does not work when LINKS is a subset of pairs, because for each i in ORIG it tries to sum Trans[i,j] over every j in DEST, while Trans[i,j] is defined only for pairs (i,j) in LINKS. If we try it, we get an error message like this:

```
error processing constraint Supply['GARY']:
        invalid subscript Trans['GARY','FRA']
```

What we want to say is that for each origin i, the sum should be over all destinations j such that (i,j) is an allowed link. This statement can be transcribed directly to AMPL, by adding a condition to the indexing expression after sum:

```
subject to Supply {i in ORIG}:
    sum {j in DEST: (i,j) in LINKS} Trans[i,j] = supply[i];
```

Rather than requiring this somewhat awkward form, however, AMPL lets us drop the j in DEST from the indexing expression to produce the following more concise constraint:

```
subject to Supply {i in ORIG}:
    sum {(i,j) in LINKS} Trans[i,j] = supply[i];
```

Because {(i,j) in LINKS} appears in a context where i has already been defined, AMPL interprets this indexing expression as the set of all j such that (i,j) is in LINKS. The demand constraint is handled similarly, and the entire revised version of the model is shown in Figure 6-2a. A small representative collection of data for this model is shown in Figure 6-2b; AMPL offers a variety of convenient ways to specify the membership of compound sets and the data indexed over them, as explained in Chapter 9.

You can see from Figure 6-2a that the indexing expression

```
{(i,j) in LINKS}
```

means something different in each of the three places where it appears. Its membership can be understood in terms of a table like this:

	FRA	DET	LAN	WIN	STL	FRE	LAF
GARY		x	x		x		x
CLEV	x	x	x	x	x		x
PITT	x			x	x	x	

The rows represent origins and the columns destinations, while each pair in the set is marked by an x. A table for {ORIG,DEST} would be completely filled in with x's, while the table shown depicts {LINKS} for the "sparse" subset of pairs defined by the data in Figure 6-2b.

At a point where i and j are not currently defined, such as in the objective

```
minimize Total_Cost:
    sum {(i,j) in LINKS} cost[i,j] * Trans[i,j];
```

```
set ORIG;    # origins
set DEST;    # destinations

set LINKS within {ORIG,DEST};

param supply {ORIG} >= 0;    # amounts available at origins
param demand {DEST} >= 0;    # amounts required at destinations

    check: sum {i in ORIG} supply[i] = sum {j in DEST} demand[j];

param cost {LINKS} >= 0;     # shipment costs per unit
var Trans {LINKS} >= 0;      # units to be shipped

minimize Total_Cost:
    sum {(i,j) in LINKS} cost[i,j] * Trans[i,j];

subject to Supply {i in ORIG}:
    sum {(i,j) in LINKS} Trans[i,j] = supply[i];

subject to Demand {j in DEST}:
    sum {(i,j) in LINKS} Trans[i,j] = demand[j];
```

Figure 6-2a: Transportation model with selected pairs (`transp3.mod`).

```
param: ORIG:  supply :=
    GARY  1400     CLEV  2600     PITT  2900 ;

param: DEST:  demand :=
    FRA    900    DET  1200    LAN   600    WIN  400
    STL   1700    FRE  1100    LAF  1000 ;

param: LINKS:  cost :=
    GARY DET 14    GARY LAN 11    GARY STL 16    GARY LAF  8
    CLEV FRA 27    CLEV DET  9    CLEV LAN 12    CLEV WIN  9
    CLEV STL 26    CLEV LAF 17
    PITT FRA 24    PITT WIN 13    PITT STL 28    PITT FRE 99 ;
```

Figure 6-2b: Data for transportation model (`transp3.dat`).

the indexing expression `{(i,j) in LINKS}` represents all the pairs in this table. But at a point where `i` has already been defined, such as in the `Supply` constraint

```
    subject to Supply {i in ORIG}:
        sum {(i,j) in LINKS} Trans[i,j] = supply[i];
```

the expression `{(i,j) in LINKS}` is associated with just the row of the table corresponding to `i`. You can think of it as taking a one-dimensional "slice" through the table in the row corresponding to the already-defined first component. Although in this case the first component is a previously defined dummy index, the same convention applies when the first component is any expression that can be evaluated to a valid set object; we could write

```
    {("GARY",j) in LINKS}
```

for example, to represent the pairs in the first row of the table.

Similarly, where `j` has already been defined, such as in the `Demand` constraint

```
subject to Demand {j in DEST}:
    sum {(i,j) in LINKS} Trans[i,j] = demand[j];
```

the expression `{(i,j) in LINKS}` selects pairs from the column of the table corresponding to `j`. Pairs in the third column of the table could be specified by `{(i,"LAN") in LINKS}`.

6.3 Sets of longer tuples

AMPL's notation for ordered pairs extends in a natural way to triples, quadruples, or ordered lists of any length. All tuples in a set must have the same dimension. A set can't contain both pairs and triples, for example, nor can the determination as to whether a set contains pairs or triples be made according to some value in the data.

The multicommodity transportation model of Figure 4-1 offers some examples of how we can use ordered triples, and by extension longer tuples. In the original version of the model, the costs and amounts shipped are indexed over origin-destination-product triples:

```
param cost {ORIG,DEST,PROD} >= 0;
var Trans {ORIG,DEST,PROD} >= 0;
```

In the objective, `cost` and `Trans` are written with three subscripts, and the total cost is determined by summing over all triples:

```
minimize Total_Cost:
    sum {i in ORIG, j in DEST, p in PROD}
        cost[i,j,p] * Trans[i,j,p];
```

The indexing expressions are the same as before, except that they list three sets instead of two. An indexing expression that listed *k* sets would similarly denote a set of *k*-tuples.

If instead we define `LINKS` as we did in Figure 6-2a, the multicommodity declarations come out like this:

```
set LINKS within {ORIG,DEST};

param cost {LINKS,PROD} >= 0;
var Trans {LINKS,PROD} >= 0;

minimize Total_Cost:
    sum {(i,j) in LINKS, p in PROD} cost[i,j,p] * Trans[i,j,p];
```

Here we see how a set of triples can be specified as combinations from a set of pairs (`LINKS`) and a set of single members (`PROD`). Since `cost` and `Trans` are indexed over `{LINKS,PROD}`, their first two subscripts must come from a pair in `LINKS`, and their third subscript from a member of `PROD`. Sets of longer tuples can be built up in an analogous way.

As a final possibility, it may be that only certain combinations of origins, destinations, and products are workable. Then it makes sense to define a set that contains only the triples of allowed combinations:

```
set ROUTES within {ORIG,DEST,PROD};
```

The costs and amounts shipped are indexed over this set:

```
param cost {ROUTES} >= 0;
var Trans {ROUTES} >= 0;
```

and in the objective, the total cost is a sum over all triples in this set:

```
minimize Total_Cost:
    sum {(i,j,p) in ROUTES} cost[i,j,p] * Trans[i,j,p];
```

Individual triples are written, by analogy with pairs, as a parenthesized and comma-separated list `(i,j,p)`. Longer lists specify longer tuples.

In the three constraints of this model, the summations must be taken over three different slices through the set ROUTES:

```
subject to Supply {i in ORIG, p in PROD}:
    sum {(i,j,p) in ROUTES} Trans[i,j,p] = supply[i,p];

subject to Demand {j in DEST, p in PROD}:
    sum {(i,j,p) in ROUTES} Trans[i,j,p] = demand[j,p];

subject to Multi {i in ORIG, j in DEST}:
    sum {(i,j,p) in ROUTES} Trans[i,j,p] <= limit[i,j];
```

In the Supply constraint, for instance, indices i and p are defined before the sum, so `{(i,j,p) in ROUTES}` refers to all j such that `(i,j,p)` is a triple in ROUTES. AMPL allows comparable slices through any set of tuples, in any number of dimensions and any combination of coordinates.

When you declare a high-dimensional set such as ROUTES, a phrase like within {ORIG,DEST,PROD} may specify a set with a huge number of members. With 10 origins, 100 destinations and 100 products, for instance, this set potentially has 100,000 members. Fortunately, AMPL does not create this set when it processes the declaration, but merely checks that each tuple in the data for ROUTES has its first component in ORIG, its second in DEST, and its third in PROD. The set ROUTES can thus be handled efficiently so long as it does not itself contain a huge number of triples.

When using high-dimensional sets in other contexts, you may have to be more careful that you do not inadvertently force AMPL to generate a large set of tuples. As an example, consider how you could constrain the volume of all products shipped out of each origin to be less than some amount. You might write either

```
subject to Supply_All {i in ORIG}:
    sum {j in DEST, p in PROD: (i,j,p) in ROUTES}
        Trans[i,j,p] <= supply_all[i];
```

or, using the more compact slice notation,

```
subject to Supply_All {i in ORIG}:
   sum {(i,j,p) in ROUTES} Trans[i,j,p] <= supply_all[i];
```

In the first case, AMPL explicitly generates the set `{j in DEST, p in PROD}` and checks for membership of `(i,j,p)` in `ROUTES`, while in the second case it is able to use a more efficient approach to finding all `(i,j,p)` from `ROUTES` that have a given `i`. In our small examples this may not seem critical, but for problems of realistic size the slice version may be the only one that can be processed in a reasonable amount of time and space.

6.4 Operations on sets of tuples

Operations on compound sets are, as much as possible, the same as the operations introduced for simple sets in Chapter 5. Sets of pairs, triples, or longer tuples can be combined with `union`, `inter`, `diff`, and `symdiff`; can be tested by `in` and `within`; and can be counted with `card`. Dimensions of operands must match appropriately, so for example you may not form the union of a set of pairs with a set of triples. Also, compound sets in AMPL cannot be declared as `ordered` or `circular`, and hence also cannot be arguments to functions like `first` and `next` that take only ordered sets.

Another set operator, `cross`, gives the set of all pairs of its arguments — the cross or Cartesian product. Thus the set expression

```
ORIG cross DEST
```

represents the same set as the indexing expression `{ORIG,DEST}`, and

```
ORIG cross DEST cross PROD
```

is the same as `{ORIG,DEST,PROD}`.

Our examples so far have been constructed so that every compound set has a domain within a cross product of previously specified simple sets; `LINKS` lies within `ORIG cross DEST`, for example, and `ROUTES` within `ORIG cross DEST cross PROD`. This practice helps to produce clear and correct models. Nevertheless, if you find it inconvenient to specify the domains as part of the data, you may define them instead within the model. AMPL provides an iterated `setof` operator for this purpose, as in the following example:

```
set ROUTES dimen 3;
set PROD = setof {(i,j,p) in ROUTES} p;
set LINKS = setof {(i,j,p) in ROUTES} (i,j);
```

Like an iterated `sum` operator, `setof` is followed by an indexing expression and an argument, which can be any expression that evaluates to a legal set member. The argument is evaluated for each member of the indexing set, and the results are combined into a new set that is returned by the operator. Duplicate members are ignored. Thus these expressions for PROD and LINKS give the sets of all objects p and pairs (i,j) such that there is some member (i,j,p) in ROUTES.

As with simple sets, membership in a compound set may be restricted by a logical condition at the end of an indexing expression. For example, the multicommodity transportation model could define

```
set DEMAND = {j in DEST, p in PROD: demand[j,p] > 0};
```

so that DEMAND contains only those pairs (j,p) with positive demand for product p at destination j. As another example, suppose that we also wanted to model transfers of the products from one origin to another. We could simply define

```
set TRANSF = {ORIG,ORIG};
```

to specify the set of all pairs of members from ORIG. But this set would include pairs like ("PITT","PITT"); to specify the set of all pairs of *different* members from ORIG, a condition must be added:

```
set TRANSF = {i1 in ORIG, i2 in ORIG: i1 <> i2};
```

This is another case where two different dummy indices, i1 and i2, need to be defined to run over the same set; the condition selects those pairs where i1 is not equal to i2.

If a set is ordered, the condition within an indexing expression can also refer to the ordering. We could declare

```
set ORIG ordered;
set TRANSF = {i1 in ORIG, i2 in ORIG: ord(i1) < ord(i2)};
```

to define a "triangular" set of pairs from ORIG that does not contain any pair and its reverse. For example, TRANSF would contain either of the pairs ("PITT","CLEV") or ("CLEV","PITT"), depending on which came first in ORIG, but it would not contain both.

Sets of numbers can be treated in a similar way, since they are naturally ordered. Suppose that we want to accommodate inventories of different ages in the multiperiod production model of Figure 4-4, by declaring:

```
set PROD;      # products
param T > 0;   # number of weeks
param A > 0;   # maximum age of inventory

var Inv {PROD,0..T,0..A} >= 0;          # tons inventoried
```

Depending on how initial inventories are handled, we might have to include a constraint that no inventory in period t can be more than t weeks old:

```
subject to Too_Old
    {p in PROD, t in 1..T, a in 1..A: a > t}: Inv[p,t,a] = 0;
```

In this case, there is a simpler way to write the indexing expression:

```
subject to Too_Old
    {p in PROD, t in 1..T, a in t+1..A}: Inv[p,t,a] = 0;
```

Here the dummy index defined by `t in 1..T` is immediately used in the phrase `a in t+1..A`. In this and other cases where an indexing expression specifies two or more sets, the comma-separated phrases are evaluated from left to right. Any dummy index defined in one phrase is available for use in all subsequent phrases.

6.5 Indexed collections of sets

Although declarations of individual sets are most common in AMPL models, sets may also be declared in collections indexed over other sets. The principles are much the same as for indexed collections of parameters, variables or constraints.

As an example of how indexed collections of sets can be useful, let us extend the multiperiod production model of Figure 4-4 to recognize different market areas for each product. We begin by declaring:

```
set PROD;
set AREA {PROD};
```

This says that for each member p of PROD, there is to be a set `AREA[p]`; its members will denote the market areas in which product p is sold.

The market demands, expected sales revenues and amounts to be sold should be indexed over areas as well as products and weeks:

```
param market {p in PROD, AREA[p], 1..T} >= 0;
param revenue {p in PROD, AREA[p], 1..T} >= 0;
var Sell {p in PROD, a in AREA[p], t in 1..T}
                    >= 0, <= market[p,a,t];
```

In the declarations for `market` and `revenue`, we define only the dummy index p that is needed to specify the set `AREA[p]`, but for the `Sell` variables we need to define all the dummy indices, so that they can be used to specify the upper bound `market[p,a,t]`. This is another example in which an index defined by one phrase of an indexing expression is used by a subsequent phrase; for each p from the set PROD, a runs over a different set `AREA[p]`.

In the objective, the expression `revenue[p,t] * Sell[p,t]` from Figure 4-4 must be replaced by a sum of revenues over all areas for product p:

```
maximize Total_Profit:
   sum {p in PROD, t in 1..T}
      (sum {a in AREA[p]} revenue[p,a,t]*Sell[p,a,t] -
         prodcost[p]*Make[p,t] - invcost[p]*Inv[p,t]);
```

The only other change is in the `Balance` constraints, where `Sell[p,t]` is similarly replaced by a summation:

```
subject to Balance {p in PROD, t in 1..T}:
   Make[p,t] + Inv[p,t-1]
      = sum {a in AREA[p]} Sell[p,a,t] + Inv[p,t];
```

```
set PROD;           # products
set AREA {PROD};    # market areas for each product
param T > 0;        # number of weeks

param rate {PROD} > 0;              # tons per hour produced
param inv0 {PROD} >= 0;             # initial inventory
param avail {1..T} >= 0;            # hours available in week
param market {p in PROD, AREA[p], 1..T} >= 0;
                                    # limit on tons sold in week

param prodcost {PROD} >= 0;        # cost per ton produced
param invcost {PROD} >= 0;         # carrying cost/ton of inventory
param revenue {p in PROD, AREA[p], 1..T} >= 0;
                                    # revenue per ton sold

var Make {PROD,1..T} >= 0;          # tons produced
var Inv {PROD,0..T} >= 0;           # tons inventoried
var Sell {p in PROD, a in AREA[p], t in 1..T}    # tons sold
                    >= 0, <= market[p,a,t];
maximize Total_Profit:
   sum {p in PROD, t in 1..T}
      (sum {a in AREA[p]} revenue[p,a,t]*Sell[p,a,t] -
         prodcost[p]*Make[p,t] - invcost[p]*Inv[p,t]);

            # Total revenue less costs for all products in all weeks
subject to Time {t in 1..T}:
   sum {p in PROD} (1/rate[p]) * Make[p,t] <= avail[t];

            # Total of hours used by all products
            # may not exceed hours available, in each week
subject to Init_Inv {p in PROD}:  Inv[p,0] = inv0[p];

            # Initial inventory must equal given value
subject to Balance {p in PROD, t in 1..T}:
   Make[p,t] + Inv[p,t-1]
      = sum {a in AREA[p]} Sell[p,a,t] + Inv[p,t];

            # Tons produced and taken from inventory
            # must equal tons sold and put into inventory
```

Figure 6-3: Multiperiod production with indexed sets (steelT3.mod).

The complete model is shown in Figure 6-3.

In the data for this model, each set within the indexed collection AREA is specified like an ordinary set:

```
      set PROD := bands coils;
      set AREA[bands] := east north ;
      set AREA[coils] := east west export ;
```

The parameters revenue and market are now indexed over three sets, so their data values are specified in a series of tables. Since the indexing is over a different set

```
param T := 4;

set PROD := bands coils;
set AREA[bands]  := east north ;
set AREA[coils]  := east west export ;

param avail :=  1 40  2 40  3 32  4 40 ;

param rate :=  bands  200  coils  140 ;
param inv0 :=  bands   10  coils    0 ;

param prodcost := bands 10    coils 11 ;
param invcost  := bands  2.5  coils  3 ;

param revenue :=

   [bands,*,*]:    1       2       3       4 :=
      east      25.0    26.0    27.0    27.0
      north     26.5    27.5    28.0    28.5

   [coils,*,*]:    1       2       3       4 :=
      east        30      35      37      39
      west        29      32      33      35
      export      25      25      25      28 ;

param market :=

   [bands,*,*]:    1       2       3       4 :=
      east      2000    2000    1500    2000
      north     4000    4000    2500    4500

   [coils,*,*]:    1       2       3       4 :=
      east      1000     800    1000    1100
      west      2000    1200    2000    2300
      export    1000     500     500     800 ;
```

Figure 6-4: Data for multiperiod production with indexed sets (steelT3.dat).

AREA[p] for each product p, the values are most conveniently arranged as one table for each product, as shown in Figure 6-4. (Chapter 9 explains the general rules behind this arrangement.)

We could instead have written this model with a set PRODAREA of pairs, such that product p will be sold in area a if and only if (p,a) is a member of PRODAREA. Our formulation in terms of PROD and AREA[p] seems preferable, however, because it emphasizes the hierarchical relationship between products and areas. Although the model must refer in many places to the set of all areas selling one product, it never refers to the set of all products sold in one area.

As a contrasting example, we can consider how the multicommodity transportation model might use indexed collections of sets. As shown in Figure 6-5, for each product we define a set of origins where that product is supplied, a set of destinations where the product is demanded, and a set of links that represent possible shipments of the product:

```
set ORIG;    # origins
set DEST;    # destinations
set PROD;    # products

set orig {PROD} within ORIG;
set dest {PROD} within DEST;
set links {p in PROD} = orig[p] cross dest[p];

param supply {p in PROD, orig[p]} >= 0; # available at origins
param demand {p in PROD, dest[p]} >= 0; # required at destinations
   check {p in PROD}: sum {i in orig[p]} supply[p,i]
                            = sum {j in dest[p]} demand[p,j];

param limit {ORIG,DEST} >= 0;

param cost {p in PROD, links[p]} >= 0;  # shipment costs per unit
var Trans {p in PROD, links[p]} >= 0;   # units to be shipped

minimize Total_Cost:
   sum {p in PROD, (i,j) in links[p]} cost[p,i,j] * Trans[p,i,j];

subject to Supply {p in PROD, i in orig[p]}:
   sum {j in dest[p]} Trans[p,i,j] = supply[p,i];

subject to Demand {p in PROD, j in dest[p]}:
   sum {i in orig[p]} Trans[p,i,j] = demand[p,j];

subject to Multi {i in ORIG, j in DEST}:
   sum {p in PROD: (i,j) in links[p]} Trans[p,i,j] <= limit[i,j];
```

Figure 6-5: Multicommodity transportation with indexed sets (multic.mod).

```
set orig {PROD} within ORIG;
set dest {PROD} within DEST;
set links {p in PROD} = orig[p] cross dest[p];
```

The declaration of links demonstrates that it is possible to have an indexed collection of compound sets, and that an indexed collection may be defined through set operations from other indexed collections. In addition to the operations previously mentioned, there are iterated union and intersection operators that apply to sets in the same way that an iterated sum applies to numbers. For example, the expressions

```
union {p in PROD} orig[p]
inter {p in PROD} orig[p]
```

represent the subset of origins that supply at least one product, and the subset of origins that supply all products.

The hierarchical relationship based on products that was observed in Figure 6-3 is seen in most of Figure 6-5 as well. The model repeatedly deals with the sets of all origins, destinations, and links associated with a particular product. The only exception comes in the last constraint, where the summation must be over all products shipped via a particular link:

```
subject to Multi {i in ORIG, j in DEST}:
  sum {p in PROD: (i,j) in links[p]} Trans[p,i,j] <= limit[i,j];
```

Here it is necessary, following sum, to use a somewhat awkward indexing expression to describe a set that does not match the hierarchical organization.

In general, almost any model that can be written with indexed collections of sets can also be written with sets of tuples. As our examples suggest, indexed collections are most suitable for entities such as products and areas that have a hierarchical relationship. Sets of tuples are preferable, on the other hand, in dealing with entities like origins and destinations that are related symmetrically.

Exercises

6-1. Return to the production and transportation model of Figures 4-6 and 4-7. Using the display command, together with indexing expressions as demonstrated in Section 6.4, you can determine the membership of a variety of compound sets; for example, you can use

```
ampl: display {j in DEST, p in PROD: demand[j,p] > 500};
set {j in DEST, p in PROD: demand[j,p] > 500}  :=
(DET,coils)    (STL,bands)    (STL,coils)    (FRE,coils);
```

to show the set of all combinations of products and destinations where the demand is greater than 500.

(a) Use display to determine the membership of the following sets, which depend only on the data:

– All combinations of origins and products for which the production rate is greater than 150 tons per hour.

– All combinations of origins, destinations and products for which there is a shipping cost of $\leq$ $10 per ton.

– All combinations of origins and destinations for which the shipping cost of coils is $\leq$ $10 per ton.

– All combinations of origins and products for which the production cost per hour is less than $30,000.

– All combinations of origins, destinations and products for which the transportation cost is more than 15% of the production cost.

– All combinations of origins, destinations and products for which the transportation cost is more than 15% but less than 25% of the production cost.

(b) Use display to determine the membership of the following sets, which depend on the optimal solution as well as on the data:

– All combinations of origins and products for which there is production of at least 1000 tons.

– All combinations of origins, destinations and products for which there is a nonzero amount shipped.

– All combinations of origins and products for which more than 10 hours are used in production.

– All combinations of origins and products such that the product accounts for more than 25% of the hours available at the origin.

– All combinations of origins and products such that the total amount of the product shipped
from the origin is at least 1000 tons.

6-2. This exercise resembles the previous one, but asks about the ordered-pair version of the
transportation model in Figure 6-2.

(a) Use `display` and indexing expressions to determine the membership of the following sets:

– Origin-destination links that have a transportation cost less than $10 per ton.

– Destinations that can be served by `GARY`.

– Origins that can serve `FRE`.

– Links that are used for transportation in the optimal solution.

– Links that are used for transportation from `CLEV` in the optimal solution.

– Destinations to which the total cost of shipping, from all origins, exceeds $20,000.

(b) Use the `display` command and the `setof` operator to determine the membership of the following sets:

– Destinations that have a shipping cost of more than 20 from any origin.

– All destination-origin pairs `(j,i)` such that the link from `i` to `j` is used in the optimal solution.

6-3. Use `display` and appropriate set expressions to determine the membership of the following
sets from the multiperiod production model of Figures 6-3 and 6-4:

– All market areas served with any of the products.

– All combinations of products, areas and weeks such that the amount actually sold in the optimal
solution equals the maximum that can be sold.

– All combinations of products and weeks such that the total sold in all areas is greater than or
equal to 6000 tons.

6-4. To try the following experiment, first enter these declarations:

```
ampl: set Q = {1..10,1..10,1..10,1..10,1..10,1..10};
ampl: set S within Q;
ampl: data;
ampl: set S := 1 2 3 3 4 5  2 3 4 4 5 6  3 4 5 5 6 7  4 5 6 7 8 9 ;
```

(a) Now try the following two commands:

```
display S;
display {(a,b,c,d,e,f) in Q: (a,b,c,d,e,f) in S};
```

The two expressions in these commands represent the same set, but do you get the same speed of
response from AMPL? Explain the cause of the difference.

(b) Predict the result of the command `display Q`.

6-5. This exercise asks you to reformulate the diet model of Figure 2-1 in a variety of ways, using
compound sets.

(a) Reformulate the diet model so that it uses a declaration

```
set GIVE within {NUTR,FOOD};
```

to define a subset of pairs `(i,j)` such that nutrient `i` can be found in food `j`.

(b) Reformulate the diet model so that it uses a declaration

```
set FN {NUTR} within FOOD;
```

to define, for each nutrient `i`, the set `FN[i]` of all foods that can supply that nutrient.

(c) Reformulate the diet model so that it uses a declaration

```
set NF {FOOD} within NUTR;
```

to define, for each food `j`, the set `NF[j]` of all nutrients supplied by that food. Explain why you find this formulation more or less natural and convenient than the one in (b).

6-6. Re-read the suggestions in Section 6.3, and complete the following reformulations of the multicommodity transportation model:

(a) Use a subset `LINKS` of origin-destination pairs.

(b) Use a subset `ROUTES` of origin-destination-product triples.

(c) Use a subset `MARKETS` of destination-product pairs, with the property that product `p` can be sold at destination `j` if and only if `(j,p)` is in the subset.

6-7. Carry through the following two suggestions from Section 6.4 for enhancements to the multi-commodity transportation problem of Figure 4-1.

(a) Add a declaration

```
set DEMAND = {j in DEST, p in PROD: demand[j,p] > 0};
```

and index the variables over `{ORIG,DEMAND}`, so that variables are defined only where they might be needed to meet demand. Make all of the necessary changes in the rest of the model to use this set.

(b) Add the declarations

```
set LINKS within {ORIG,DEST};
set TRANSF = {i1 in ORIG, i2 in ORIG: i1 <> i2};
```

Define variables over `LINKS` to represent shipments to destinations, and over `TRANSF` to represent shipments between origins. The constraint at each origin now must say that total shipments out — to other origins as well as to destinations — must equal supply plus shipments in from other origins. Complete the formulation for this case.

6-8. Reformulate the model from Exercise 3-3(b) so that it uses a set `LINK1` of allowable plant-mill shipment pairs, and a set `LINK2` of allowable mill-factory shipment pairs.

6-9. As chairman of the program committee for a prestigious scientific conference, you must assign submitted papers to volunteer referees. To do so in the most effective way, you can formulate an LP model along the lines of the assignment model discussed in Chapter 3, but with a few extra twists.

After looking through the papers and the list of referees, you can compile the following data:

```
set Papers;
set Referees;
set Categories;

set PaperKind within {Papers,Categories};
set Willing within {Referees,Categories};
```

The contents of the first two sets are self-evident, while the third set contains subject categories into which papers may be classified. The set `PaperKind` contains a pair `(p,c)` if paper `p` falls into category c; in general, a paper can fit into several categories. The set `Willing` contains a pair (r,c) if referee r is willing to handle papers in category c.

(a) What is the dimension of the set

```
{(r,c) in Willing, (p,c) in PaperKind}
```

and what is the significance of the tuples contained in this set?

(b) Based on your answer to (a), explain why the declaration

```
set CanHandle = setof {(r,c) in Willing, (p,c) in PaperKind} (r,p);
```

gives the set of pairs (r,p) such that referee r can be assigned paper p.

Your model could use parameters `ppref` and variables `Review` indexed over `CanHandle`; `ppref[r,p]` would be the preference of referee r for paper p, and `Review[r,p]` would be 1 if referee r were assigned paper p, or 0 otherwise. Assuming higher preferences are better, write out the declarations for these components and for an objective function to maximize the sum of preferences of all assignments.

(c) Unfortunately, you don't have the referees' preferences for individual papers, since they haven't seen any papers yet. What you have are their preferences for different categories:

```
param cpref {Willing} integer >= 0, <= 5;
```

Explain why it would make sense to replace `ppref[r,p]` in your objective by

```
max {(r,c) in Willing: (p,c) in PaperKind} cpref[r,c]
```

(d) Finally, you must define the following parameters that indicate how much work is to be done:

```
param nreferees integer > 0;        # referees needed per paper
param minwork integer > 0;          # min papers to each referee
param maxwork integer > minwork;    # max papers to each referee
```

Formulate the appropriate assignment constraints. Complete the model, by formulating constraints that each paper must have the required number of referees, and that each referee must be assigned an acceptable number of papers.

7

Parameters and Expressions

A large optimization model invariably uses many numerical values. As we have explained before, only a concise symbolic description of these values need appear in an AMPL model, while the explicit data values are given in separate data statements, to be described in Chapter 9.

In AMPL a single named numerical value is called a *parameter*. Although some parameters are defined as individual scalar values, most occur in vectors or matrices or other collections of numerical values indexed over sets. We will thus loosely refer to an indexed collection of parameters as ''a parameter'' when the meaning is clear. To begin this chapter, Section 7.1 describes the rules for declaring parameters and for referring to them in an AMPL model.

Parameters and other numerical values are the building blocks of the expressions that make up a model's objective and constraints. Sections 7.2 and 7.3 describe arithmetic expressions, which have a numerical value, and logical expressions, which evaluate to true or false. Along with the standard unary and binary operators of conventional algebraic notation, AMPL provides iterated operators like sum and prod, and a conditional (if-then-else) operator that chooses between two expressions, depending on the truth of a third expression.

The expressions in objectives and constraints necessarily involve variables, whose declaration and use will be discussed in Chapter 8. There are several common uses for expressions that involve only sets and parameters, however. Section 7.4 describes how logical expressions are used to test the validity of data, either directly in a parameter declaration, or separately in a check statement. Section 7.5 introduces features for defining new parameters through arithmetic expressions in previously declared parameters and sets, and 7.6 describes randomly-generated parameters.

Although the key purpose of parameters is to represent numerical values, they can also represent logical values or arbitrary strings. These possibilities are covered in Sections 7.7 and 7.8, respectively. AMPL provides a range of operators for strings, but as they are most often used in AMPL commands and programming rather than in models, we defer their introduction to Section 13.7.

7.1 Parameter declarations

A parameter declaration describes certain data required by a model, and indicates how the model will refer to data values in subsequent expressions.

The simplest parameter declaration consists of the keyword `param` and a name:

```
param T;
```

At any point after this declaration, `T` can be used to refer to a numerical value.

More often, the name in a parameter declaration is followed by an indexing expression:

```
param avail {1..T};
param demand {DEST,PROD};
param revenue {p in PROD, AREA[p], 1..T};
```

One parameter is defined for each member of the set specified by the indexing expression. Thus a parameter is uniquely determined by its name and its associated set member; throughout the rest of the model, you would refer to this parameter by writing the name and bracketed ''subscripts'':

```
avail[i]
demand[j,p]
revenue[p,a,t]
```

If the indexing is over a simple set of objects as described in Chapter 5, there is one subscript. If the indexing is over a set of pairs, triples, or longer tuples as described in Chapter 6, there must be a corresponding pair, triple, or longer list of subscripts separated by commas. The subscripts can be any expressions, so long as they evaluate to members of the underlying index set.

An unindexed parameter is a scalar value, but a parameter indexed over a simple set has the characteristics of a vector or an array; when the indexing is over a sequence of integers, say

```
param avail {1..T};
```

the individual subscripted parameters are `avail[1]`, `avail[2]`, ..., `avail[T]`, and there is an obvious analogy to the vectors of linear algebra or the arrays of a programming language like Fortran or C. AMPL's concept of a vector is more general, however, since parameters may also be indexed over sets of strings, which need not even be ordered. Indexing over sets of strings is best suited for parameters that correspond to places, products and other entities for which no numbering is especially natural. Indexing over sequences of numbers is more appropriate for parameters that correspond to weeks, stages, and the like, which by their nature tend to be ordered and numbered; even for these, you may prefer to use ordered sets of strings as described in Section 5.6.

A parameter indexed over a set of pairs is like a two-dimensional array or matrix. If the indexing is over all pairs from two sets, as in

```
set ORIG;
set DEST;
param cost {ORIG,DEST};
```

then there is a parameter cost[i,j] for every combination of i from ORIG and j from DEST, and the analogy to a matrix is strongest — although again the subscripts are more likely to be strings than numbers. If the indexing is over a subset of pairs, however:

```
set ORIG;
set DEST;
set LINKS within {ORIG,DEST};
param cost {LINKS};
```

then cost[i,j] exists only for those i from ORIG and j from DEST such that (i,j) is a member of LINKS. In this case, you can think of cost as being a ''sparse'' matrix.

Similar comments apply to parameters indexed over triples and longer tuples, which resemble arrays of higher dimension in programming languages.

7.2 Arithmetic expressions

Arithmetic expressions in AMPL are much the same as in other computer languages. Literal numbers consist of an optional sign preceding a sequence of digits, which may or may not include a decimal point (for example, -17 or 2.71828 or +.3). At the end of a literal there may also be an exponent, consisting of the letter d, D, e, or E and an optional sign followed by digits (1e30 or 7.66439D-07).

Literals, parameters, and variables are combined into expressions by the standard operations of addition (+), subtraction (-), multiplication (*), division (/), and exponentiation (^). The familiar conventions of arithmetic apply. Exponentiation has higher precedence than multiplication and division, which have higher precedence than addition and subtraction; successive operations of the same precedence group to the left, except for exponentiation, which groups to the right. Parentheses may be used to change the order of evaluation.

Arithmetic expressions may also use the div operator, which returns the truncated quotient when its left operand is divided by its right operand; the mod operator, which computes the remainder; and the less operator, which returns its left operand minus its right operand if the result is positive, or zero otherwise. For purposes of precedence and grouping, AMPL treats div and mod like division, and less like subtraction.

A list of arithmetic operators (and logical operators, to be described shortly) is given in Table 7-1. As much as possible, AMPL follows common programming languages in its choice of operator symbols, such as * for multiplication and / for division. There is sometimes more than one standard, however, as with exponentiation, where some languages use ^ while others use **. In this and other cases, AMPL provides alternate forms. Table 7-1 shows the more common forms to the left, and the alternatives (if any)

Usual style	alternative style	type of operands	type of result
if-then-else		logical, arithmetic	arithmetic
or	||	logical	logical
exists forall		logical	logical
and	&&	logical	logical
not (unary)	!	logical	logical
< <= = <> > >=	< <= == != > >=	arithmetic	logical
in not in		object, set	logical
+ - less		arithmetic	arithmetic
sum prod min max		arithmetic	arithmetic
* / div mod		arithmetic	arithmetic
+ - (unary)		arithmetic	arithmetic
^	**	arithmetic	arithmetic

Exponentiation and if-then-else are right-associative; the other operators are left-associative. The logical operand of if-then-else appears after if, and the arithmetic operands after then and (optionally) else.

Table 7-1: Arithmetic and logical operators, in increasing precedence.

to the right; you can mix them as you like, but your models will be easier to read and understand if you are consistent in your choices.

Another way to build arithmetic expressions is by applying functions to other expressions. A function reference consists of a name followed by a parenthesized argument or comma-separated list of arguments; an arithmetic argument can be any arithmetic expression. Here are a few examples, which compute the minimum, absolute value, and square root of their arguments, respectively:

```
min(T,20)
abs(sum {i in ORIG} supply[i] - sum {j in DEST} demand[j])
sqrt((tan[j]-tan[k])^2)
```

Table 7-2 lists the built-in arithmetic functions that are typically found in models. Except for min and max, the names of any of these functions may be redefined, but their original meanings will become inaccessible. For example, a model may declare a parameter named tan as in the last example above, but then it cannot also refer to the function tan.

The set functions card and ord, which were described in Chapter 5, also produce an arithmetic result. In addition, AMPL provides several ''rounding'' functions (Section 11.3) and a variety of random-number functions (Section 7.6 below). A mechanism for ''importing'' functions defined by your own programs is described in Appendix A.22.

abs (x)	absolute value, $\lvert x \rvert$
acos (x)	inverse cosine, $\cos^{-1}(x)$
acosh (x)	inverse hyperbolic cosine, $\cosh^{-1}(x)$
asin (x)	inverse sine, $\sin^{-1}(x)$
asinh (x)	inverse hyperbolic sine, $\sinh^{-1}(x)$
atan (x)	inverse tangent, $\tan^{-1}(x)$
atan2 (y, x)	inverse tangent, $\tan^{-1}(y/x)$
atanh (x)	inverse hyperbolic tangent, $\tanh^{-1}(x)$
cos (x)	cosine
cosh (x)	hyperbolic cosine
exp (x)	exponential, e^x
log (x)	natural logarithm, $\log_e(x)$
log10 (x)	common logarithm, $\log_{10}(x)$
max $(x, y, \ldots)$	maximum (2 or more arguments)
min $(x, y, \ldots)$	minimum (2 or more arguments)
sin (x)	sine
sinh (x)	hyperbolic sine
sqrt (x)	square root
tan (x)	tangent
tanh (x)	hyperbolic tangent

Table 7-2: Built-in arithmetic functions for use in models.

Finally, the indexed operators such as Σ and Π from algebraic notation are generalized in AMPL by expressions for iterating operations over sets. In particular, most large-scale linear programming models contain iterated summations:

```
sum {i in ORIG} supply[i]
```

The keyword sum may be followed by any indexing expression. The subsequent arithmetic expression is evaluated once for each member of the index set, and all the resulting values are added. Thus the sum above, from the transportation model of Figure 3-1a, represents the total supply available, at all origins. The sum operator has lower precedence than *, so the objective of the same model can be written

```
sum {i in ORIG, j in DEST} cost[i,j] * Trans[i,j]
```

to represent the total of cost[i,j] * Trans[i,j] over all combinations of origins and destinations. The precedence of sum is higher than that of + or –, however, so for the objective of the multiperiod production model in Figure 6-3 we must write

```
sum {p in PROD, t in 1..T}
   (sum {a in AREA[p]} revenue[p,a,t]*Sell[p,a,t] -
      prodcost[p]*Make[p,t] - invcost[p]*Inv[p,t]);
```

The outer sum applies to the entire parenthesized expression following it, while the inner sum applies only to the term revenue[p,a,t] * Sell[p,a,t].

Other iterated arithmetic operators are `prod` for multiplication, `min` for minimum, and `max` for maximum. As an example, we could use

```
max {i in ORIG} supply[i]
```

to describe the greatest supply available at any origin.

Bear in mind that, while an AMPL arithmetic function or operator may be applied to variables as well as to parameters or to numeric members of sets, most operations on variables are not linear. AMPL's requirements for arithmetic expressions in a linear program are described in Section 8.2. Some of the nonlinear functions of variables that can be handled by certain solvers are discussed in Chapter 18.

7.3 Logical and conditional expressions

The values of arithmetic expressions can be tested against each other by comparison operators:

=	equal to
<>	not equal to
<	less than
<=	less than or equal to
>	greater than
>=	greater than or equal to

The result of a comparison is either "true" or "false". Thus $T > 1$ is true if the parameter T has a value greater than 1, and is false otherwise; and

```
sum {i in ORIG} supply[i] = sum {j in DEST} demand[j]
```

is true if and only if total supply equals total demand.

Comparisons are one example of AMPL's logical expressions, which evaluate to true or false. Set membership tests using `in` and `within`, described in Section 5.4, are another example. More complex logical expressions can be built up with logical operators. The `and` operator returns true if and only if both its operands are true, while `or` returns true if and only if at least one of its operands is true; the unary operator `not` returns false for true and true for false. Thus the expression

```
T >= 0 and T <= 10
```

is only true if T lies in the interval $[0, 10]$, while the following from Section 5.5,

```
i in MAXREQ or n_min[i] > 0
```

is true if i is a member of MAXREQ, or `n_min[i]` is positive, or both. Where several operators are used together, any comparison, membership or arithmetic operator has higher precedence than the logical operators; `and` has higher precedence than `or`, while `not` has higher precedence than either. Thus the expression

```
not i in MAXREQ or n_min[i] > 0 and n_min[i] <= 10
```

is interpreted as

```
(not (i in MAXREQ)) or ((n_min[i] > 0) and (n_min[i] <= 10))
```

Alternatively, the `not in` operator could be used:

```
i not in MAXREQ or n_min[i] > 0 and n_min[i] <= 10
```

The precedences are summarized in Table 7-1, which also gives alternative forms.

Like + and *, the operators `or` and `and` have iterated versions. The iterated `or` is denoted by `exists`, and the iterated `and` by `forall`. For example, the expression

```
exists {i in ORIG} demand[i] > 10
```

is true if and only if at least one origin has a demand greater than 10, while

```
forall {i in ORIG} demand[i] > 10
```

is true if and only if every origin has demand greater than 10.

Another use for a logical expression is as an operand to the conditional or `if-then-else` operator, which returns one of two different arithmetic values depending on whether the logical expression is true or false. Consider the two collections of inventory balance constraints in the multiperiod production model of Figure 5-3:

```
subject to Balance0 {p in PROD}:
   Make[p,first(WEEKS)] + inv0[p]
      = Sell[p,first(WEEKS)] + Inv[p,first(WEEKS)];
subject to Balance {p in PROD, t in WEEKS: ord(t) > 1}:
   Make[p,t] + Inv[p,prev(t)] = Sell[p,t] + Inv[p,t];
```

The `Balance0` constraints are basically the `Balance` constraints with `t` set to `first(WEEKS)`. The only difference is in the second term, which represents the previous week's inventory; it is given as `inv0[p]` for the first week (in the `Balance0` constraints) but is represented by the variable `Inv[p,prev(t)]` for subsequent weeks (in the `Balance` constraints). We would like to combine these constraints into one declaration, by having a term that takes the value `inv0[p]` when `t` is the first week, and takes the value `Inv[p,prev(t)]` otherwise. Such a term is written in AMPL as:

```
if t = first(WEEKS) then inv0[p] else Inv[p,prev(t)]
```

Placing this expression into the constraint declaration, we can write

```
subject to Balance {p in PROD, t in WEEKS}:
   Make[p,t] +
      (if t = first(WEEKS) then inv0[p] else Inv[p,prev(t)])
         = Sell[p,t] + Inv[p,t];
```

This form communicates the inventory balance constraints more concisely and directly than two separate declarations.

The general form of a conditional expression is

```
if a then b else c
```

where *a* is a logical expression. If *a* evaluates to true, the conditional expression takes the value of *b*; if *a* is false, the expression takes the value of *c*. If *c* is zero, the `else` *c* part can be dropped. Most often *b* and *c* are arithmetic expressions, but they can also be string or set expressions, so long as both are expressions of the same kind. Because `then` and `else` have lower precedence than any other operators, a conditional expression needs to be parenthesized (as in the example above) unless it occurs at the end of a statement.

AMPL also has an `if-then-else` for use in programming; like the conditional statements in many programming languages, it executes one or another block of statements depending on the truth of some logical expression. We describe it with other AMPL programming features in Chapter 13. The `if-then-else` that we have described here is not a statement, but rather an expression whose value is conditionally determined. It therefore belongs inside a declaration, in a place where an expression would normally be evaluated.

7.4 Restrictions on parameters

If `T` is intended to represent the number of weeks in a multiperiod model, it should be an integer and greater than 1. By including these conditions in `T`'s declaration,

```
param T > 1 integer;
```

you instruct AMPL to reject your data if you inadvertently set `T` to 1:

```
error processing param T:
        failed check: param T = 1
                is not > 1;
```

or to 2.5:

```
error processing param T:
        failed check: param T = 2.5
                is not an integer;
```

AMPL will not send your problem instance to a solver as long as any errors of this kind remain.

In the declaration of an indexed collection of parameters, a simple restriction such as `integer` or `>= 0` applies to every parameter defined. Our examples often use this option to specify that vectors and arrays are nonnegative:

```
param demand {DEST,PROD} >= 0;
```

If you include dummy indices in the indexing expression, however, you can use them to specify a different restriction for each parameter:

```
param f_min {FOOD} >= 0;
param f_max {j in FOOD} >= f_min[j];
```

The effect of these declarations is to define a pair of parameters f_max[j] >= f_min[j] for every j in the set FOOD.

A restriction phrase for a parameter declaration may be the word integer or binary or a comparison operator followed by an arithmetic expression. While integer restricts a parameter to integral (whole-number) values, binary restricts it to zero or one. The arithmetic expression may refer to sets and parameters previously defined in the model, and to dummy indices defined by the current declaration. There may be several restriction phrases in the same declaration, in which case they may optionally be separated by commas.

In special circumstances, a restriction phrase may even refer to the parameter in whose declaration it appears. Some multiperiod production models, for example, are defined in terms of a parameter cumulative_market[p,t] that represents the cumulative demand for product p in weeks 1 through t. Since cumulative demand does not decrease, you might try to write a restriction phrase like this:

```
param cumulative_market {p in PROD, t in 1..T}
   >= cumulative_market[p,t-1];   # ERROR
```

For the parameters cumulative_market[p,1], however, the restriction phrase will refer to cumulative_market[p,0], which is undefined; AMPL will reject the declaration with an error message. What you need here again is a conditional expression that handles the first period specially:

```
param cumulative_market {p in PROD, t in 1..T}
   >= if t = 1 then 0 else cumulative_market[p,t-1];
```

The same thing could be written a little more compactly as

```
param cumulative_market {p in PROD, t in 1..T}
   >= if t > 1 then cumulative_market[p,t-1];
```

since "else 0" is assumed. Almost always, some form of if-then-else expression is needed to make this kind of self-reference possible.

As you might suspect from this last example, sometimes it is desirable to place a more complex restriction on the model's data than can be expressed by a restriction phrase within a declaration. This is the purpose of the check statement. For example, in the transportation model of Figure 3-1a, total supply must equal total demand:

```
check: sum {i in ORIG} supply[i] = sum {j in DEST} demand[j];
```

The multicommodity version, in Figure 4-1, uses an indexed check to say that total supply must equal total demand for each product:

```
check {p in PROD}:
   sum {i in ORIG} supply[i,p] = sum {j in DEST} demand[j,p];
```

Here the restriction is tested once for each member p of PROD. If the check fails for any member, AMPL prints an error message and rejects all of the data.

You can think of the check statement as specifying a kind of constraint, but only on the data. The restriction clause is a logical expression, which may use any previously defined sets and parameters as well as dummy indices defined in the statement's indexing expression. After the data values have been read, the logical expression must evaluate to true; if an indexing expression has been specified, the logical expression is evaluated separately for each assignment of set members to the dummy indices, and must be true for each.

We strongly recommend the use of restriction phrases and `check` statements to validate a model's data. These features will help you to catch data errors at an early stage, when they are easy to fix. Data errors not caught will, at best, cause errors in the generation of the variables and constraints, so that you will get some kind of error message from AMPL. In other cases, data errors lead to the generation of an incorrect linear program. If you are fortunate, the incorrect LP will have a meaningless optimal solution, so that — possibly after a good deal of effort — you will be able to work backward to find the error in the data. At worst, the incorrect LP will have a plausible solution, and the error will go undetected.

7.5 Computed parameters

It is seldom possible to arrange that the data values available to a model are precisely the coefficient values required by the objective and constraints. Even in the simple production model of Figure 1-4, for example, we wrote the constraint as

```
sum {p in PROD} (1/rate[p]) * Make[p] <= avail;
```

because production rates were given in tons per hour, while the coefficient of `Make[p]` had to be in hours per ton. Any parameter expression may be used in the constraints and objective, but the expressions are best kept simple. When more complex expressions are needed, the model is usually easier to understand if new, computed parameters are defined in terms of the data parameters.

The declaration of a computed parameter has an assignment phrase, which resembles the restriction phrase described in the previous section except for the use of an = operator to indicate that the parameter is being set equal to a certain expression, rather than merely being restricted by an inequality. As a first example, suppose that the data values provided to the multicommodity transportation model of Figure 4-1 consist of the total demand for each product, together with each destination's share of demand. The destinations' shares are percentages between zero and 100, but their sum over all destinations might not exactly equal 100%, because of rounding and approximation. Thus we declare data parameters to represent the shares, and a computed parameter equal to their sum:

```
param share {DEST} >= 0, <= 100;
param tot_sh = sum {j in DEST} share[j];
```

We can then declare a data parameter to represent total demands, and a computed parameter that equals demand at each destination:

```
param tot_dem {PROD} >= 0;
param demand {j in DEST, p in PROD}
        = share[j] * tot_dem[p] / tot_sh;
```

The division by `tot_sh` acts as a correction factor for a sum not equal to 100%. Once demand has been defined in this way, the model can use it as in Figure 4-1:

```
subject to Demand {j in DEST, p in PROD}:
   sum {i in ORIG} Trans[i,j,p] = demand[j,p];
```

We could avoid computed parameters by substituting the formulas for `tot_sh` and `demand[j,p]` directly into this constraint:

```
subject to Demand {j in DEST, p in PROD}:
   sum {i in ORIG} Trans[i,j,p]
        = share[j] * tot_dem[p] / sum {k in DEST} share[k];
```

This alternative makes the model a little shorter, but the computation of the demand and the structure of the constraint are both harder to follow.

As another example, consider a scenario for the multiperiod production model (Figure 4-4) in which minimum inventories are computed. Specifically, suppose that the inventory of product p for week t must be at least a certain fraction of `market[p,t+1]`, the maximum that can be sold in the following week. We thus use the following declarations for the data to be supplied:

```
param frac > 0;
param market {PROD,1..T+1} >= 0;
```

and then declare

```
param mininv {p in PROD, t in 0..T} = frac * market[p,t+1];
var Inv {p in PROD, t in 0..T} >= mininv[p,t];
```

to define and use parameters `mininv[p,t]` that represent the minimum inventory of product p for week t. AMPL keeps all = definitions of parameters up to date throughout a session. Thus for example if you change the value of `frac` the values of all the `mininv` parameters automatically change accordingly.

If you define a computed parameter as in the examples above, then you cannot also specify a data value for it. An attempt to do so will result in an error message:

```
mininv was defined in the model
context:  param  >>> mininv <<<  :=  bands 2   3000
```

However, there is an alternative way in which you can define an initial value for a parameter but allow it to be changed later.

If you define a parameter using the `default` operator in place of =, then the parameter is initialized rather than defined. Its value is taken from the value of the expression to the right of the `default` operator, but does not change if the expression's value later changes. Initial values can be overridden by data statements, and they also may be changed by subsequent assignment statements. This feature is most useful for writing AMPL scripts that update certain values repeatedly, as shown in Section 13.2.

If you define a parameter using the operator `default` in place of `=`, then you can specify values in data statements to override the ones that would otherwise be computed. For instance, by declaring

```
param mininv {p in PROD, t in 0..T}
   default frac * market[p,t+1];
```

you can allow a few exceptional minimum inventories to be specified as part of the data for the model, either in a list:

```
param mininv :=
        bands 2   3000
        coils 2   2000
        coils 3   2000 ;
```

or in a table:

```
param market:     1      2      3      4 :=
        bands       .   3000      .      .
        coils       .   2000   2000      . ;
```

(AMPL uses ''`.`'' in a data statement to indicate an omitted entry, as explained in Chapter 9 and A.12.2.)

The expression that gives the default value of a parameter is evaluated only when the parameter's value is first needed, such as when an objective or constraint that uses the parameter is processed by a `solve` command.

In most `=` and `default` phrases, the operator is followed by an arithmetic expression in previously defined sets and parameters (but not variables) and currently defined dummy indices. Some parameters in an indexed collection may be given a computed or default value in terms of others in the same collection, however. As an example, you can smooth out some of the variation in the minimum inventories by defining the `mininv` parameter to be a running average like this:

```
param mininv {p in PROD, t in 0..T} =
   if t = 0 then inv0[p]
       else 0.5 * (mininv[p,t-1] + frac * market[p,t+1]);
```

The values of `mininv` for week 0 are set explicitly to the initial inventories, while the values for each subsequent week `t` are defined in terms of the previous week's values. AMPL permits any ''recursive'' definition of this kind, but will signal an error if it detects a circular reference that causes a parameter's value to depend directly or indirectly on itself.

You can use the phrases defined in this section together with the restriction phrases of the previous section, to further check the values that are computed. For example the declaration

```
param mininv {p in PROD, t in 0..T}
   = frac * market[p,t+1], >= 0;
```

will cause an error to be signaled if the computed value of any of the `mininv` parameters is negative. This check is triggered whenever an AMPL session uses `mininv` for any purpose.

7.6 Randomly generated parameters

When you're testing out a model, especially in the early stages of development, you may find it convenient to let randomly generated data stand in for actual data to be obtained later. Randomly generated parameters can also be useful in experimenting with alternative model formulations or solvers.

Randomly generated parameters are like the computed parameters introduced in the preceding section, except that their defining expressions are made random by use of AMPL's built-in random number generation functions listed in Table A-3. As an example of the simplest case, the individual parameter `avail` representing hours available in `steel.mod` may be defined to equal a random function:

```
param avail_mean > 0;
param avail_variance > 0, < avail_mean / 2;

param avail = max(Normal(avail_mean, avail_variance), 0);
```

Adding some indexing gives a multi-stage version of this model:

```
param avail {STAGE} =
    max(Normal(avail_mean, avail_variance), 0);
```

For each stage s, this gives `avail[s]` a different random value from the same random distribution. To specify stage-dependent random distributions, you would add indexing to the mean and variance parameters as well:

```
param avail_mean {STAGE} > 0;
param avail_variance {s in STAGE} > 0, < avail_mean[s] / 2;

param avail {s in STAGE} =
    max(Normal(avail_mean[s], avail_variance[s]), 0);
```

The `max(..., 0)` expression is included to handle the rare case in which the normal distribution with a positive mean returns a negative value.

More general ways of randomly computing parameters arise naturally from the preceding section's examples. In the multicommodity transportation problem, you can define random shares of demand:

```
param share {DEST} = Uniform(0,100);
param tot_sh = sum {j in DEST} share[j];

param tot_dem {PROD} >= 0;
param demand {j in DEST, p in PROD}
        = share[j] * tot_dem[p] / tot_sh;
```

Parameters `tot_sh` and `demand` then also become random, because they are defined in terms of random parameters. In the multiperiod production model, you can define the demand quantities `market[p,t]` in terms of an initial value and a random amount of increase per period:

```
param market1 {PROD} >= 0;
param max_incr {PROD} >= 0;

param market {p in PROD, t in 1..T+1} =
    if t = 1 then market1[p]
        else Uniform(0,max_incr) * market[p,t-1];
```

A recursive definition of this kind provides a way of generating simple random processes over time.

All of the AMPL random functions are based on a uniform random number generator with a very long period. When you start AMPL or give a `reset` command, however, the generator is reset and the ''random'' values are the same as before. You can request different values by changing the AMPL option `randseed` to some integer other than its default value of 1; the command for this purpose is

```
option randseed n;
```

where *n* is some integer value. Nonzero values give sequences that repeat each time AMPL is reset. A value of 0 requests AMPL to pick a seed based on the current value of the system clock, resulting (for practical purposes) in a different seed at each reset.

AMPL's `reset data` command, when applied to a randomly computed parameter, also causes a new sample of random values to be determined. The use of this command is discussed in Section 11.3.

7.7 Logical parameters

Although parameters normally represent numeric values, they can optionally be used to stand for true-false values or for character strings.

The current version of AMPL does not support a full-fledged ''logical'' type of parameter that would stand for only the values true and false, but a parameter of type `binary` may be used to the same effect. As an illustration, we describe an application of the preceding inventory example to consumer goods. Certain products in each week may be specially promoted, in which case they require a higher inventory fraction. Using parameters of type `binary`, we can represent this situation by the following declarations:

```
param fr_reg > 0;        # regular inventory fraction
param fr_pro > fr_reg;   # fraction for promoted items

param promote {PROD,1..T+1} binary;
param market {PROD,1..T+1} >= 0;
```

The binary parameters `promote[p,t]` are 0 when there is no promotion, and 1 when there is a promotion. Thus we can define the minimum-inventory parameters by use of an `if-then-else` expression as follows:

```
param mininv {p in PROD, t in 0..T} =
    (if promote[p,t] = 1 then fr_pro else fr_reg)
        * market[p,t+1];
```

We can also say the same thing more concisely:

```
param mininv {p in PROD, t in 0..T} =
    (if promote[p,t] then fr_pro else fr_reg) * market[p,t+1];
```

When an arithmetic expression like `promote[p,t]` appears where a logical expression is required, AMPL interprets any nonzero value as true, and zero as false. You do need to exercise a little caution to avoid being tripped up by this implicit conversion. For example, in Section 7.4 we used the expression

```
if t = 1 then 0 else cumulative_market[p,t-1]
```

If you accidentally write

```
if t then 0 else cumulative_market[p,t-1]     # DIFFERENT
```

it's perfectly legal, but it doesn't mean what you intended.

7.8 Symbolic parameters

You may permit a parameter to represent character string values, by including the keyword `symbolic` in its declaration. A symbolic parameter's values may be strings or numbers, just like a set's members, but the string values may not participate in arithmetic.

A major use of symbolic parameters is to designate individual set members that are to be treated specially. For example, in a model of traffic flow, there is a set of intersections, two of whose members are designated as the entrance and exit. Symbolic parameters can be used to represent these two members:

```
set INTER;

param entr symbolic in INTER;
param exit symbolic in INTER, <> entr;
```

In the data statements, an appropriate string is assigned to each symbolic parameter:

```
set INTER := a b c d e f g ;

param entr := a ;
param exit := g ;
```

These parameters are subsequently used in defining the objective and constraints; the complete model is developed in Section 15.2.

Another use of symbolic parameters is to associate descriptive strings with set members. Consider for example the set of ''origins'' in the transportation model of Figure 3-1a. When we introduced this set at the beginning of Chapter 3, we described each originating city by means of a 4-character string and a longer descriptive string. The short strings became the members of the AMPL set ORIG, while the longer strings played no further role. To make both available, we could declare

```
set ORIG;
param orig_name {ORIG} symbolic;
param supply {ORIG} >= 0;
```

Then in the data we could specify

```
param: ORIG:   orig_name                  supply :=
       GARY    "Gary, Indiana"            1400
       CLEV    "Cleveland, Ohio"          2600
       PITT    "Pittsburgh, Pennsylvania" 2900 ;
```

Since the long strings do not have the form of AMPL names, they do need to be quoted. They still play no role in the model or the resulting linear program, but they can be retrieved for documentary purposes by the display and printf commands described in Chapter 12.

Just as there are arithmetic and logical operators and functions, there are AMPL string operators and functions for working with string values. These features are mostly used in AMPL command scripts rather than in models, so we defer their description to Section 13.7.

Exercises

7-1. Show how the multicommodity transportation model of Figure 4-1 could be modified so that it applies the following restrictions to the data. Use either a restriction phrase in a set or param declaration, or a check statement, whichever is appropriate.

- No city is a member of both ORIG and DEST.
- The number of cities in DEST must be greater than the number in ORIG.
- Demand does not exceed 1000 at any one city in DEST.
- Total supply for each product at all origins must equal total demand for that product at all destinations.
- Total supply for all products at all origins must equal total demand for all products at all destinations.
- Total supply of all products at an origin must not exceed total capacity for all shipments from that origin.
- Total demand for all products at a destination must not exceed total capacity for all shipments to that destination.

7-2. Show how the multiperiod production model of Figure 4-4 could be modified so that it applies the following restrictions to the data.

- The number of weeks is a positive integer greater than 1.

- The initial inventory of a product does not exceed the total market demand for that product over all weeks.

- The inventory cost for a product is never more than 10% of the expected revenue for that product in any one week.

- The number of hours in a week is between 24 and 40, and does not change by more than 8 hours from one week to the next.

- For each product, the expected revenue never decreases from one week to the next.

7-3. The solutions to the following exercises involve the use of an `if-then-else` operator to formulate a constraint.

(a) In the example of the constraint `Balance` in Section 7.3, we used an expression beginning

```
if t = first(WEEKS) then ...
```

Find an equivalent expression that uses the function `ord(t)`.

(b) Combine the `Diet_Min` and `Diet_Max` constraints of Figure 5-1's diet model into one constraint declaration.

(c) In the multicommodity transportation model of Figure 4-1, imagine that there is more demand at the destinations than we can meet from the supply produced at the origins. To make up the difference, a limited number of additional tons can be purchased (rather than manufactured) for shipment at certain origins.

To model this situation, suppose that we declare a subset of origins,

```
set BUY_ORIG within ORIG;
```

where the additional tons can be bought. The relevant data values and decision variables could be indexed over this subset:

```
param buy_supply {BUY_ORIG,PROD} >= 0;  # available for purchase
param buy_cost {BUY_ORIG,PROD} > 0;     # purchase cost per ton

var Buy {i in BUY_ORIG, p in PROD} >= 0, <= buy_supply[i,p];
                                        # amount to buy
```

Revise the objective function to include the purchase costs. Revise the `Supply` constraints to say that, for each origin and each product, total tons shipped out must equal tons of supply from production plus (if applicable) tons purchased.

(d) Formulate the same model as in (c), but with `BUY_ORIG` being the set of pairs `(i,p)` such that product `p` can be bought at origin `i`.

7-4. This exercise is concerned with the following sets and parameters from Figure 4-1:

```
set ORIG;   # origins
set DEST;   # destinations
set PROD;   # products

param supply {ORIG,PROD} >= 0;
param demand {DEST,PROD} >= 0;
```

(a) Write `param` declarations, using the = operator, to compute parameters having the following definitions:

- `prod_supply[p]` is the total supply of product `p` at all origins.

- `dest_demand[j]` is the total demand for all products at destination `j`.
- `true_limit[i,j,p]` is the largest quantity of product `p` that can be shipped from `i` to `j` — that is, the largest value that does not exceed `limit[i,j]`, or the supply of `p` at `i`, or the demand for `p` at `j`.
- `max_supply[p]` is the largest supply of product `p` available at any origin.
- `max_diff[p]` is the largest difference, over all combinations of origins and destinations, between the supply and demand for product `p`.

(b) Write `set` declarations, using the `=` operator, to compute these sets:

- Products `p` whose demand is at least 500 at some destination `j`.
- Products `p` whose demand is at least 250 at all destinations `j`.
- Products `p` whose demand is equal to 500 at some destination `j`.

7-5. AMPL parameters can be defined to contain many kinds of series, especially by using recursive definitions. For example, we can make `s[j]` equal the sum of the first `j` integers, for `j` from 1 to some given limit `N`, by writing

```
param N;
param s {j in 1..N} = sum {jj in 1..j} jj;
```

or, using a formula for the sum,

```
param s {j in 1..N} = j * (j+1) / 2;
```

or, using a recursive definition,

```
param s {j in 1..N} = if j = 1 then 1 else s[j-1] + j;
```

This exercise asks you to play with some other possibilities.

(a) Define `fact[n]` to be `n` factorial, the product of the first `n` integers. Give both a recursive and a nonrecursive definition as above.

(b) The Fibonacci numbers are defined mathematically by $f_0 = f_1 = 1$ and $f_n = f_{n-1} + f_{n-2}$. Using a recursive declaration, define `fib[n]` in AMPL to equal the n-th Fibonacci number.

Use another AMPL declaration to verify that the n-th Fibonacci number equals the closest integer to $(\frac{1}{2} + \frac{1}{2}\sqrt{5})^n / \sqrt{5}$.

(c) Here's another recursive definition, called Ackermann's function, for positive integers i and j:

$$A(i, 0) = i + 1$$
$$A(0, j + 1) = A(1, j)$$
$$A(i + 1, j + 1) = A(A(i, j + 1), j)$$

Using a recursive declaration, define `ack[i,j]` in AMPL so that it will equal $A(i, j)$. Use `display` to print `ack[0,0]`, `ack[1,1]`, `ack[2,2]` and so forth. What difficulty do you encounter?

(d) What are the values `odd[i]` defined by the following odd declaration?

```
param odd {i in 1..N} =
   if i = 1 then 3 else
      min {j in odd[i-1]+2 .. odd[i-1]*2 by 2:
           not exists {k in 1 .. i-1} j mod odd[k] = 0} j;
```

Once you've figured it out, create a simpler and more efficient declaration that gives a set rather than an array of these numbers.

(e) A ''tree'' consists of a collection of nodes, one of which we designate as the ''root''. Each node except the root has a unique predecessor node in the tree, such that if you work backwards from a node to its predecessor, then to its predecessor's predecessor, and so forth, you always eventually reach the root. A tree can be drawn like this, with the root at the left and an arrow from each node to its successors:

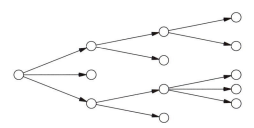

We can store the structure of a tree in AMPL sets and parameters as follows:

```
set NODES;
param Root symbolic in NODES;
param pred {i in NODES diff {Root}} symbolic in NODES diff {i};
```

Every node `i`, except `Root`, has a predecessor `pred[i]`.

The depth of a node is the number of predecessors that you encounter on tracing back to the root; the depth of the root is 0. Give an AMPL definition for `depth[i]` that correctly computes the depth of each node `i`. To check your answer, apply your definition to AMPL data for the tree depicted above; after reading in the data, use `display` to view the parameter `depth`.

An error in the data could give a tree plus a disconnected cycle, like this:

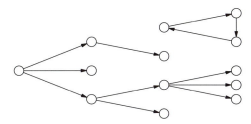

If you enter such data, what will happen when you try to display `depth`?

8

Linear Programs: Variables, Objectives and Constraints

The best-known kind of optimization model, which has served for all of our examples so far, is the linear program. The variables of a linear program take values from some continuous range; the objective and constraints must use only linear functions of the variables. Previous chapters have described these requirements informally or implicitly; here we will be more specific.

Linear programs are particularly important because they accurately represent many practical applications of optimization. The simplicity of linear functions makes linear models easy to formulate, interpret, and analyze. They are also easy to solve; if you can express your problem as a linear program, even in thousands of constraints and variables, then you can be confident of finding an optimal solution accurately and quickly.

This chapter describes how variables are declared, defines the expressions that AMPL recognizes as being linear in the variables, and gives the rules for declaring linear objectives and constraints. Much of the material on variables, objectives and constraints is basic to other AMPL models as well, and will be used in later chapters.

Because AMPL is fundamentally an algebraic modeling language, this chapter concentrates on features for expressing linear programs in terms of algebraic objectives and constraints. For linear programs that have certain special structures, such as networks, AMPL offers alternative notations that may make models easier to write, read and solve. Such special structures are among the topics of Chapters 15 through 17.

8.1 Variables

The variables of a linear program have much in common with its numerical parameters. Both are symbols that stand for numbers, and that may be used in arithmetic expressions. Parameter values are supplied by the modeler or computed from other values, while the values of variables are determined by an optimizing algorithm (as implemented in one of the packages that we refer to as solvers).

Syntactically, variable declarations are the same as the parameter declarations defined in Chapter 7, except that they begin with the keyword `var` rather than `param`. The meaning of qualifying phrases within the declaration may be different, however, when these phrases are applied to variables rather than to parameters.

Phrases beginning with `>=` or `<=` are by far the most common in declarations of variables for linear programs. They have appeared in all of our examples, beginning with the production model of Figure 1-4:

```
var Make {p in PROD} >= 0, <= market[p];
```

This declaration creates an indexed collection of variables `Make[p]`, one for each member `p` of the set `PROD`; the rules in this respect are exactly the same as for parameters. The effect of the two qualifying phrases is to impose a restriction, or constraint, on the permissible values of the variables. Specifically, `>= 0` implies that all of the variables `Make[p]` must be assigned nonnegative values by the optimizing algorithm, while the phrase `<= market[p]` says that, for each product p, the value given to `Make[p]` may not exceed the value of the parameter `market[p]`.

In general, either `>=` or `<=` may be followed by any arithmetic expression in previously defined sets and parameters and currently defined dummy indices. Most linear programs are formulated in such a way that every variable must be nonnegative; an AMPL variable declaration can specify nonnegativity either directly by `>= 0`, or indirectly as in the diet model of Figure 5-1:

```
param f_min {FOOD} >= 0;
param f_max {j in FOOD} >= f_min[j];

var Buy {j in FOOD} >= f_min[j], <= f_max[j];
```

The values following `>=` and `<=` are lower and upper *bounds* on the variables. Because these bounds represent a kind of constraint, they could just as well be imposed by the constraint declarations described later in this chapter. By placing bounds in the `var` declaration instead, you may be able to make the model shorter or clearer, although you will not make the optimal solution any different or easier to find. Some solvers do treat bounds specially in order to speed up their algorithms, but with AMPL all bounds are identified automatically, no matter how they are expressed in your model.

Variable declarations may not use the comparison operators `<`, `>` or `<>` in qualifying phrases. For linear programming it makes no sense to constrain a variable to be, say, < 3, since it could always be chosen as 2.99999... or as close to 3 as you like.

An `=` phrase in a variable declaration gives rise to a definition, as in a parameter declaration. Because a variable is being declared, however, the expression to the right of the `=` operator may contain previously declared variables as well as sets and parameters. For example, instead of writing the complicated objective from the multi-period production model of Figure 6-3 (`steelT3.mod`) as

```
maximize Total_Profit:
    sum {p in PROD, t in 1..T}
       (sum {a in AREA[p]} revenue[p,a,t]*Sell[p,a,t] -
          prodcost[p]*Make[p,t] - invcost[p]*Inv[p,t]);
```

you could instead define variables to represent the total revenues, production costs, and inventory costs:

```
var Total_Revenue =
    sum {p in PROD, t in 1..T}
       sum {a in AREA[p]} revenue[p,a,t] * Sell[p,a,t];
var Total_Prod_Cost =
    sum {p in PROD, t in 1..T} prodcost[p] * Make[p,t];
var Total_Inv_Cost =
    sum {p in PROD, t in 1..T} invcost[p] * Inv[p,t];
```

The objective would then be the sum of these three defined variables:

```
maximize Total_Profit:
    Total_Revenue - Total_Prod_Cost - Total_Inv_Cost;
```

The structure of the objective is clearer this way. Also the defined variables are conveniently available to a `display` statement to show how the three main components of profit compare:

```
ampl: display Total_Revenue, Total_Prod_Cost, Total_Inv_Cost;
Total_Revenue = 801385
Total_Prod_Cost = 285643
Total_Inv_Cost = 1221
```

Declarations of defined variables like these do not give rise to additional constraints in the resulting problem instance. Rather, the linear expression to the right of the = is substituted for every occurrence of the defined variable in the objective and constraints. Defined variables are even more useful for nonlinear programming, where the substitution may be only implicit, so we will return to this topic in Chapter 18.

If the expression to the right of the = operator contains no variables, then you are merely defining variables to be fixed to values given by the data. In that case you should use a `param` declaration instead. On the other hand, if you only want to fix some variables temporarily while developing or analyzing a model, then you should leave the declarations unchanged and instead fix them with the `fix` command described in Section 11.4.

A := or `default` phrase in a variable declaration gives *initial* values to the indicated variables. Variables not assigned an initial value by := can also be assigned initial values from a data file. Initial values of variables are normally changed — ideally to optimal values — when a solver is invoked. Thus the main purpose of initial values of variables is to give the solver a good starting solution. Solvers for linear programming can seldom make good use of a starting solution, however, so we defer further discussion of this topic to Chapter 18 on nonlinear programming.

Finally, variables may be declared as `integer` so that they must take whole number values in any optimal solution, or as `binary` so that they may only take the values 0 and

1. Models that contain any such variables are integer programs, which are the topic of Chapter 20.

8.2 Linear expressions

An arithmetic expression is *linear* in a given variable if, for every unit increase or decrease in the variable, the value of the expression increases or decreases by some fixed amount. An expression that is linear in all its variables is called a linear expression. (Strictly speaking, these are *affine* expressions, and a linear expression is an affine expression with constant term zero. For simplicity, we will ignore this distinction.)

AMPL recognizes as a linear expression any sum of terms of the form

> *constant-expr*
> *variable-ref*
> (*constant-expr*) * *variable-ref*

provided that each *constant-expr* is an arithmetic expression that contains no variables, while *var-ref* is a reference (possibly subscripted) to a variable. The parentheses around the *constant-expr* may be omitted if the result is the same according to the rules of operator precedence (Table A-1). The following examples, from the constraints in the multi-period production model of Figure 6-3, are all linear expressions under this definition:

```
avail[t]
Make[p,t] + Inv[p,t-1]
sum {p in PROD} (1/rate[p]) * Make[p,t]
sum {a in AREA[p]} Sell[p,a,t] + Inv[p,t]
```

The model's objective,

```
sum {p in PROD, t in 1..T}
   (sum {a in AREA[p]} revenue[p,a,t] * Sell[p,a,t] -
      prodcost[p] * Make[p,t] - invcost[p] * Inv[p,t])
```

is also linear because subtraction of a term is the addition of its negative, and a sum of sums is itself a sum.

Various kinds of expressions are equivalent to a sum of terms of the forms above, and are also recognized as linear by AMPL. Division by an arithmetic expression is equivalent to multiplication by its inverse, so

```
(1/rate[p]) * Make[p,t]
```

may be written in a linear program as

```
Make[p,t] / rate[p]
```

The order of multiplications is irrelevant, so the *variable-ref* need not come at the end of a term; for instance,

```
revenue[p,a,t] * Sell[p,a,t]
```

is equivalent to

```
Sell[p,a,t]  *  revenue[p,a,t]
```

As an example combining these principles, imagine that `revenue[p,a,t]` is in dollars per metric ton, while `Sell` remains in tons. If we define conversion factors

```
param mt_t = 0.90718474;    # metric tons per ton
param t_mt = 1 / mt_t;      # tons per metric ton
```

then both

```
sum {a in AREA[p]} mt_t * revenue[p,a,t] * Sell[p,a,t]
```

and

```
sum {a in AREA[p]} revenue[p,a,t] * Sell[p,a,t] / t_mt
```

are linear expressions for total revenue.

To continue our example, if costs are also in dollars per metric ton, the objective could be written as

```
mt_t * sum {p in PROD, t in 1..T}
    (sum {a in AREA[p]} revenue[p,a,t] * Sell[p,a,t] -
        prodcost[p] * Make[p,t] - invcost[p] * Inv[p,t])
```

or as

```
sum {p in PROD, t in 1..T}
    (sum {a in AREA[p]} revenue[p,a,t] * Sell[p,a,t] -
        prodcost[p] * Make[p,t] - invcost[p] * Inv[p,t]) / t_mt
```

Multiplication and division distribute over any summation to yield an equivalent linear sum of terms. Notice that in the first form, `mt_t` multiplies the entire `sum {p in PROD, t in 1..T}`, while in the second `t_mt` divides only the summand that follows `sum {p in PROD, t in 1..T}`, because the `/` operator has higher precedence than the `sum` operator. In these examples the effect is the same, however.

Finally, an `if-then-else` operator produces a linear result if the expressions following `then` and `else` are both linear and no variables appear in the logical expression between `if` and `else`. The following example appeared in a constraint in Section 7.3:

```
Make[j,t] +
    (if t = first(WEEKS) then inv0[j] else Inv[j,prev(t)])
```

The variables in a linear expression may not appear as the operands to any other operators, or in the arguments to any functions. This rule applies to iterated operators like `max`, `min`, `abs`, `forall`, and `exists`, as well as `^` and standard numerical functions like `sqrt`, `log`, and `cos`.

To summarize, a linear expression may be any sum of terms in the forms

> *constant-expr*
> *var-ref*
> (*constant-expr*) * (*linear-expr*)
> (*linear-expr*) * (*constant-expr*)
> (*linear-expr*) / (*constant-expr*)
> if *logical-expr* then *linear-expr* else *linear-expr*

where *constant-expr* is any arithmetic expression that contains no references to variables, and *linear-expr* is any other (simpler) linear expression. Parentheses may be omitted if the result is the same by the rules of operator precedence in Table A-1. AMPL automatically performs the transformations that convert any such expression to a simple sum of linear terms.

8.3 Objectives

The declaration of an objective function consists of one of the keywords `minimize` or `maximize`, a name, a colon, and a linear expression in previously defined sets, parameters and variables. We have seen examples such as

```
minimize Total_Cost: sum {j in FOOD} cost[j] * Buy[j];
```

and

```
maximize Total_Profit:
    sum {p in PROD, t in 1..T}
        (sum {a in AREA[p]} revenue[p,a,t] * Sell[p,a,t] -
            prodcost[p] * Make[p,t] - invcost[p] * Inv[p,t]);
```

The name of the objective plays no further role in the model, with the exception of certain "columnwise" declarations to be introduced in Chapters 15 and 16. Within AMPL commands, the objective's name refers to its value. Thus for example after solving a feasible instance of the Figure 2-1 diet model we could issue the command

```
ampl: display {j in FOOD} 100 * cost[j] * Buy[j] / Total_Cost;
100*cost[j]*Buy[j]/Total_Cost [*] :=
BEEF   14.4845
 CHK    4.38762
FISH    3.8794
 HAM   24.4792
 MCH   16.0089
 MTL   16.8559
 SPG   15.6862
 TUR    4.21822
;
```

to show the percentage of the total cost spent on each food.

Although a particular linear program must have one objective function, a model may contain more than one objective declaration. Moreover, any `minimize` or `maximize` declaration may define an indexed collection of objective functions, by including an indexing expression after the objective name. In these cases, you may issue an `objective` command, before typing `solve`, to indicate which objective is to be optimized.

As an example, recall that when trying to solve the model of Figure 2-1 with the data of Figure 2-2, we found that no solution could satisfy all of the constraints; we subsequently increased the sodium (NA) limit to 50000 to make a feasible solution possible. It is reasonable to ask: How much of an increase in the sodium limit is really necessary to permit a feasible solution? For this purpose we can introduce a new objective equal to the total sodium in the diet:

```
minimize Total_NA: sum {j in FOOD} amt["NA",j] * Buy[j];
```

(We create this objective only for sodium, because we have no reason to minimize most of the other nutrients.) We can solve the linear program for total cost as before, since AMPL chooses the model's first objective by default:

```
ampl: model diet.mod;
ampl: data diet2a.dat;
ampl: display n_max["NA"];
n_max['NA'] = 50000

ampl: minimize Total_NA: sum {j in FOOD} amt["NA",j] * Buy[j];
ampl: solve;
MINOS 5.5: optimal solution found.
13 iterations, objective 118.0594032
Objective = Total_Cost
```

The solver tells us the minimum cost, and we can also use `display` to look at the total sodium, even though it's not currently being minimized:

```
ampl: display Total_NA;
Total_NA = 50000
```

Next we can use the `objective` command to switch the objective to minimization of total sodium. The `solve` command then re-optimizes with this alternative objective, and we display `Total_Cost` to determine the resulting cost:

```
ampl: objective Total_NA;

ampl: solve;
MINOS 5.5: optimal solution found.
1 iterations, objective 48186

ampl: display Total_Cost;
Total_Cost = 123.627
```

We see that sodium can be brought down by about 1800, though the cost is forced up by about $5.50 as a result. (Healthier diets are in general more expensive, because they force the solution away from the one that minimizes costs.)

As another example, here's how we could experiment with different optimal solutions for the office assignment problem of Figure 3-2. First we solve the original problem:

```
ampl: model transp.mod; data assign.dat; solve;
CPLEX 8.0.0: optimal solution; objective 28
24 dual simplex iterations (0 in phase I)

ampl: option display_1col 1000, omit_zero_rows 1;
ampl: option display_eps .000001;

ampl: display Total_Cost,
ampl?    {i in ORIG, j in DEST} cost[i,j] * Trans[i,j];
Total_Cost = 28

cost[i,j]*Trans[i,j] :=
Coullard   C118    6
Daskin     D241    4
Hazen      C246    1
Hopp       D237    1
Iravani    C138    2
Linetsky   C250    3
Mehrotra   D239    2
Nelson     C140    4
Smilowitz  M233    1
Tamhane    C251    3
White      M239    1
;
```

To keep the objective value at this optimal level while we experiment, we add a constraint that fixes the expression for the objective equal to the current value, 28:

```
ampl: subject to Stay_Optimal:
ampl?    sum {i in ORIG, j in DEST}
ampl?       cost[i,j] * Trans[i,j] = 28;
```

Next, recall that cost[i,j] is the ranking that person i has given to office j, while Trans[i,j] is set to 1 if it's optimal to put person i in office j, or 0 otherwise. Thus

```
sum {j in DEST} cost[i,j] * Trans[i,j]
```

always equals the ranking of person i for the office to which i is assigned. We use this expression to declare a new objective function:

```
ampl: minimize Pref_of {i in ORIG}:
ampl?    sum {j in DEST} cost[i,j] * Trans[i,j];
```

This statement creates, for each person i, an objective Pref_of[i] that minimizes the ranking of i for the room that i is assigned. Then we can select any one person and optimize his or her ranking in the assignment:

```
ampl: objective Pref_of["Coullard"];
ampl: solve;
CPLEX 8.0.0: optimal solution; objective 3
3 simplex iterations (0 in phase I)
```

Looking at the new assignment, we see that the original objective is unchanged, and that the selected individual's situation is in fact improved, although of course at the expense of others:

```
ampl: display Total_Cost,
ampl?    {i in ORIG, j in DEST} cost[i,j] * Trans[i,j];
Total_Cost = 28

cost[i,j]*Trans[i,j] :=
Coullard   D241   3
Daskin     D237   1
Hazen      C246   1
Hopp       C251   5
Iravani    C138   2
Linetsky   C250   3
Mehrotra   D239   2
Nelson     C140   4
Smilowitz  M233   1
Tamhane    C118   5
White      M239   1
;
```

We were able to make this change because there are several optimal solutions to the original total-ranking objective. A solver arbitrarily returns one of these, but by use of a second objective we can force it toward others.

8.4 Constraints

The simplest kind of constraint declaration begins with the keywords `subject to`, a name, and a colon. Even the `subject to` is optional; AMPL assumes that any declaration not beginning with a keyword is a constraint. Following the colon is an algebraic description of the constraint, in terms of previously defined sets, parameters and variables. Thus in the production model introduced in Figure 1-4, we have the following constraint imposed by limited processing time:

```
subject to Time:
    sum {p in PROD} (1/rate[p]) * Make[p] <= avail;
```

The name of a constraint, like the name of an objective, is not used anywhere else in an algebraic model, though it figures in alternative "columnwise" formulations (Chapter 16) and is used in the AMPL command environment to specify the constraint's dual value and other associated quantities (Chapter 14).

Most of the constraints in large linear programming models are defined as indexed collections, by giving an indexing expression after the constraint name. The constraint `Time`, for example, is generalized in subsequent examples to say that the production time may not exceed the time available in each processing stage `s` (Figure 1-6a):

```
subject to Time {s in STAGE}:
    sum {p in PROD} (1/rate[p,s]) * Make[p] <= avail[s];
```

or in each week `t` (Figure 4-4):

```
subject to Time {t in 1..T}:
   sum {p in PROD} (1/rate[p]) * Make[p,t] <= avail[t];
```

Another constraint from the latter example says that production, sales and inventories must balance for each product p in each week t:

```
subject to Balance {p in PROD, t in 1..T}:
   Make[p,t] + Inv[p,t-1] = Sell[p,t] + Inv[p,t];
```

A constraint declaration can specify any valid indexing expression, which defines a set (as explained in Chapters 5 and 6); there is one constraint for each member of this set. The constraint name can be subscripted, so that Time[1] or Balance[p,t+1] refers to a particular constraint from an indexed collection.

The indexing expression in a constraint declaration should specify a dummy index (like s, t and p in the preceding examples) for each dimension of the indexing set. Then when the constraint corresponding to a particular indexing-set member is processed by AMPL, the dummy indices take their values from that member. This use of dummy indices is what permits a single constraint expression to represent many constraints; the indexing expression is AMPL's translation of a phrase such as "for all products p and weeks $t = 1$ to T" that might be seen in an algebraic statement of the model.

By using more complex indexing expressions, you can specify more precisely the constraints to be included in a model. Consider, for example, the following variation on the production time constraint:

```
subject to Time {t in 1..T: avail[t] > 0}:
   sum {p in PROD} (1/rate[p]) * Make[p,t] <= avail[t];
```

This says that if avail[t] is specified as zero in the data for any week t, it is to be interpreted as meaning "no constraint on time available in week t" rather than "limit of zero on time available in week t". In the simpler case where there is just one Time constraint not indexed over weeks, you can specify an analogous conditional definition as follows:

```
subject to Time {if avail > 0}:
   sum {p in PROD} (1/rate[p]) * Make[p] <= avail;
```

The pseudo-indexing expression {if avail > 0} causes one constraint, named Time, to be generated if the condition avail > 0 is true, and no constraint at all to be generated if the condition is false. (The same notation can be used to conditionally define other model components.)

AMPL's algebraic description of a constraint may consist of any two linear expressions separated by an equality or inequality operator:

linear-expr   <=   *linear-expr*
linear-expr   =   *linear-expr*
linear-expr   >=   *linear-expr*

While it is customary in mathematical descriptions of linear programming to place all terms containing variables to the left of the operator and all other terms to the right (as in constraint Time), AMPL imposes no such requirement (as seen in constraint Balance).

Convenience and readability should determine what terms you place on each side of the operator. AMPL takes care of canonicalizing constraints, such as by combining linear terms involving the same variable and moving variables from one side of a constraint to the other. The `expand` command described in Section 1.4 shows the canonical forms of the constraints.

AMPL also allows double-inequality constraints such as the following from the diet model of Figure 2-1:

```
subject to Diet {i in NUTR}:
    n_min[i] <= sum {j in FOOD} amt[i,j] * Buy[j] <= n_max[i];
```

This says that the middle expression, the amount of nutrient `i` supplied by all foods, must be greater than or equal to `n_min[i]` and also less than or equal to `n_max[i]`. The permissible forms for a constraint of this kind are

const-expr `<=` *linear-expr* `<=` *const-expr*
const-expr `>=` *linear-expr* `>=` *const-expr*

where each *const-expr* must contain no variables. The effect is to give upper and lower bounds on the value of the *linear-expr*. If your model requires variables in the left-hand or right-hand *const-expr*, you must define two different constraints in separate declarations.

For most applications of linear programming, you need not worry about the form of the constraints. If you simply write the constraints in the most convenient way, they will be recognized as proper linear constraints according to the rules in this chapter. There do exist situations, however, in which your choice of formulation will determine whether AMPL recognizes your model as linear. Imagine that we want to further constrain the production model so that no product p may represent more than a certain fraction of total production. We define a parameter `max_frac` to represent the limiting fraction; the constraint then says that production of p divided by total production must be less than or equal to `max_frac`:

```
subject to Limit {p in PROD}:
    Make[p] / sum {q in PROD} Make[q] <= max_frac;
```

This is not a linear constraint to AMPL, because its left-hand expression contains a division by a sum of variables. But if we rewrite it as

```
subject to Limit {p in PROD}:
    Make[p] <= max_frac * sum {q in PROD} Make[q];
```

then AMPL does recognize it as linear.

AMPL simplifies constraints as it prepares the model and data for handing to a solver. For example, it may eliminate variables fixed at a value, combine single-variable constraints with the simple bounds on the variables, or drop constraints that are implied by other constraints. You can normally ignore this presolve phase, but there are ways to observe its effects and modify its actions, as explained in Section 14.1.

Exercises

8-1. In the diet model of Figure 5-1, add a `:=` phrase to the `var` declaration (as explained in Section 8.1) to initialize each variable to a value midway between its lower and upper bounds.

Read this model into AMPL along with the data from Figure 5-2. Using `display` commands, determine which constraints (if any) the initial solution fails to satisfy, and what total cost this solution gives. Is the total cost more or less than the optimal total cost?

8-2. This exercise asks you to reformulate various kinds of constraints to make them linear.

(a) The following constraint says that the inventory `Inv[p,t]` for product `p` in any period `t` must not exceed the smallest one-period production `Make[p,t]` of product `p`:

```
subject to Inv_Limit {p in PROD, t in 1..T}:
    Inv[p,t] <= min {tt in 1..T} Make[p,tt];
```

This constraint is not recognized as linear by AMPL, because it applies the `min` operator to variables. Formulate a linear constraint that has the same effect.

(b) The following constraint says that the change in total inventories from one period to the next may not exceed a certain parameter `max_change`:

```
subject to Max_Change {t in 1..T}:
    abs(sum {p in PROD} Inv[p,t-1] - sum {p in PROD} Inv[p,t])
        <= max_change;
```

This constraint is not linear because it applies the `abs` function to an expression involving variables. Formulate a linear constraint that has the same effect.

(c) The following constraint says that the ratio of total production to total inventory in a period may not exceed `max_inv_ratio`:

```
subject to Max_Inv_Ratio {t in 1..T}:
    (sum {p in PROD} Inv[p,t]) / (sum {p in PROD} Make[p,t])
        <= max_inv_ratio;
```

This constraint is not linear because it divides one sum of variables by another. Formulate a linear constraint that has the same effect.

(d) What can you say about formulation of an alternative linear constraint for the following cases?

– In (a), `min` is replaced by `max`.

– In (b), `<= max_change` is replaced by `>= min_change`.

– In (c), the parameter `max_inv_ratio` is replaced by a new variable, `Ratio[t]`.

8-3. This exercise deals with some more possibilities for using more than one objective function in a diet model. Here we consider the model of Figure 5-1, together with the data from Figure 5-2.

Suppose that the costs are indexed over stores as well as foods:

```
set STORE:
param cost {STORE,FOOD} > 0;
```

A separate objective function may then be defined for each store:

```
minimize Total_Cost {s in STORE}:
    sum {j in FOOD} cost[s,j] * Buy[j];
```

Consider the following data for three stores:

```
set STORE := "A&P" JEWEL VONS ;

param cost:   BEEF   CHK   FISH   HAM   MCH   MTL   SPG   TUR :=
       "A&P"  3.19  2.59  2.29  2.89  1.89  1.99  1.99  2.49
       JEWEL  3.09  2.79  2.29  2.59  1.59  1.99  2.09  2.30
        VONS  2.59  2.99  2.49  2.69  1.99  2.29  2.00  2.69 ;
```

Using the `objective` command, find the lowest-cost diet for each store. Which store offers the lowest total cost?

Consider now an additional objective that represents total packages purchased, regardless of cost:

```
minimize Total_Number:
    sum {j in FOOD} Buy[j];
```

What is the minimum value of this objective? What are the costs at the three stores when this objective is minimized? Explain why you would expect these costs to be higher than the costs computed in (a).

8-4. This exercise relates to the assignment example of Section 8.3.

(a) What is the best-ranking office that you can assign to each individual, given that the total of the rankings must stay at the optimal value of 28? How many different optimal assignments do there seem to be, and which individuals get different offices in different assignments?

(b) Modify the assignment example so that it will find the best-ranking office that you can assign to each individual, given that the total of the rankings may increase from 28, but may not exceed 30.

(c) After making the modification suggested in (b), the person in charge of assigning offices has tried again to minimize the objective `Pref_of["Coullard"]`. This time, the reported solution is as follows:

```
ampl: display Total_Cost,
ampl?    {i in ORIG, j in DEST} cost[i,j]*Trans[i,j];
Total_Cost = 30

cost[i,j]*Trans[i,j] :=
Coullard   M239    1
Daskin     D241    4
Hazen      C246    1
Hopp       C251    2.5
Hopp       D237    0.5
Iravani    C138    2
Linetsky   C250    3
Mehrotra   D239    2
Nelson     C140    4
Smilowitz  M233    1
Tamhane    C118    5
White      C251    2.5
White      D237    1.5
;
```

Coullard is now assigned her first choice, but what is the difficulty with the overall solution? Why doesn't it give a useful resolution to the assignment problem as we have stated it?

8-5. Return to the assignment version of the transportation model, in Figures 3-1a and 3-2.

(a) Add parameters `worst[i]` for each `i` in ORIG, and constraints saying that `Trans[i,j]` must equal 0 for every combination of `i` in ORIG and `j` in DEST such that `cost[i,j]` is greater than `worst[i]`. (See the constraint `Time` in Section 8.4 for a similar example.) In the assignment interpretation of this model, what do the new constraints mean?

(b) Use the model from (a) to show that there is an optimal solution, with the objective equal to 28, in which no one gets an office worse than their fifth choice.

(c) Use the model from (a) to show that at least one person must get an office worse than fourth choice.

(d) Use the model from (a) to show that if you give Nelson his first choice, without any restrictions on the other individuals' choices, the objective cannot be made smaller than 31. Determine similarly how small the objective can be made if each other individual is given first choice.

9

Specifying Data

As we emphasize throughout this book, there is a distinction between an AMPL model for an optimization problem, and the data values that define a particular instance of the problem. Chapters 5 through 8 focused on the declarations of sets, parameters, variables, objectives and constraints that are necessary to describe models. In this chapter and the next, we take a closer look at the statements that specify the data.

Examples of AMPL *data statements* appear in almost every chapter. These statements offer several formats for lists and tables of set and parameter values. Some formats are most naturally created and maintained in a text editing or word processing environment, while others are easy to generate from programs like database systems and spreadsheets. The `display` command (Chapter 12) also produces output in these formats. Wherever possible, similar syntax and concepts are used for both sets and parameters.

This chapter first explains how AMPL's `data` command is used, in conjunction with data statements, to read data values from files such as those whose names end in `.dat` throughout our examples. Options to the `data` command also allow or force selected sets and parameters to be read again.

Subsequent sections describe data statements, first for lists and then for tables of set and parameter data, followed by brief sections on initial values for variables, values for indexed collections of sets, and default values. A summary of data statement formats appears in Section A.12.

A final section describes the `read` command, which reads *unformatted* lists of values into sets and parameters. Chapter 10 is devoted to AMPL's features for data stored in relational database tables.

9.1 Formatted data: the `data` command

Declarations like `param` and `var`, and commands like `solve` and `display`, are executed in *model mode*, the standard mode for most modeling activity. But model mode is inconvenient for reading long lists of set and parameter values. Instead AMPL reads its data statements in a *data mode* that is initiated by the `data` command. In its most common use, this command consists of the keyword `data` followed by the name of a file. For example,

    ampl: *data diet.dat;*

reads data from a file named `diet.dat`. Filenames containing spaces, semicolons, or nonprinting characters must be enclosed in quotes.

While reading in data mode, AMPL treats white space, that is, any sequence of space, tab, and "newline" characters, as a single space. Commas separating strings or numbers are also ignored. Judicious use of these separators can help to arrange data into easy-to-read lists and tables; our examples use a combination of spaces and newlines. If data statements are produced as output from other data management software and sent directly to AMPL, however, then you may ignore visual appearance and use whatever format is convenient.

Data files often contain numerous character strings, representing set members or the values of symbolic parameters. Thus in data mode AMPL does not, in general, require strings to be enclosed in quotes. Strings that include any character other than letters, digits, underscores, period, + and – must be quoted, however, as in the case of A&P. You may use a pair of either single quotes (`'A&P'`) or double quotes (`"A&P"`), unless the string contains a quote, in which case the other kind of quote must surround it (`"DOMINICK'S"`) or the surrounding quote must be doubled within it (`'DOMINICK''S'`).

A string that looks like a number (for example `"+1"` or `"3e4"`) must also be quoted, to distinguish it from a set member or parameter value that is actually a number. Numbers that have the same internal representation are considered to be the same, so that for example `2`, `2.00`, `2.e0` and `0.02E+2` all denote the same set member.

When AMPL finishes reading a file in data mode, it normally reverts to whatever mode it was in before the `data` command was executed. Hence a data file can itself contain `data` commands that read data from other files. If the last data statement in a data file lacks its terminating semicolon, however, then data mode persists regardless of the previous mode.

A `data` command with no filename puts AMPL into data mode, so subsequent input is taken as data statements:

    ampl: *model dietu.mod;*
    ampl: *data;*
    ampl data: *set MINREQ := A B1 B2 C CAL;*
    ampl data: *set MAXREQ := A NA CAL;*
    ampl data: *display NUTR;*
    set NUTR := A B1 B2 C CAL NA;

    ampl:

AMPL leaves data mode when it sees any statement (like `display`) that does not begin with a keyword (like `set` or `param`) that begins a `data` statement. The `model` command, with or without filename, also causes a return to model mode.

Model components may be assigned values from any number of data files, by using multiple `data` commands. Regardless of the number of files, AMPL checks that no component is assigned a value more than once, and duplicate assignments are flagged as errors. In some situations, however, it is convenient to be able to change the data by issuing new `data` statements; for example, after solving for one scenario of a model, you may want to modify some of the data by reading a new data file that corresponds to a second scenario. The data values in the new file would normally be treated as erroneous duplicates, but you can tell AMPL to accept them by first giving a `reset data` or `update data` command. These alternatives are described in Section 11.3, along with the use of `reset data` to resample randomly-computed parameters, and of `let` to directly assign new set or parameter values.

9.2 Data in lists

For an unindexed (scalar) parameter, a data statement assigns one value:

```
param avail := 40;
```

Most of a typical model's parameters are indexed over sets, however, and their values are specified in a variety of lists and tables that are introduced in this section and the next, respectively.

We start with sets of simple one-dimensional objects, and the one-dimensional collections of parameters indexed over them. We then turn to two-dimensional sets and parameters, for which we have the additional option of organizing the data into "slices". The options for two dimensions are then shown to generalize readily to higher dimensions, for which we present some three-dimensional examples. Finally, we show how data statements for a set and the parameters indexed over it can be combined to provide a more concise and convenient representation.

Lists of one-dimensional sets and parameters

For a parameter indexed over a one-dimensional set like

```
set PROD;
param rate {PROD} > 0;
```

the specification of the set can be simply a listing of its members:

```
set PROD := bands coils plate ;
```

and the parameter's specification may be virtually the same except for the addition of a value after each set member:

```
param rate := bands 200  coils 140  plate 160 ;
```

The parameter specification could equally well be written

```
param rate :=
        bands  200
        coils  140
        plate  160 ;
```

since extra spaces and line breaks are ignored.

If a one-dimensional set has been declared with the attribute `ordered` or `circular` (Section 5.6), then the ordering of its members is taken from the data statement that defines it. For example, we specified

```
set WEEKS := 27sep 04oct 11oct 18oct ;
```

as the membership of the ordered set `WEEKS` in Figure 5-4.

Members of a set must all be different; AMPL will warn of duplicates:

```
duplicate member coils for set PROD
context:  set PROD := bands coils plate coils  >>> ; <<<
```

Also a parameter may not be given more than one value for each member of the set over which it is indexed. A violation of this rule provokes a similar message:

```
rate['bands'] already defined
context:  param rate := bands 200 bands 160  >>> ; <<<
```

The context bracketed by >>> and <<< isn't the exact point of the error, but the message makes the situation clear.

A set may be specified as empty by giving an empty list of members; simply put the semicolon right after the : = operator. A parameter indexed over an empty set has no data associated with it.

Lists of two-dimensional sets and parameters

The extension of data lists to the two-dimensional case is largely straightforward, but with each set member denoted by a pair of objects. As an example, consider the following sets from Figure 6-2a:

```
set ORIG;    # origins
set DEST;    # destinations

set LINKS within {ORIG,DEST};   # transportation links
```

The members of `ORIG` and `DEST` can be given as for any one-dimensional sets:

```
set ORIG := GARY CLEV PITT ;
set DEST := FRA DET LAN WIN STL FRE LAF ;
```

Then the membership of `LINKS` may be specified as a list of tuples such as you would find in a model's indexing expressions,

```
set LINKS :=
   (GARY,DET) (GARY,LAN) (GARY,STL) (GARY,LAF) (CLEV,FRA)
   (CLEV,DET) (CLEV,LAN) (CLEV,WIN) (CLEV,STL) (CLEV,LAF)
   (PITT,FRA) (PITT,WIN) (PITT,STL) (PITT,FRE) ;
```

or as a list of pairs, without the parentheses and commas:

```
set LINKS :=
     GARY DET    GARY LAN    GARY STL    GARY LAF
     CLEV FRA    CLEV DET    CLEV LAN    CLEV WIN
     CLEV STL    CLEV LAF    PITT FRA    PITT WIN
     PITT STL    PITT FRE ;
```

The order of members within each pair is significant — the first must be from ORIG, and the second from DEST — but the pairs themselves may appear in any order.

An alternative, more concise way to describe this set of pairs is to list all second components that go with each first component:

```
set LINKS :=
     (GARY,*) DET LAN STL LAF
     (CLEV,*) FRA DET LAN WIN STL LAF
     (PITT,*) FRA WIN STL FRE ;
```

It is also easy to list all first components that go with each second component:

```
set LINKS :=
     (*,FRA) CLEV PITT    (*,DET) GARY CLEV    (*,LAN) GARY CLEV
     (*,WIN) CLEV PITT    (*,LAF) GARY CLEV    (*,FRE) PITT
     (*,STL) GARY CLEV PITT ;
```

An expression such as (GARY,*) or (*,FRA), resembling a pair but with a component replaced by a *, is a data *template*. Each template is followed by a list, whose entries are substituted for the * to generate pairs; these pairs together make up a *slice* through the dimension of the set where the * appears. A tuple without any *'s, like (GARY,DET), is in effect a template that specifies only itself, so it is not followed by any values. At the other extreme, in the table that consists of pairs alone,

```
set LINKS :=
     GARY DET    GARY LAN    GARY STL    GARY LAF
     CLEV FRA    CLEV DET    CLEV LAN    CLEV WIN
     CLEV STL    CLEV LAF    PITT FRA    PITT WIN
     PITT STL    PITT FRE ;
```

a default template (*,*) applies to all entries.

For a parameter indexed over a two-dimensional set, the AMPL list formats are again derived from those for sets by placing parameter values after the set members. Thus if we have the parameter cost indexed over the set LINKS:

```
param cost {LINKS} >= 0;
```

then the set data statement for LINKS is extended to become the following param data statement for cost:

```
param cost :=
     GARY DET 14   GARY LAN 11   GARY STL 16   GARY LAF  8
     CLEV FRA 27   CLEV DET  9   CLEV LAN 12   CLEV WIN  9
     CLEV STL 26   CLEV LAF 17   PITT FRA 24   PITT WIN 13
     PITT STL 28   PITT FRE 99 ;
```

Lists of slices through a set extend similarly, by placing a parameter value after each implied set member. Thus, corresponding to our concise data statement for LINKS:

```
set LINKS :=
   (GARY,*) DET LAN STL LAF
   (CLEV,*) FRA DET LAN WIN STL LAF
   (PITT,*) FRA WIN STL FRE ;
```

there is the following statement for the values of cost:

```
param cost :=
   [GARY,*] DET 14   LAN 11   STL 16   LAF  8
   [CLEV,*] FRA 27   DET  9   LAN 12   WIN  9   STL 26   LAF 17
   [PITT,*] FRA 24   WIN 13   STL 28   FRE 99 ;
```

The templates are given in brackets to distinguish them from the set templates in parentheses, but they work in the same way. Thus a template such as [GARY, *] indicates that the ensuing entries will be for values of cost that have a first index of GARY, and an entry such as DET 14 gives cost["GARY", "DET"] a value of 14.

All of the above applies just as well to the use of templates that slice on the first dimension, so that for instance you could also specify parameter cost by:

```
param cost :=
   [*,FRA] CLEV 27   PITT 24
   [*,DET] GARY 14   CLEV  9
   [*,LAN] GARY 11   CLEV 12
   [*,WIN] CLEV  9   PITT 13
   [*,STL] GARY 16   CLEV 26   PITT 28
   [*,FRE] PITT 99
   [*,LAF] GARY  8   CLEV 17
```

You can even think of the list-of-pairs example,

```
param cost :=
   GARY DET 14   GARY LAN 11   GARY STL 16   GARY LAF  8
   . . .
```

as also being a case of this form, corresponding to the default template [*, *].

Lists of higher-dimensional sets and parameters

The concepts underlying data lists for two-dimensional sets and parameters extend straightforwardly to higher-dimensional cases. The only difference of any note is that nontrivial slices may be made through more than one dimension. Hence we confine the presentation here to some illustrative examples in three dimensions, followed by a sketch of the general rules for the AMPL data list format that are given in Section A.12.

We take our example from Section 6.3, where we suggest a version of the multicommodity transportation model that defines a set of triples and costs indexed over them:

```
set ROUTES within {ORIG,DEST,PROD};
param cost {ROUTES} >= 0;
```

Suppose that ORIG and DEST are as above, that PROD only has members bands and coils, and that ROUTES has as members certain triples from {ORIG, DEST, PROD}. Then the membership of ROUTES can be given most simply by a list of triples, either

```
set ROUTES :=
   (GARY,LAN,coils)  (GARY,STL,coils)  (GARY,LAF,coils)
   (CLEV,FRA,bands)  (CLEV,FRA,coils)  (CLEV,DET,bands)
   (CLEV,DET,coils)  (CLEV,LAN,bands)  (CLEV,LAN,coils)
   (CLEV,WIN,coils)  (CLEV,STL,bands)  (CLEV,STL,coils)
   (CLEV,LAF,bands)  (PITT,FRA,bands)  (PITT,WIN,bands)
   (PITT,STL,bands)  (PITT,FRE,bands)  (PITT,FRE,coils)  ;
```

or

```
set ROUTES :=
   GARY LAN coils    GARY STL coils    GARY LAF coils
   CLEV FRA bands    CLEV FRA coils    CLEV DET bands
   CLEV DET coils    CLEV LAN bands    CLEV LAN coils
   CLEV WIN coils    CLEV STL bands    CLEV STL coils
   CLEV LAF bands    PITT FRA bands    PITT WIN bands
   PITT STL bands    PITT FRE bands    PITT FRE coils  ;
```

Using templates as before, but with three items in each template, we can break the specification into slices through one dimension by placing one * in each template. In the following example, we slice through the second dimension:

```
set ROUTES :=
   (CLEV,*,bands) FRA DET LAN STL LAF
   (PITT,*,bands) FRA WIN STL FRE

   (GARY,*,coils) LAN STL LAF
   (CLEV,*,coils) FRA DET LAN WIN STL
   (PITT,*,coils) FRE ;
```

Because the set contains no members with origin GARY and product bands, the template (GARY, *, bands) is omitted.

When the set's dimension is more than two, the slices can also be through more than one dimension. A slice through two dimensions, in particular, naturally involves placing two *'s in each template. Here we slice through both the first and third dimensions:

```
set ROUTES :=
   (*,FRA,*)  CLEV bands  CLEV coils  PITT bands
   (*,DET,*)  CLEV bands  CLEV coils
   (*,LAN,*)  GARY coils  CLEV bands  CLEV coils
   (*,WIN,*)  CLEV coils  PITT bands
   (*,STL,*)  GARY coils  CLEV bands  CLEV coils  PITT bands
   (*,FRE,*)  PITT bands  PITT coils
   (*,LAF,*)  GARY coils  CLEV bands ;
```

Since these templates have two *'s, they must be followed by pairs of components, which are substituted from left to right to generate the set members. For instance the template (*, FRA, *) followed by CLEV bands specifies that (CLEV, FRA, bands) is a member of the set.

Any of the above forms suffices for giving the values of parameter `cost` as well. We could write

```
param cost :=
   [CLEV,*,bands] FRA 27   DET   9   LAN 12   STL 26   LAF 17
   [PITT,*,bands] FRA 24   WIN 13   STL 28   FRE 99

   [GARY,*,coils] LAN 11   STL 16   LAF   8
   [CLEV,*,coils] FRA 23   DET   8   LAN 10   WIN   9   STL 21
   [PITT,*,coils] FRE 81 ;
```

or

```
param cost :=
   [*,*,bands]   CLEV FRA 27   CLEV DET   9   CLEV LAN 12
                 CLEV STL 26   CLEV LAF 17   PITT FRA 24
                 PITT WIN 13   PITT STL 28   PITT FRE 99

   [*,*,coils]   GARY LAN 11   GARY STL 16   GARY LAF   8
                 CLEV FRA 23   CLEV DET   8   CLEV LAN 10
                 CLEV WIN   9   CLEV STL 21   PITT FRE 81
```

or

```
param cost :=
   CLEV DET bands   9   CLEV DET coils   8   CLEV FRA bands 27
   CLEV FRA coils 23   CLEV LAF bands 17   CLEV LAN bands 12
   CLEV LAN coils 10   CLEV STL bands 26   CLEV STL coils 21
   CLEV WIN coils   9   GARY LAF coils   8   GARY LAN coils 11
   GARY STL coils 16   PITT FRA bands 24   PITT FRE bands 99
   PITT FRE coils 81   PITT STL bands 28   PITT WIN bands 13 ;
```

By placing the `*`'s in different positions within the templates, we can slice one-dimensionally in any of three different ways, or two-dimensionally in any of three different ways. (The template `[*,*,*]` would specify a three-dimensional list like

```
param cost :=
   CLEV DET bands   9   CLEV DET coils   8   CLEV FRA bands 27
   ...
```

as already shown above.)

More generally, a template for an *n*-dimensional set or parameter in list form must have *n* entries. Each entry is either a legal set member or a `*`. Templates for sets are enclosed in parentheses (like the tuples in set-expressions) and templates for parameters are enclosed in brackets (like the subscripts of parameters). Following a template is a series of items, each item consisting of one set member for each `*`, and additionally one parameter value in the case of a parameter template. Each item defines an *n*-tuple, by substituting its set members for the `*`s in the template; either this tuple is added to the set being specified, or the parameter indexed by this tuple is assigned the value in the item.

A template applies to all items between it and the next template (or the end of the data statement). Templates having different numbers of `*`s may even be used together in the same data statement, so long as each parameter is assigned a value only once. Where no template appears, a template of all *s is assumed.

Combined lists of sets and parameters

When we give data statements for a set and a parameter indexed over it, like

```
set PROD := bands coils plate ;
param rate := bands 200   coils 140   plate 160 ;
```

we are specifying the set's members twice. AMPL lets us avoid this duplication by including the set's name in the `param` data statement:

```
param: PROD: rate := bands 200   coils 140   plate 160 ;
```

AMPL uses this statement to determine both the membership of PROD and the values of rate.

Another common redundancy occurs when we need to supply data for several parameters indexed over the same set, such as rate, profit and market all indexed over PROD in Figure 1-4a. Rather than write a separate data statement for each parameter,

```
param rate    := bands  200   coils  140   plate  160 ;
param profit := bands   25   coils   30   plate   29 ;
param market := bands 6000   coils 4000   plate 3500 ;
```

we can combine these statements into one by listing all three parameter names after the keyword `param`:

```
param: rate profit market :=
   bands 200 25 6000   coils 140 30 4000   plate 160 29 3500 ;
```

Since AMPL ignores extra spaces and line breaks, we have the option of rearranging this information into an easier-to-read table:

```
param:     rate  profit  market :=
   bands    200    25     6000
   coils    140    30     4000
   plate    160    29     3500 ;
```

Either way, we still have the option of adding the indexing set's name to the statement,

```
param: PROD:   rate  profit  market :=
       bands    200    25     6000
       coils    140    30     4000
       plate    160    29     3500 ;
```

so that the specifications of the set and all three parameters are combined.

The same rules apply to lists of any higher-dimensional sets and the parameters indexed over them. Thus for our two-dimensional example LINKS we could write

```
param: LINKS: cost :=
   GARY DET 14   GARY LAN 11   GARY STL 16   GARY LAF   8
   CLEV FRA 27   CLEV DET  9   CLEV LAN 12   CLEV WIN   9
   CLEV STL 26   CLEV LAF 17   PITT FRA 24   PITT WIN  13
   PITT STL 28   PITT FRE 99 ;
```

to specify the membership of LINKS and the values of the parameter cost indexed over it, or

```
param: LINKS: cost    limit :=
   GARY DET        14    1000
   GARY LAN        11     800
   GARY STL        16    1200
   GARY LAF         8    1100
   CLEV FRA        27    1200
   CLEV DET         9     600
   CLEV LAN        12     900
   CLEV WIN         9     950
   CLEV STL        26    1000
   CLEV LAF        17     800
   PITT FRA        24    1500
   PITT WIN        13    1400
   PITT STL        28    1500
   PITT FRE        99    1200 ;
```

to specify the values of cost and limit together. The same options apply when templates are used, making possible further alternatives such as

```
param: LINKS: cost :=
   [GARY,*] DET 14   LAN 11   STL 16   LAF   8
   [CLEV,*] FRA 27   DET  9   LAN 12   WIN   9   STL 26   LAF 17
   [PITT,*] FRA 24   WIN 13   STL 28   FRE 99 ;
```

and

```
param:   LINKS:   cost   limit :=
   [GARY,*] DET     14    1000
            LAN     11     800
            STL     16    1200
            LAF      8    1100
   [CLEV,*] FRA     27    1200
            DET      9     600
            LAN     12     900
            WIN      9     950
            STL     26    1000
            LAF     17     800
   [PITT,*] FRA     24    1500
            WIN     13    1400
            STL     28    1500
            FRE     99    1200 ;
```

Here the membership of the indexing set is specified along with the two parameters; for example, the template [GARY,*] followed by the set member DET and the values 14 and `1000` indicates that `(GARY,DET)` is to be added to the set `LINKS`, that `cost[GARY,DET]` has the value `14`, and that `limit[GARY,DET]` has the value `1000`.

As our illustrations suggest, the key to the interpretation of a `param` statement that provides values for several parameters or for a set and parameters is in the first line, which consists of `param` followed by a colon, then optionally the name of an indexing set followed by a colon, then by a list of parameter names terminated by the `:=` assignment operator. Each subsequent item in the list consists of a number of set members equal to the number of `*`s in the most recent template and then a number of parameter values equal to the number of parameters listed in the first line.

Normally the parameters listed in the first line of a `param` statement are all indexed over the same set. This need not be the case, however, as seen in the case of Figure 5-1. For this variation on the diet model, the nutrient restrictions are given by

```
set MINREQ;
set MAXREQ;

param n_min {MINREQ} >= 0;
param n_max {MAXREQ} >= 0;
```

so that `n_min` and `n_max` are indexed over sets of nutrients that may overlap but that are not likely to be the same.

Our sample data for this model specifies:

```
set MINREQ := A B1 B2 C CAL ;
set MAXREQ := A NA CAL ;

param:     n_min   n_max  :=
    A        700    20000
    C        700       .
    B1         0       .
    B2         0       .
    NA         .    50000
    CAL    16000    24000  ;
```

Each period or dot (`.`) indicates to AMPL that no value is being given for the corresponding parameter and index. For example, since `MINREQ` does not contain a member `NA`, the parameter `n_min[NA]` is not defined; consequently a `.` is given as the entry for `NA` and `n_min` in the data statement. We cannot simply leave a space for this entry, because AMPL will take it to be `50000`: data mode processing ignores all extra spaces. Nor should we put a zero in this entry; in that case we will get a message like

```
error processing param n_min:
        invalid subscript n_min['NA'] discarded.
```

when AMPL first tries to access `n_min`, usually at the first `solve`.

When we name a set in the first line of a `param` statement, the set must not yet have a value. If the specification of parameter data in Figure 5-1 had been given as

```
param: NUTR:   n_min   n_max :=
         A         700   20000
         C         700       .
         B1          0       .
         B2          0       .
         NA          .   50000
         CAL     16000   24000 ;
```

AMPL would have generated the error message

```
dietu.dat, line 16 (offset 366):
        NUTR was defined in the model
context:  param: NUTR >>> : <<<   n_min  n_max :=
```

because the declaration of NUTR in the model,

```
set NUTR = MINREQ union MAXREQ;
```

defines it already as the union of MINREQ and MAXREQ.

9.3 Data in tables

The table format of data, with indices running along the left and top edges and values corresponding to pairs of indices, can be more concise or easier to read than the list format described in the previous section. Here we describe tables first for two-dimensional parameters and then for slices from higher-dimensional ones. We also show how the corresponding multidimensional sets can be specified in tables that have entries of + or − rather than parameter value entries.

AMPL also supports a convenient extension of the table format, in which more than two indices may appear along the left and top edge. The rules for specifying such tables are provided near the end of this section.

Two-dimensional tables

Data values for a parameter indexed over two sets, such as the shipping cost data from the transportation model of Figure 3-1a:

```
set ORIG;
set DEST;
param cost {ORIG,DEST} >= 0;
```

are very naturally specified in a table (Figure 3-1b):

```
param cost:   FRA  DET  LAN  WIN  STL  FRE  LAF :=
         GARY   39   14   11   14   16   82    8
         CLEV   27    9   12    9   26   95   17
         PITT   24   14   17   13   28   99   20 ;
```

The row labels give the first index and the column labels the second index, so that for example `cost["GARY","FRA"]` is set to 39. To enable AMPL to recognize this as a table, a colon must follow the parameter name, while the `:=` operator follows the list of column labels.

For larger index sets, the columns of tables become impossible to view within the width of a single screen or page. To deal with this situation, AMPL offers several alternatives, which we illustrate on the small table above.

When only one of the index sets is uncomfortably large, the table may be transposed so that the column labels correspond to the smaller set:

```
param cost (tr):
          GARY CLEV PITT :=
    FRA    39   27   24
    DET    14    9   14
    LAN    11   12   17
    WIN    14    9   13
    STL    16   26   28
    FRE    82   95   99
    LAF     8   17   20 ;
```

The notation (`tr`) after the parameter name indicates a transposed table, in which the column labels give the first index and the row labels the second index. When both of the index sets are large, either the table or its transpose may be divided up in some way. Since line breaks are ignored, each row may be divided across several lines:

```
param cost:   FRA   DET   LAN   WIN
              STL   FRE   LAF        :=
       GARY    39    14    11    14
               16    82     8
       CLEV    27     9    12     9
               26    95    17
       PITT    24    14    17    13
               28    99    20        ;
```

Or the table may be divided columnwise into several smaller ones:

```
param cost:   FRA   DET   LAN   WIN :=
       GARY    39    14    11    14
       CLEV    27     9    12     9
       PITT    24    14    17    13

          :   STL   FRE   LAF  :=
       GARY    16    82     8
       CLEV    26    95    17
       PITT    28    99    20 ;
```

A colon indicates the start of each new sub-table; in this example, each has the same row labels, but a different subset of the column labels.

In the alternative formulation of this model presented in Figure 6-2a, `cost` is not indexed over all combinations of members of `ORIG` and `DEST`, but over a subset of pairs from these sets:

```
set LINKS within {ORIG,DEST};
param cost {LINKS} >= 0;
```

As we have seen in Section 9.2, the membership of LINKS can be given concisely by a list of pairs:

```
set LINKS :=
    (GARY,*) DET LAN STL LAF
    (CLEV,*) FRA DET LAN WIN STL LAF
    (PITT,*) FRA WIN STL FRE ;
```

Rather than being given in a similar list, the values of cost can be given in a table like this:

```
param cost:   FRA   DET   LAN   WIN   STL   FRE   LAF :=
    GARY        .    14    11     .    16     .     8
    CLEV       27     9    12     9    26     .    17
    PITT       24     .     .    13    28    99     . ;
```

A cost value is given for all pairs that exist in LINKS, while a dot (.) serves as a place-holder for pairs that are not in LINKS. The dot can appear in any AMPL table to indicate "no value specified here".

The set LINKS may itself be given by a table that is analogous to the one for cost:

```
set LINKS:   FRA   DET   LAN   WIN   STL   FRE   LAF :=
    GARY       -     +     +     -     +     -     +
    CLEV       +     +     +     +     +     -     +
    PITT       +     -     -     +     +     +     - ;
```

A + indicates a pair that is a member of the set, and a – indicates a pair that is not a member. Any of AMPL's table formats for specifying parameters can be used for sets in this way.

Two-dimensional slices of higher-dimensional data

To provide data for parameters of more than two dimensions, we can specify the values in two-dimensional slices that are represented as tables. The rules for using slices are much the same as for lists. As an example, consider again the three-dimensional parameter cost defined by

```
set ROUTES within {ORIG,DEST,PROD};
param cost {ROUTES} >= 0;
```

The values for this parameter that we specified in list format in the previous section as

```
param cost :=
    [*,*,bands]  CLEV FRA 27   CLEV DET  9   CLEV LAN 12
                 CLEV STL 26   CLEV LAF 17   PITT FRA 24
                 PITT WIN 13   PITT STL 28   PITT FRE 99

    [*,*,coils]  GARY LAN 11   GARY STL 16   GARY LAF  8
                 CLEV FRA 23   CLEV DET  8   CLEV LAN 10
                 CLEV WIN  9   CLEV STL 21   PITT FRE 81
```

can instead be written in table format as

```
param cost :=

[*,*,bands]: FRA   DET   LAN   WIN   STL   FRE   LAF :=
       CLEV   27    9    12     .    26     .    17
       PITT   24     .     .    13    28    99     .

[*,*,coils]: FRA   DET   LAN   WIN   STL   FRE   LAF :=
       GARY    .     .    11     .    16     .     8
       CLEV   23     8    10     9    21     .     .
       PITT    .     .     .     .     .    81     . ;
```

Since we are working with two-dimensional tables, there must be two *'s in the templates. A table value's row label is substituted for the first *, and its column label for the second, unless the opposite is specified by (tr) right after the template. You can omit any rows or columns that would have no significant entries, such as the row for GARY in the [*,*,bands] table above.

As before, a dot in the table for any slice indicates a tuple that is not a member of the table.

An analogous table to specify the set ROUTES can be constructed by putting a + where each number appears:

```
set ROUTES :=

(*,*,bands): FRA DET LAN WIN STL FRE LAF :=
       CLEV   +   +   +   -   +   -   +
       PITT   +   -   -   +   +   +   -

(*,*,coils): FRA DET LAN WIN STL FRE LAF :=
       GARY   -   -   +   -   +   -   +
       CLEV   +   +   +   +   +   -   -
       PITT   -   -   -   -   -   +   - ;
```

Since the templates are now set templates rather than parameter templates, they are enclosed in parentheses rather than brackets.

Higher-dimensional tables

By putting more than one index to the left of each row or at the top of each column, you can describe multidimensional data in a single table rather than a series of slices. We'll continue with the three-dimensional cost data to illustrate some of the wide variety of possibilities.

By putting the first two indices, from sets ORIG and DEST, to the left, with the third index from set PROD at the top, we produce the following three-dimensional table of the costs:

```
param cost: bands coils :=
    CLEV FRA    27     23
    CLEV DET     8      8
    CLEV LAN    12     10
    CLEV WIN     .      9
    CLEV STL    26     21
    CLEV LAF    17      .
    PITT FRA    24      .
    PITT WIN    13      .
    PITT STL    28      .
    PITT FRE    99     81
    GARY LAN     .     11
    GARY STL     .     16
    GARY LAF     .      8 ;
```

Putting only the first index to the left, and the second and third at the top, we arrive instead at the following table, which for convenience we break into two pieces:

```
param cost:   FRA    DET    LAN    WIN    STL    FRE    LAF
          : bands  bands  bands  bands  bands  bands  bands :=
     CLEV    27      9     12      .     26      .     17
     PITT    24      .      .     13     28     99      .

          :   FRA    DET    LAN    WIN    STL    FRE    LAF
          : coils  coils  coils  coils  coils  coils  coils :=
     GARY     .      .     11      .     16      .      8
     CLEV    23      8     10      9     21      .      .
     PITT     .      .      .      .      .     81      . ;
```

In general a colon must precede each of the table heading lines, while a : = is placed only after the last heading line.

The indices are taken in the order that they appear, first at the left and then at the top, if no indication is given to the contrary. As with other tables, you can add the indicator (tr) to transpose the table, so that the indices are still taken in order but first from the top and then from the left:

```
param cost (tr):  CLEV  CLEV  CLEV  CLEV  CLEV  CLEV
              :    FRA   DET   LAN   WIN   STL   LAF :=
        bands       27     8    12     .    26    17
        coils       23     8    10     9    21     .

              :   PITT  PITT  PITT  PITT  GARY  GARY  GARY
              :    FRA   WIN   STL   FRE   LAN   STL   LAF :=
        bands       24    13    28    99     .     .     .
        coils        .     .     .    81    11    16     8 ;
```

Templates can also be used to specify more precisely what goes where. For multidimensional tables the template has two symbols in it, * to indicate those indices that appear at the left and : to indicate those that appear at the top. For example the template [*,:,*] gives a representation in which the first and third indices are at the left and the second is at the top:

```
param cost :=
   [*,:,*] :  FRA   DET   LAN   WIN   STL   FRE   LAF :=
   CLEV bands   27     9    12     .    26     .    17
   CLEV coils   23     8    10     9    21     .     .
   PITT bands   24     .     .    13    28    99     .
   PITT coils    .     .     .     .     .    81     .
   GARY coils    .     .    11     .    16     .     8 ;
```

The ordering of the indices is always preserved in tables of this kind. The third index is never correctly placed before the first, for example, no matter what transposition or templates are employed.

For parameters of four or more dimensions, the ideas of slicing and multidimensional tables can be applied together provide an especially broad choice of table formats. If cost were indexed over ORIG, DEST, PROD, and 1..T, for instance, then the templates [*,:,bands,*] and [*,:,coils,*] could be used to specify two slices through the third index, each specified by a multidimensional table with two indices at the left and one at the top.

Choice of format

The arrangement of slices to represent multidimensional data has no effect on how the data values are used in the model, so you can choose the most convenient format. For the cost parameter above, it may be appealing to slice along the third dimension, so that the data values are organized into one shipping-cost table for each product. Alternatively, placing all of the origin-product pairs at the left gives a particularly concise representation. As another example, consider the revenue parameter from Figure 6-3:

```
set PROD;          # products
set AREA {PROD};   # market areas for each product
param T > 0;       # number of weeks

param revenue {p in PROD, AREA[p], 1..T} >= 0;
```

Because the index set AREA[p] is potentially different for each product p, slices through the first (PROD) dimension are most attractive. In the sample data from Figure 6-4, they look like this:

```
param T := 4 ;
set PROD := bands coils ;
set AREA[bands] := east north ;
set AREA[coils] := east west export ;

param revenue :=
   [bands,*,*]:    1      2      3      4    :=
        east      25.0   26.0   27.0   27.0
        north     26.5   27.5   28.0   28.5
   [coils,*,*]:    1      2      3      4    :=
        east      30     35     37     39
        west      29     32     33     35
        export    25     25     25     28 ;
```

We have a separate revenue table for each product p, with market areas from AREA[p] labeling the rows, and weeks from 1..T labeling the columns.

9.4 Other features of data statements

Additional features of the AMPL data format are provided to handle special situations. We describe here the data statements that specify default values for parameters, that define the membership of individual sets within an indexed collection of sets, and that assign initial values to variables.

Default values

Data statements must provide values for exactly the parameters in your model. You will receive an error message if you give a value for a nonexistent parameter:

```
error processing param cost:
        invalid subscript cost['PITT','DET','coils'] discarded.
```

or if you fail to give a value for a parameter that does exist:

```
error processing objective Total_Cost:
        no value for cost['CLEV','LAN','coils']
```

The error message appears the first time that AMPL tries to use the offending parameter, usually after you type solve.

If the same value would appear many times in a data statement, you can avoid specifying it repeatedly by including a default phrase that provides the value to be used when no explicit value is given. For example, suppose that the parameter cost above is indexed over all possible triples:

```
set ORIG;
set DEST;
set PROD;

param cost {ORIG,DEST,PROD} >= 0;
```

but that a very high cost is assigned to routes that should not be used. This can be expressed as

```
param cost  default 9999  :=
 [*,*,bands]: FRA  DET  LAN  WIN  STL  FRE  LAF :=
       CLEV  27    9   12    .   26    .   17
       PITT  24    .    .   13   28   99    .
 [*,*,coils]: FRA  DET  LAN  WIN  STL  FRE  LAF :=
       GARY   .    .   11    .   16    .    8
       CLEV  23    8   10    9   21    .    .
       PITT   .    .    .    .    .   81    . ;
```

Missing parameters like `cost["GARY","FRA","bands"]`, as well as those explicitly marked ''omitted'' by use of a dot (like `cost["GARY","FRA","coils"]`), are given the value 9999. In total, 24 values of 9999 are assigned.

The `default` feature is especially useful when you want all parameters of an indexed collection to be assigned the same value. For instance, in Figure 3-2, we apply a transportation model to an assignment problem by setting all supplies and demands to 1. The model declares

```
param supply {ORIG} >= 0;
param demand {DEST} >= 0;
```

but in the data we give only a default value:

```
param supply default 1 ;
param demand default 1 ;
```

Since no other values are specified, the default of 1 is automatically assigned to every element of `supply` and `demand`.

As explained in Chapter 7, a parameter declaration in the model may include a `default` expression. This offers an alternative way to specify a single default value:

```
param cost {ORIG,DEST,PROD} >= 0, default 9999;
```

If you just want to avoid storing a lot of 9999's in a data file, however, it is better to put the `default` phrase in the data statement. The `default` phrase should go in the model when you want the default value to depend in some way on other data. For instance, a different arbitrarily large cost could be given for each product by specifying:

```
param huge_cost {PROD} > 0;
param cost {ORIG, DEST, p in PROD} >= 0, default huge_cost[p];
```

A discussion of `default`'s relation to the = phrase in `param` statements is given in Section 7.5.

Indexed collections of sets

For an indexed collection of sets, separate data statements specify the members of each set in the collection. In the example of Figure 6-3, for example, the sets named `AREA` are indexed by the set `PROD`:

```
set PROD;          # products
set AREA {PROD};   # market areas for each product
```

The membership of these sets is given in Figure 6-4 by:

```
set PROD := bands coils ;
set AREA[bands] := east north ;
set AREA[coils] := east west export ;
```

Any of the data statement formats for a set may be used with indexed collections of sets. The only difference is that the set name following the keyword `set` is subscripted.

As for other sets, you may specify one or more members of an indexed collection to be empty, by giving an empty list of elements. If you want to provide a data statement only for those members of an indexed collection that are not empty, define the empty set as the default value in the model:

```
set AREA {PROD} default {};
```

Otherwise you will be warned about any set whose data statement is not provided.

Initial values for variables

You may optionally assign initial values to the variables of a model, using any of the options for assigning values to parameters. A variable's name stands for its value, and a constraint's name stands for the associated dual variable's value. (See Section 12.5 for a short explanation of dual variables.)

Any `param` data statement may specify initial values for variables. The variable or constraint name is simply used in place of a parameter name, in any of the formats described by the previous sections of this chapter. To help clarify the intent, the keyword `var` may be substituted for `param` at the start of a data statement. For example, the following data table gives initial values for the variable `Trans` of Figure 3-1a:

```
var Trans:   FRA   DET   LAN   WIN   STL   FRE   LAF  :=
       GARY  100   100   800   100   100   500   200
       CLEV  900   100   100   500   500   200   200
       PITT  100   900   100   500   100   900   200 ;
```

As another example, in the model of Figure 1-4, a single table can give values for the parameters `rate`, `profit` and `market`, and initial values for the variables `Make`:

```
param:     rate   profit   market   Make :=
    bands   200     25       6000    3000
    coils   140     30       4000    2500
    plate   160     29       3500    1500 ;
```

All of the previously described features for default values also apply to variables.

Initial values of variables (as well as the values of expressions involving these initial values) may be viewed before you type `solve`, using the `display`, `print` or `printf` commands described in Sections 12.1 through 12.4. Initial values are also optionally passed to the solver, as explained in Section 14.1 and A.18.1. After a solution is returned, the variables no longer have their initial values, but even then you can refer to the initial values by placing an appropriate suffix after the variable's name, as shown in Section A.11.

The most common use of initial values is to give a good starting guess to a solver for nonlinear optimization, which is discussed in Chapter 18.

9.5 Reading unformatted data: the `read` command

The `read` command provides a particularly simple way of getting values into AMPL, given that the values you need are listed in a regular order in a file. The file must be *unformatted* in the sense that it contains nothing except the values to be read — no set or parameter names, no colons or : = operators.

In its simplest form, `read` specifies a list of parameters and a file from which their values are to be read. The values in the file are assigned to the entries in the list in the order that they appear. For example, if you want to read the number of weeks and the hours available each week for our simple production model (Figure 4-4),

```
param T > 0;
param avail {1..T} >= 0;
```

from a file `week_data.txt` containing

```
4
40 40 32 40
```

then you can give the command

```
read T, avail[1], avail[2], avail[3], avail[4] <week_data.txt;
```

Or you can use an indexing expression to say the same thing more concisely and generally:

```
read T, {t in 1..T} avail[t] <week_data.txt;
```

The notation < *filename* specifies the name of a file for reading. (Analogously, > indicates writing to a file; see A.15.)

In general, the `read` command has the form

```
read item-list < filename ;
```

with the *item-list* being a comma-separated list of items that may each be any of the following:

> *parameter*
> { *indexing* } *parameter*
> { *indexing* } (*item-list*)

The first two are used in our example above, while the third allows for the same indexing to be applied to several items. Using the same production example, to read in values for

```
param prodcost {PROD} >= 0;
param invcost {PROD} >= 0;
param revenue {PROD,1..T} >= 0;
```

from a file organized by parameters, you could read each parameter separately:

```
read {p in PROD} prodcost[p] < cost_data;
read {p in PROD} invcost[p] < cost_data;
read {p in PROD, t in 1..T} revenue[p,t] < cost_data;
```

reading from file `cost_data` first all the production costs, then all the inventory costs, and then all the revenues.

If the data were organized by product instead, you could say

```
read {p in PROD}
    (prodcost[p], invcost[p], {t in 1..T} revenue[p,t])
        <cost_data;
```

to read the production and inventory costs and the revenues for the first product, then for the second product, and so forth.

A parenthesized *item-list* may itself contain parenthesized *item-lists*, so that if you also want to read

```
param market {PROD,1..T} >= 0;
```

from the same file at the same time, you could say

```
read {p in PROD} (prodcost[p], invcost[p],
    {t in 1..T} (revenue[p,t], market[p,t])) <cost_data;
```

in which case for each product you would read the two costs as before, and then for each week the product's revenue and market demand.

As our descriptions suggest, the form of a `read` statement's *item-list* depends on how the data values are ordered in the file. When you are reading data indexed over sets of strings that, like `PROD`, are not inherently ordered, then the order in which values are read is the order in which AMPL is internally representing them. If the members of the set came directly from a `set` data statement, then the ordering will be the same as in the data statement. Otherwise, it is best to put an `ordered` or `ordered by` phrase in the model's `set` declaration to ensure that the ordering is always what you expect; see Section 5.6 for more about ordered sets.

An alternative that avoids knowing the order of the members in a set is to specify them explicitly in the file that is read. As an example, consider how you might use a `read` statement rather than a data statement to get the values from the `cost` parameter of Section 9.4 that was defined as

```
param cost {ORIG,DEST,PROD} >= 0, default 9999;
```

You could set up the `read` statement as follows:

```
param ntriples integer;
param ic symbolic in ORIG;
param jc symbolic in DEST;
param kc symbolic in PROD;

read ntriples, {1..ntriples}
    (ic, jc, kc, cost[ic,jc,kc]) <cost_data;
```

The corresponding file `cost_data` must begin something like this:

```
18
CLEV FRA bands 27
PITT FRA bands 24
CLEV FRA coils 23
...
```

with 15 more entries needed to give all 18 data values shown in the Section 9.4 example.

Strings in a file for the `read` command that include any character other than letters, digits, underscores, period, + and – must be quoted, just as for data mode. However, the `read` statement itself is interpreted in model mode, so if the statement refers to any particular string, as in, say,

```
read {t in 1..T} revenue ["bands",t];
```

that string must be quoted. The filename following < need not be quoted unless it contains spaces, semicolons, or nonprinting characters.

If a `read` statement contains no < *filename*, values are read from the current input stream. Thus if you have typed the `read` command at an AMPL prompt, you can type the values at subsequent prompts until all of the listed items have been assigned values. For example:

```
ampl: read T, {t in 1..T} avail[t];
ampl? 4
ampl? 40 40 32 40
ampl: display avail;
avail [*] :=
1 40   2 40   3 32   4 40
;
```

The prompt changes from `ampl?` back to `ampl:` when all the needed input has been read.

The filename ''–'' (a literal minus sign) is taken as the standard input of the AMPL process; this is useful for providing input interactively.

Further uses of `read` within AMPL scripts, to read values directly from script files or to prompt users for values at the command line, are described in Chapter 13.

All of our examples assume that underlying sets such as ORIG and PROD have already been assigned values, through data statements as described earlier in this chapter, or through other means such as database access or assignment to be described in later chapters. Thus the `read` statement would normally supplement rather than replace other input commands. It is particularly useful in handling long files of data that are generated for certain parameters by programs outside of AMPL.

Exercises

9-1. Section 9.2 gave a variety of data statements for a three-dimensional set, ROUTES. Construct some other alternatives for this set as follows:

(a) Use templates that look like `(CLEV,FRA,*)`.

(b) Use templates that look like `(*,*,bands)`, with the list format.

(c) Use templates that look like `(CLEV,*,*)`, with the table format.

(d) Specify some of the set's members using templates with one `*`, and some using templates with two `*`'s.

9-2. Rewrite the production model data of Figure 5-4 so that it consists of just three data statements arranged as follows:

The set `PROD` and parameters `rate`, `inv0`, `prodcost` and `invcost` are given in one table.

The set `WEEKS` and parameter `avail` are given in one table.

The parameters `revenue` and `market` are given in one table.

9-3. For the assignment problem whose data is depicted in Figure 3-2, suppose that the only information you receive about people's preferences for offices is as follows:

Coullard	M239 M233 D241 D237 D239
Daskin	D237 M233 M239 D241 D239 C246 C140
Hazen	C246 D237 M233 M239 C250 C251 D239
Hopp	D237 M233 M239 D241 C251 C250
Iravani	D237 C138 C118 D241 D239
Linetsky	M233 M239 C250 C251 C246 D237
Mehrotra	D237 D239 M239 M233 D241 C118 C251
Nelson	D237 M233 M239
Smilowitz	M233 M239 D239 D241 C251 C250 D237
Tamhane	M239 M233 C251 C250 C118 C138 D237
White	M239 M233 D237 C246

This means that, for example, Coullard's first choice is M239, her second choice is M233, and so on through her fifth choice, D239, but she hasn't given any preference for the other offices.

To use this information with the transportation model of Figure 3-1a as explained in Chapter 3, you must set `cost["Coullard","M239"]` to 1, `cost["Coullard","M233"]` to 2, and so forth. For an office not ranked, such as C246, you can set `cost["Coullard","C246"]` to 99, to indicate that it is a highly undesirable assignment.

(a) Using the list format and a `default` phrase, convert the information above to an appropriate AMPL data statement for the parameter `cost`.

(b) Do the same, but with a table format.

9-4. Sections 9.2 and 9.3 gave a variety of data statements for a three-dimensional parameter, `cost`, indexed over the set `ROUTES` of triples. Construct some other alternatives for this parameter as follows:

(a) Use templates that look like `[CLEV,FRA,*]`.

(b) Use templates that look like `[*,*,bands]`, employing the list format.

(c) Use templates that look like `[CLEV,*,*]`, employing the table format.

(d) Specify some of the parameter values using templates with one `*`, and some using templates with two `*`'s.

9-5. For the three-dimensional parameter `revenue` of Figure 6-4, construct alternative data statements as follows:

(a) Use templates that look like [*,east,*], employing the table format.

(b) Use templates that look like [*,*,1], employing the table format.

(c) Use templates that look like [bands,*,1].

9-6. Given the following declarations,

```
set ORIG;
set DEST;
var Trans {ORIG, DEST} >= 0;
```

how could you use a data statement to assign an initial value of 300 to all of the Trans variables?

10

Database Access

The structure of indexed data in AMPL has much in common with the structure of the relational tables widely used in database applications. The AMPL `table` declaration lets you take advantage of this similarity to define explicit connections between sets, parameters, variables, and expressions in AMPL, and relational database tables maintained by other software. The `read table` and `write table` commands subsequently use these connections to import data values into AMPL and to export data and solution values from AMPL.

The relational tables read and written by AMPL reside in files whose names and locations you specify as part of the `table` declaration. To work with these files, AMPL relies on *table handlers*, which are add-ons that can be loaded as needed. Handlers may be provided by the vendors of solvers or database software. AMPL has built-in handlers for two simple relational table formats useful for experimentation, and the AMPL web site provides a handler that works with the widely available ODBC interface.

This chapter begins by showing how AMPL entities can be put into correspondence with the columns of relational tables, and how the same correspondences can be described and implemented by use of AMPL's `table` declaration. Subsequent sections present basic features for reading and writing external relational tables, additional rules for handling complications that arise when reading and writing the same table, and mechanisms for writing a series of tables or columns and for reading spreadsheet data. The final section briefly describes some standard and built-in handlers.

10.1 General principles of data correspondence

Consider the following declarations from `diet.mod` in Chapter 2, defining the set `FOOD` and three parameters indexed over it:

```
set FOOD;
param cost {FOOD} > 0;
param f_min {FOOD} >= 0;
param f_max {j in FOOD} >= f_min[j];
```

A *relational table* giving values for these components has four *columns*:

```
FOOD      cost      f_min     f_max
BEEF      3.19      2         10
CHK       2.59      2         10
FISH      2.29      2         10
HAM       2.89      2         10
MCH       1.89      2         10
MTL       1.99      2         10
SPG       1.99      2         10
TUR       2.49      2         10
```

The column headed FOOD lists the members of the AMPL set also named FOOD. This is the table's *key* column; entries in a key column must be unique, like a set's members, so that each key value identifies exactly one row. The column headed cost gives the values of the like-named parameter indexed over set FOOD; here the value of cost["BEEF"] is specified as 3.19, cost["CHK"] as 2.59, and so forth. The remaining two columns give values for the other two parameters indexed over FOOD.

The table has eight *rows* of data, one for each set member. Thus each row contains all of the table's data corresponding to one member — one food, in this example.

In the context of database software, the table rows are often viewed as data *records,* and the columns as *fields* within each record. Thus a data entry form has one entry field for each column. A form for the diet example (from Microsoft Access) might look like Figure 10-1. Data records, one for each table row, can be entered or viewed one at a time by using the controls at the bottom of the form.

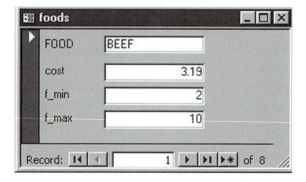

Figure 10-1: Access data entry form.

Parameters are not the only entities indexed over the set FOOD in this example. There are also the variables:

```
var Buy {j in FOOD} >= f_min[j], <= f_max[j];
```

and assorted result expressions that may be displayed:

```
ampl: model diet.mod;
ampl: data diet2a.dat;

ampl: solve;
MINOS 5.5: optimal solution found.
13 iterations, objective 118.0594032

ampl: display Buy, Buy.rc, {j in FOOD} Buy[j]/f_max[j];
:         Buy         Buy.rc     Buy[j]/f_max[j]     :=
BEEF     5.36061      8.88178e-16    0.536061
CHK      2            1.18884        0.2
FISH     2            1.14441        0.2
HAM      10          -0.302651       1
MCH      10          -0.551151       1
MTL      10          -1.3289         1
SPG      9.30605      0              0.930605
TUR      2            2.73162        0.2
;
```

All of these can be included in the relational table for values indexed over FOOD:

FOOD	cost	f_min	f_max	Buy	BuyRC	BuyFrac
BEEF	3.19	2	10	5.36061	8.88178e-16	0.536061
CHK	2.59	2	10	2	1.18884	0.2
FISH	2.29	2	10	2	1.14441	0.2
HAM	2.89	2	10	10	-0.302651	1
MCH	1.89	2	10	10	-0.551151	1
MTL	1.99	2	10	10	-1.3289	1
SPG	1.99	2	10	9.30605	0	0.930605
TUR	2.49	2	10	2	2.73162	0.2

Where the first four columns would typically be read into AMPL from a database, the last three are results that would be written back from AMPL to the database. We have invented the column headings BuyRC and BuyFrac, because the AMPL expressions for the quantities in those columns are typically not valid column headings in database management systems. The table declaration provides for input/output and naming distinctions such as these, as subsequent sections will show.

Other entities of diet.mod are indexed over the set NUTR of nutrients: parameters n_min and n_max, dual prices and other values associated with constraint Diet, and expressions involving these. Since nutrients are entirely distinct from foods, however, the values indexed over nutrients go into a separate relational table from the one for foods. It might look like this:

```
NUTR    n_min    n_max    NutrDual
A         700    20000    0
B1        700    20000    0
B2        700    20000    0.404585
C         700    20000    0
CAL     16000    24000    0
NA          0    50000    -0.00306905
```

As this example suggests, any model having more than one indexing set will require more than one relational table to hold its data and results. Databases that consist of multiple tables are a standard feature of relational data management, to be found in all but the simplest "flat file" database packages.

Entities indexed over the same higher-dimensional set have a similar correspondence to a relational table, but with one key column for each dimension. In the case of Chapter 4's steelT.mod, for example, the following parameters and variables are indexed over the same two-dimensional set of product-time pairs:

```
set PROD;       # products
param T > 0;    # number of weeks

param market {PROD,1..T} >= 0;
param revenue {PROD,1..T} >= 0;
var Make {PROD,1..T} >= 0;
var Sell {p in PROD, t in 1..T} >= 0, <= market[p,t];
```

A corresponding relational table thus has two key columns, one containing members of PROD and the other members of 1..T, and then a column of values for each parameter and variable. Here's an example, corresponding to the data in steelT.dat:

```
PROD    TIME    market    revenue    Make    Sell
bands    1       6000       25       5990    6000
bands    2       6000       26       6000    6000
bands    3       4000       27       1400    1400
bands    4       6500       27       2000    2000
coils    1       4000       30       1407     307
coils    2       2500       35       1400    2500
coils    3       3500       37       3500    3500
coils    4       4200       39       4200    4200
```

Each ordered pair of items in the two key columns is unique in this table, just as these pairs are unique in the set {PROD,1..T}. The market column of the table implies, for example, that market["bands",1] is 6000 and that market["coils",3] is 3500. From the first row, we can also see that revenue["bands",1] is 25, Make["bands",1] is 5990, and Sell["bands",1] is 6000. Again various names from the AMPL model are used as column headings, except for TIME, which must be invented to stand for the expression 1..T. As in the previous example, the column headings can be any identifiers acceptable to the database software, and the table declaration will take care of the correspondences to AMPL names (as explained below).

AMPL entities that have sufficiently similar indexing generally fit into the same relational table. We could extend the `steelT.mod` table, for instance, by adding a column for values of

```
var Inv {PROD,0..T} >= 0;
```

The table would then have the following layout:

```
PROD    TIME    market  revenue   Make    Sell     Inv
bands   0          .        .       .       .        10
bands   1        6000      25     5990    6000        0
bands   2        6000      26     6000    6000        0
bands   3        4000      27     1400    1400        0
bands   4        6500      27     2000    2000        0
coils   0          .        .       .       .         0
coils   1        4000      30     1407     307     1100
coils   2        2500      35     1400    2500        0
coils   3        3500      37     3500    3500        0
coils   4        4200      39     4200    4200        0
```

We use ''.'' here to mark table entries that correspond to values not defined by the model and data. There is no `market["bands",0]` in the data for this model, for example, although there does exist a value for `Inv["bands",0]` in the results. Database packages vary in their handling of ''missing'' entries of this sort.

Parameters and variables may also be indexed over a set of pairs that is read as data rather than being constructed from one-dimensional sets. For instance, in the example of `transp3.mod` from Chapter 3, we have:

```
set LINKS within {ORIG,DEST};
param cost {LINKS} >= 0;    # shipment costs per unit
var Trans {LINKS} >= 0;     # actual units to be shipped
```

A corresponding relational table has two key columns corresponding to the two components of the indexing set `LINKS`, plus a column each for the parameter and variable that are indexed over `LINKS`:

```
ORIG    DEST    cost    Trans
GARY    DET      14        0
GARY    LAF       8      600
GARY    LAN      11        0
GARY    STL      16      800
CLEV    DET       9     1200
CLEV    FRA      27        0
CLEV    LAF      17      400
CLEV    LAN      12      600
CLEV    STL      26        0
CLEV    WIN       9      400
PITT    FRA      24      900
PITT    FRE      99     1100
PITT    STL      28      900
PITT    WIN      13        0
```

The structure here is the same as in the previous example. There is a row in the table only for each origin-destination pair that is actually in the set LINKS, however, rather than for every possible origin-destination pair.

10.2 Examples of table-handling statements

To transfer information between an AMPL model and a relational table, we begin with a table declaration that establishes the correspondence between them. Certain details of this declaration depend on the software being used to create and maintain the table. In the case of the four-column table of diet data defined above, some of the possibilities are as follows:

- For a Microsoft Access table in a database file diet.mdb:

```
table Foods IN "ODBC" "diet.mdb":
    FOOD <- [FOOD], cost, f_min, f_max;
```

- For a Microsoft Excel range from a workbook file diet.xls:

```
table Foods IN "ODBC" "diet.xls":
    FOOD <- [FOOD], cost, f_min, f_max;
```

- For an ASCII text table in file Foods.tab:

```
table Foods IN:
    FOOD <- [FOOD], cost, f_min, f_max;
```

Each table declaration has two parts. Before the colon, the declaration provides general information. First comes the table name — Foods in the examples above — which will be the name by which the table is known within AMPL. The keyword IN states that the default for all non-key table columns will be read-only; AMPL will read values *in* from these columns and will not write out to them.

Details for locating the table in an external database file are provided by the character strings such as "ODBC" and "diet.mdb", with the AMPL table name (Foods) providing a default where needed:

- For Microsoft Access, the table is to be read from database file diet.mdb using AMPL's ODBC handler. The table's name within the database file is taken to be Foods by default.

- For Microsoft Excel, the table is to be read from spreadsheet file diet.xls using AMPL's ODBC handler. The spreadsheet range containing the table is taken to be Foods by default.

- Where no details are given, the table is read by default from the ASCII text file Foods.tab using AMPL's built-in text table handler.

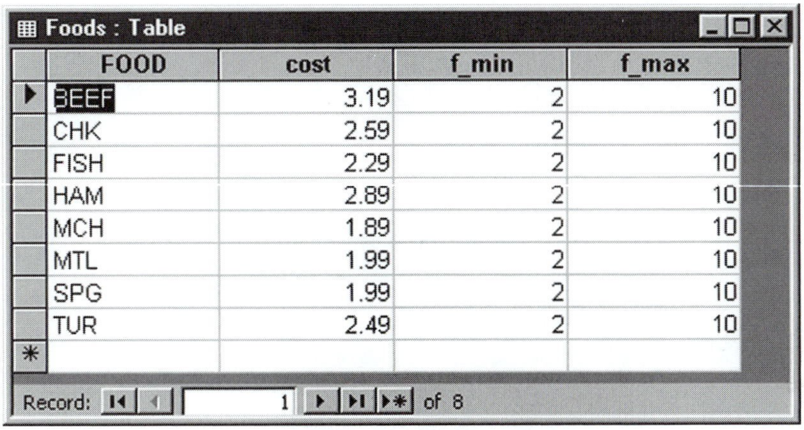

Figure 10-2: Access relational table.

In general, the format of the character strings in the `table` declaration depends upon the table handler being used. The strings required by the handlers used in our examples are described briefly in Section 10.7, and in detail in online documentation for specific table handlers.

After the colon, the `table` declaration gives the details of the correspondence between AMPL entities and relational table columns. The four comma-separated entries correspond to four columns in the table, starting with the key column distinguished by surrounding brackets `[...]`. In this example, the names of the table columns (`FOOD`, `cost`, `f_min`, `f_max`) are the same as the names of the corresponding AMPL components. The expression `FOOD <- [FOOD]` indicates that the entries in the key column `FOOD` are to be copied into AMPL to define the members of the set `FOOD`.

The `table` declaration only defines a correspondence. To read values from columns of a relational table into AMPL sets and parameters, it is necessary to give an explicit

```
read table
```

command.

Thus, if the data values were in an Access relational table like Figure 10-2, the `table` declaration for Access could be used together with the `read table` command to read the members of `FOOD` and values of `cost`, `f_min` and `f_max` into the corresponding AMPL set and parameters:

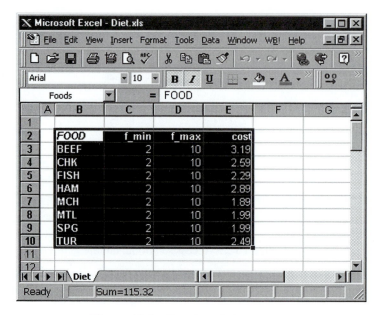

Figure 10-3: Excel worksheet range.

```
ampl: model diet.mod;
ampl: table Foods IN "ODBC" "diet.mdb":
ampl?    FOOD <- [FOOD], cost, f_min, f_max;
ampl: read table Foods;
ampl: display cost, f_min, f_max;
:      cost f_min f_max     :=
BEEF   3.19    2    10
CHK    2.59    2    10
FISH   2.29    2    10
HAM    2.89    2    10
MCH    1.89    2    10
MTL    1.99    2    10
SPG    1.99    2    10
TUR    2.49    2    10
;
```

(The display command confirms that the database values were read as intended.) If the data values were instead in an Excel worksheet range like Figure 10-3, the values would be read in the same way, but using the table declaration for Excel:

```
ampl: model diet.mod;
ampl: table Foods IN "ODBC" "diet.xls":
ampl?    FOOD <- [FOOD], cost, f_min, f_max;
ampl: read table Foods;
```

And if the values were in a file `Foods.tab` containing a text table like this:

```
ampl.tab 1 3
FOOD     cost    f_min    f_max
BEEF     3.19    2        10
CHK      2.59    2        10
FISH     2.29    2        10
HAM      2.89    2        10
MCH      1.89    2        10
MTL      1.99    2        10
SPG      1.99    2        10
TUR      2.49    2        10
```

the declaration for a text table would be used:

```
ampl: model diet.mod;
ampl: table Foods IN: FOOD <- [FOOD], cost, f_min, f_max;
ampl: read table Foods;
```

Because the AMPL table name `Foods` is the same in all three of these examples, the `read table` command is the same for all three: `read table Foods`. In general, the `read table` command only specifies the AMPL name of the table to be read. All information about what is to be read, and how it is to be handled, is taken from the named table's definition in the `table` declaration.

To create the second (7-column) relational table example of the previous section, we could use a pair of table declarations:

```
table ImportFoods IN "ODBC" "diet.mdb" "Foods":
    FOOD <- [FOOD], cost, f_min, f_max;
table ExportFoods OUT "ODBC" "diet.mdb" "Foods":
    FOOD <- [FOOD], Buy, Buy.rc ~ BuyRC,
    {j in FOOD} Buy[j]/f_max[j] ~ BuyFrac;
```

or a single table declaration combining the input and output information:

```
table Foods "ODBC" "diet.mdb": [FOOD] IN, cost IN,
    f_min IN, f_max IN, Buy OUT, Buy.rc ~ BuyRC OUT,
    {j in FOOD} Buy[j]/f_max[j] ~ BuyFrac OUT;
```

These examples show how the AMPL table name (such as `ExportFoods`) may be different from the name of the corresponding table within the external file (as indicated by the subsequent string `"Foods"`). A number of other useful options are also seen here: `IN` and `OUT` are associated with individual columns of the table, rather than with the whole table; `[FOOD] IN` is used as an abbreviation for `FOOD <- [FOOD]`; columns of the table are associated with the values of variables `Buy` and expressions `Buy.rc` and `Buy[j]/f_max[j]`; `Buy.rc ~ BuyRC` and `{j in FOOD} Buy[j]/f_max[j] ~ BuyFrac` associate an AMPL expression (to the left of the `~` operator) with a database column heading (to the right).

To write meaningful results back to the Access database, we need to read all of the diet model's data, then solve, and then give a `write table` command. Here's how it

all might look using separate `table` declarations to read and write the Access table
`Foods`:

```
ampl: model diet.mod;
ampl: table ImportFoods IN "ODBC" "diet.mdb" "Foods":
ampl?    FOOD <- [FOOD], cost, f_min, f_max;
ampl: table Nutrs IN "ODBC" "diet.mdb": NUTR <- [NUTR],
ampl?    n_min, n_max;
ampl: table Amts IN "ODBC" "diet.mdb": [NUTR, FOOD], amt;
ampl: read table ImportFoods;
ampl: read table Nutrs;
ampl: read table Amts;
ampl: solve;
ampl: table ExportFoods OUT "ODBC" "diet.mdb" "Foods":
ampl?    FOOD <- [FOOD],
ampl?    Buy, Buy.rc ~ BuyRC,
ampl?    {j in FOOD} Buy[j]/f_max[j] ~ BuyFrac;
ampl: write table ExportFoods;
```

and here is an alternative using a single declaration to both read and write `Foods`:

```
ampl: model diet.mod;
ampl: table Foods "ODBC" "diet.mdb":
ampl?    [FOOD] IN, cost IN, f_min IN, f_max IN,
ampl?    Buy OUT, Buy.rc ~ BuyRC OUT,
ampl?    {j in FOOD} Buy[j]/f_max[j] ~ BuyFrac OUT;
ampl: table Nutrs IN "ODBC" "diet.mdb":
ampl?    NUTR <- [NUTR], n_min, n_max;
ampl: table Amts IN "ODBC" "diet.mdb": [NUTR, FOOD], amt;
ampl: read table Foods;
ampl: read table Nutrs;
ampl: read table Amts;
ampl: solve;
ampl: write table Foods;
```

Either way, the Access table `Foods` would end up having three additional columns, as
seen in Figure 10-4.

The same operations are handled similarly for other types of database files. In general, the actions of a `write table` command are determined by the previously declared
AMPL table named in the command, and by the status of the external file associated with
the AMPL table through its `table` declaration. Depending on the circumstances, the
`write table` command may create a new external file or table, overwrite an existing
table, overwrite certain columns within an existing table, or append columns to an existing table.

The `table` declaration is the same for multidimensional AMPL entities, except that
there must be more than one key column specified between brackets [and]. For the
steel production example discussed previously, the correspondence to a relational table
could be set up like this:

FOOD	cost	f_min	f_max	Buy	BuyRC	BuyFrac
BEEF	3.19	2	10	5.360613811	3.05311E-16	0.536061381
CHK	2.59	2	10	2	1.18884058	0.2
FISH	2.29	2	10	2	1.144407502	0.2
HAM	2.89	2	10	10	-0.30265132	1
MCH	1.89	2	10	10	-0.5511509	1
MTL	1.99	2	10	10	-1.32890026	1
SPG	1.99	2	10	9.306052856	-2.0123E-15	0.930605286
TUR	2.49	2	10	2	2.731619778	0.2

Record: ◄◄ ◄ [1] ► ►I ►* of 8

Figure 10-4: Access relational table with output columns.

```
table SteelProd "ODBC" "steel.mdb":
   [PROD, TIME], market IN, revenue IN,
   Make OUT, Sell OUT, Inv OUT;
```

Here the key columns PROD and TIME are not specified as IN. This is because the parameters to be read in, market and revenue, are indexed in the AMPL model over the set {PROD, 1..T}, whose membership would be specified by use of other, simpler tables. The read table SteelProd command merely uses the PROD and TIME entries of each database row to determine the pair of indices (subscripts) that are to be associated with the market and revenue entries in the row.

Our transportation example also involves a relational table for two-dimensional entities, and the associated table declaration is similar:

```
table TransLinks "ODBC" "trans.xls" "Links":
   LINKS <- [ORIG, DEST], cost IN, Trans OUT;
```

The difference here is that LINKS, the AMPL set of pairs over which cost and Trans are indexed, is part of the data rather than being determined from simpler sets or parameters. Thus we write LINKS <- [ORIG, DEST], to request that pairs from the key columns be read into LINKS at the same time that the corresponding values are read into cost. This distinction is discussed further in the next section.

As you can see from even our simple examples so far, table statements tend to be cumbersome to type interactively. Instead they are usually placed in AMPL programs, or scripts, which are executed as described in Chapter 13. The read table and write table statements may be included in the scripts as well. You can define a table and then immediately read or write it, as seen in some of our examples, but a script is often more readable if the complex `table` statements are segregated from the statements that read and write the tables.

The rest of this chapter will concentrate on `table` statements. Complete sample scripts and Access or Excel files for the diet, production, and transportation examples can be obtained from the AMPL web site.

10.3 Reading data from relational tables

To use an external relational table for reading only, you should employ a `table` declaration that specifies a read/write status of `IN`. Thus it should have the general form

> `table` *table-name* `IN` *string-list*$_{opt}$:
> *key-spec*, *data-spec*, *data-spec*, ... ;

where the optional *string-list* is specific to the database type and access method being used. (In the interest of brevity, most subsequent examples do not show a *string-list*.) The *key-spec* names the key columns, and the *data-spec* gives the data columns. Data values are subsequently read from the table into AMPL entities by the command

> `read table` *table-name* ;

which determines the values to be read by referring to the `table` declaration that defined *table-name*.

Reading parameters only

To assign values from data columns to like-named AMPL parameters, it suffices to give a bracketed list of key columns and then a list of data columns. The simplest case, where there is only one key column, is exemplified by

> `table Foods IN: [FOOD], cost, f_min, f_max;`

This indicates that the relational table has four columns, comprising a key column `FOOD` and data columns `cost`, `f_min` and `f_max`. The data columns are associated with parameters `cost`, `f_min` and `f_max` in the current AMPL model. Since there is only one key column, all of these parameters must be indexed over one-dimensional sets.

When the command

> `read table Foods`

is executed, the relational table is read one row at a time. A row's entry in the key column is interpreted as a subscript to each of the parameters, and these subscripted parameters are assigned the row's entries from the associated data columns. For example, if the relational table is

FOOD	cost	f_min	f_max
BEEF	3.19	2	10
CHK	2.59	2	10
FISH	2.29	2	10
HAM	2.89	2	10
MCH	1.89	2	10
MTL	1.99	2	10
SPG	1.99	2	10
TUR	2.49	2	10

processing the first row assigns the values 3.19 to parameter cost['BEEF'], 2 to f_min['BEEF'], and 10 to f_max['BEEF']; processing the second row assigns 2.59 to cost['CHK'], 2 to f_min['CHK'], and 10 to f_max['CHK']; and so forth through the six remaining rows.

At the time that the read table command is executed, AMPL makes no assumptions about how the parameters are declared; they need not be indexed over a set named FOOD, and indeed the members of their indexing sets may not yet even be known. Only later, when AMPL first uses each parameter in some computation, does it check the entries read from key column FOOD to be sure that each is a valid subscript for that parameter.

The situation is analogous for multidimensional parameters. The name of each data column must also be the name of an AMPL parameter, and the dimension of the parameter's indexing set must equal the number of key columns. For example, when two key columns are listed within the brackets:

```
table SteelProd IN: [PROD, TIME], market, revenue;
```

the listed data columns, market and revenue, must correspond to AMPL parameters market and revenue that are indexed over two-dimensional sets.

When read table SteelProd is executed, each row's entries in the key columns are interpreted as a pair of subscripts to each of the parameters. Thus if the relational table has contents

PROD	TIME	market	revenue
bands	1	6000	25
bands	2	6000	26
bands	3	4000	27
bands	4	6500	27
coils	1	4000	30
coils	2	2500	35
coils	3	3500	37
coils	4	4200	39

processing the first row will assign 6000 to market['bands',1] and 25 to revenue['bands',1]; processing the second row will assign 6000 to market['bands',2] and 26 to revenue['bands',2]; and so forth through all eight rows. The pairs of subscripts given by the key column entries must be valid for market and revenue when the values of these parameters are first needed by AMPL, but the parameters need not be declared over sets named PROD and TIME. (In fact, in the model from which this example is taken, the parameters are indexed by {PROD, 1..T} where T is a previously defined parameter.)

Since a relational table has only one collection of key columns, AMPL applies the same subscripting to each of the parameters named by the data columns. These parameters are thus usually indexed over the same AMPL set. Parameters indexed over similar sets may also be accommodated in one database table, however, by leaving blank any entries in rows corresponding to invalid subscripts. The way in which a blank entry is indicated is specific to the database software being used.

Values of unindexed (scalar) parameters may be supplied by a relational table that has one row and no key columns, so that each data column contains precisely one value. The corresponding table declaration has an empty *key-spec*, []. For example, to read a value for the parameter T that gives the number of periods in steelT.mod, the table declaration is

```
table SteelPeriods IN: [], T;
```

and the corresponding relational table has one column, also named T, whose one entry is a positive integer.

Reading a set and parameters

It is often convenient to read the members of a set from a table's key column or columns, at the same time that parameters indexed over that set are read from the data columns. To indicate that a set should be read from a table, the *key-spec* in the table declaration is written in the form

    set-name <- [*key-col-spec*, *key-col-spec*, ...]

The <- symbol is intended as an arrow pointing in the direction that the information is moved, from the key columns to the AMPL set.

The simplest case involves reading a one-dimensional set and the parameters indexed over it, as in this example for diet.mod:

```
table Foods IN: FOOD <- [FoodName], cost, f_min, f_max;
```

When the command read table Foods is executed, all entries in the key column FoodName of the relational table are read into AMPL as members of the set FOOD, and the entries in the data columns cost, f_min and f_max are read into the like-named AMPL parameters as previously described. If the key column is named FOOD like the AMPL set, the appropriate table declaration becomes

```
table Foods IN: FOOD <- [FOOD], cost, f_min, f_max;
```

In this special case only, the *key-spec* can also be written in the abbreviated form [FOOD] IN.

An analogous syntax is employed for reading a multidimensional set along with parameters indexed over it. In the case of transp3.mod, for instance, the table declaration could be:

```
table TransLinks IN: LINKS <- [ORIG, DEST], cost;
```

When `read table TransLinks` is executed, each row of the table provides a pair of entries from key columns `ORIG` and `DEST`. All such pairs are read into AMPL as members of the two-dimensional set `LINKS`. Finally, the entries in column `cost` are read into parameter `cost` in the usual way.

As in our previous multidimensional example, the names in brackets need not correspond to sets in the AMPL model. The bracketed names serve only to identify the key columns. The name to the left of the arrow is the only one that must name a previously declared AMPL set; moreover, this set must have been declared to have the same dimension, or arity, as the number of key columns.

It makes sense to read the set `LINKS` from a relational table, because `LINKS` is specifically declared in the model in a way that leaves the corresponding data to be read separately:

```
set ORIG;
set DEST;
set LINKS within {ORIG,DEST};
param cost {LINKS} >= 0;
```

By contrast, in the similar model `transp2.mod`, `LINKS` is defined in terms of two one-dimensional sets:

```
set ORIG;
set DEST;
set LINKS = {ORIG,DEST};
param cost {LINKS} >= 0;
```

and in `transp.mod`, no named two-dimensional set is defined at all:

```
set ORIG;
set DEST;
param cost {ORIG,DEST} >= 0;
```

In these latter cases, a `table` declaration would still be needed for reading parameter `cost`, but it would not specify the reading of any associated set:

```
table TransLinks IN: [ORIG, DEST], cost;
```

Separate relational tables would instead be used to provide members for the one-dimensional sets `ORIG` and `DEST` and values for the parameters indexed over them.

When a `table` declaration specifies an AMPL set to be assigned members, its list of *data-spec*s may be empty. In that case only the key columns are read, and the only action of `read table` is to assign the key column values as members of the specified AMPL set. For instance, with the statement

```
table TransLinks IN: LINKS <- [ORIG, DEST];
```

a subsequent `read table` statement would cause just the values for the set `LINKS` to be read, from the two key columns in the corresponding database table.

Establishing correspondences

An AMPL model's set and parameter declarations do not necessarily correspond in all respects to the organization of tables in relevant databases. Where the difference is substantial, it may be necessary to use the database's query language (often SQL) to derive temporary tables that have the structure required by the model; an example is given in the discussion of the ODBC handler later in this chapter. A number of common, simple differences can be handled directly, however, through features of the `table` declaration.

Differences in naming are perhaps the most common. A `table` declaration can associate a data column with a differently named AMPL parameter by use of a *data-spec* of the form *param-name ~ data-col-name*. Thus, for example, if table `Foods` were instead defined by

```
table Foods IN:
    [FOOD], cost, f_min ~ lowerlim, f_max ~ upperlim;
```

the AMPL parameters `f_min` and `f_max` would be read from data columns `lowerlim` and `upperlim` in the relational table. (Parameter `cost` would be read from column `cost` as before.)

A similarly generalized form, *index ~ key-col-name*, can be used to associate a kind of dummy index with a key column. This index may then be used in a subscript to the optional *param-name* in one or more *data-spec*s. Such an arrangement is useful in a number of situations where the key column entries do not exactly correspond to the subscripts of the parameters that are to receive table values. Here are three common cases.

Where a numbering of some kind in the relational table is systematically different from the corresponding numbering in the AMPL model, a simple expression involving a key column *index* can translate from the one numbering scheme to the other. For example, if time periods were counted from 0 in the relational table data rather than from 1 as in the model, an adjustment could be made in the `table` declaration as follows:

```
table SteelProd IN: [p ~ PROD, t ~ TIME],
    market[p,t+1] ~ market, revenue[p,t+1] ~ revenue;
```

In the second case, where AMPL parameters have subscripts from the same sets but in different orders, key column *index*es must be employed to provide a correct index order. If `market` is indexed over {PROD, 1..T} but `revenue` is indexed over {1..T, PROD}, for example, a `table` declaration to read values for these two parameters should be written as follows:

```
table SteelProd IN: [p ~ PROD, t ~ TIME],
    market, revenue[t,p] ~ revenue;
```

Finally, where the values for an AMPL parameter are divided among several database columns, key column *index*es can be employed to describe the values to be found in each column. For instance, if the `revenue` values are given in one column for `"bands"` and in another column for `"coils"`, the corresponding `table` declaration could be written like this:

```
    table SteelProd IN: [t ~ TIME],
       revenue["bands",t] ~ revbands,
       revenue["coils",t] ~ revcoils;
```

It is tempting to try to shorten declarations of these kinds by dropping the ~ *data-col-name*, to produce, say,

```
    table SteelProd IN:
       [p ~ PROD, t ~ TIME], market, revenue[t,p];   # ERROR
```

This will usually be rejected as an error, however, because `revenue[t,p]` is not a valid name for a relational table column in most database software. Instead it is necessary to write

```
    table SteelProd IN:
       [p ~ PROD, t ~ TIME], market, revenue[t,p] ~ revenue;
```

to indicate that the AMPL parameters `revenue[t,p]` receive values from the column `revenue` of the table.

More generally, a ~ synonym will have to be used in any situation where the AMPL expression for the recipient of a column's data is not itself a valid name for a database column. The rules for valid column names tend to be the same as the rules for valid component names in AMPL models, but they can vary in details depending on the database software that is being used to create and maintain the tables.

Reading other values

In a `table` declaration used for input, an *assignable* AMPL expression may appear anywhere that a parameter name would be allowed. An expression is assignable if it can be assigned a value, such as by placing it on the left side of `:=` in a `let` command.

Variable names are assignable expressions. Thus a `table` declaration can specify columns of data to be read into variables, for purposes of evaluating a previously stored solution or providing a good initial solution for a solver.

Constraint names are also assignable expressions. Values ''read into a constraint'' are interpreted as initial dual values for some solvers, such as MINOS.

Any variable or constraint name qualified by an assignable suffix is also an assignable expression. Assignable suffixes include the predefined `.sstatus` and `.relax` as well as any user-defined suffixes. For example, if the diet problem were changed to have integer variables, the following `table` declaration could help to provide useful information for the CPLEX solver (see Section 14.3):

```
    table Foods IN: FOOD IN,
       cost, f_min, f_max, Buy, Buy.priority ~ prior;
```

An execution of `read table Foods` would supply members for set FOOD and values for parameters `cost`, `f_min` and `f_max` in the usual way, and would also assign initial values and branching priorities to the `Buy` variables.

10.4 Writing data to relational tables

To use an external relational table for writing only, you should employ a `table` declaration that specifies its read/write status to be OUT. The general form of such a declaration is

> table *table-name* OUT *string-list* :
>   *key-spec*, *data-spec*, *data-spec*, ... ;

where the optional *string-list* is specific to the database type and access method being used. (Again, most subsequent examples do not include a *string-list*.) AMPL expression values are subsequently written to the table by the command

> write table *table-name* ;

which uses the `table` declaration that defined *table-name* to determine the information to be written.

A `table` declaration for writing specifies an external file and possibly a relational table within that file, either explicitly in the *string-list* or implicitly by default rules. Normally the named external file or table is created if it does not exist, or is overwritten otherwise. To specify that instead certain columns are to be replaced or are to be added to a table, the `table` declaration must incorporate one or more *data-spec*s that have read/write status IN or INOUT, as discussed in Section 10.5. A specific table handler may also have its own more detailed rules for determining when files and tables are modified or overwritten, as explained in its documentation.

The *key-spec*s and *data-spec*s of `table` declarations for writing external tables superficially resemble those for reading. The range of AMPL expressions allowed when writing is much broader, however, including essentially all set-valued and numeric-valued expressions. Moreover, whereas the table rows to be read are those of some existing table, the rows to be written must be determined from AMPL expressions in some part of a `table` declaration. Specifically, rows to be written can be inferred either from the *data-spec*s, using the same conventions as in `display` commands, or from the *key-spec*. Each of these alternatives employs a characteristic `table` syntax as described below.

Writing rows inferred from the data specifications

If the *key-spec* is simply a bracketed list of the names of key columns,

> [*key-col-name*, *key-col-name*, ...]

the `table` declaration works much like the `display` command. It determines the external table rows to be written by taking the union of the indexing sets stated or implied in the *data-spec*s. The format of the *data-spec* list is the same as in `display`, except that all of the items listed must have the same dimension.

In the simplest case, the *data-spec*s are the names of model components indexed over the same set:

> table Foods OUT: [FoodName], f_min, Buy, f_max;

When `write table Foods` is executed, it creates a key column `FoodName` and data columns `f_min`, `Buy`, and `f_max`. Since the AMPL components corresponding to the data columns are all indexed over the AMPL set `FOOD`, one row is created for each member of `FOOD`. In a representative row, a member of `FOOD` is written to the key column `FoodName`, and the values of `f_min`, `Buy`, and `f_max` subscripted by that member are written to the like-named data columns. For the data used in the diet example, the resulting relational table would be:

FoodName	f_min	Buy	f_max
BEEF	2	5.36061	10
CHK	2	2	10
FISH	2	2	10
HAM	2	10	10
MCH	2	10	10
MTL	2	10	10
SPG	2	9.30605	10
TUR	2	2	10

Tables corresponding to higher-dimensional sets are handled analogously, with the number of bracketed key-column names listed in the *key-spec* being equal to the dimension of the items in the *data-spec*. Thus a table containing the results from `steelT.mod` could be defined as

```
table SteelProd OUT: [PROD, TIME], Make, Sell, Inv;
```

Because `Make` and `Sell` are indexed over `{PROD,1..T}`, while `Inv` is indexed over `{PROD,0..T}`, a subsequent `write table SteelProd` command would produce a table with one row for each member of the union of these sets:

PROD	TIME	Make	Sell	Inv
bands	0	.	.	10
bands	1	5990	6000	0
bands	2	6000	6000	0
bands	3	1400	1400	0
bands	4	2000	2000	0
coils	0	.	.	0
coils	1	1407	307	1100
coils	2	1400	2500	0
coils	3	3500	3500	0
coils	4	4200	4200	0

Two rows are empty in the columns for `Make` and `Sell`, because `("bands",0)` and `("coils",0)` are not members of the index sets of `Make` and `Sell`. We use a "." here to indicate the empty table entries, but the actual appearance and handling of empty entries will vary depending on the database software being used.

If this form is applied to writing suffixed variable or constraint names, such as the dual and slack values related to the constraint `Diet`:

```
table Nutrs OUT: [Nutrient],
    Diet.lslack, Diet.ldual, Diet.uslack, Diet.udual;   # ERROR
```

a subsequent `write table Nutrs` command is likely to be rejected, because names with a ''dot'' in the middle are not allowed as column names by most database software:

```
ampl: write table Nutrs;
Error executing "write table" command:
   Error writing table Nutrs with table handler ampl.odbc:
   Column 2's name "Diet.lslack" contains non-alphanumeric
      character '.'.
```

This situation requires that each AMPL expression be followed by the operator ~ and a corresponding valid column name for use in the relational table:

```
table Nutrs OUT: [Nutrient],
    Diet.lslack ~ lb_slack, Diet.ldual ~ lb_dual,
    Diet.uslack ~ ub_slack, Diet.udual ~ ub_dual;
```

This says that the values represented by `Diet.lslack` should be placed in a column named `lb_slack`, the values represented by `Diet.ldual` should be placed in a column named `lb_dual`, and so forth. With the table defined in this way, a `write table Nutrs` command produces the intended relational table:

Nutrient	lb_slack	lb_dual	ub_slack	ub_dual
A	1256.29	0	18043.7	0
B1	336.257	0	18963.7	0
B2	0	0.404585	19300	0
C	982.515	0	18317.5	0
CAL	3794.62	0	4205.38	0
NA	50000	0	0	-0.00306905

The ~ can also be used with unsuffixed names, if it is desired to assign the dabatase column a name different from the corresponding AMPL entity.

More general expressions for the values in data columns require the use of dummy indices, in the same way that they are used in the data-list of a `display` command. Since indexed AMPL expressions are rarely valid column names for a database, they should generally be followed by ~ *data-col-name* to provide a valid name for the corresponding relational table column that is to be written. To write a column `servings` containing the number of servings of each food to be bought and a column `percent` giving the amount bought as a percentage of the maximum allowed, for example, the `table` declaration could be given as either

```
table Purchases OUT: [FoodName],
    Buy ~ servings, {j in FOOD} 100*Buy[j]/f_max[j] ~ percent;
```

or

```
table Purchases OUT: [FoodName],
    {j in FOOD} (Buy[j] ~ servings,
                 100*Buy[j]/f_max[j] ~ percent);
```

Either way, since both *data-spec*s give expressions indexed over the AMPL set FOOD, the resulting table has one row for each member of that set:

```
FoodName    servings    percent
BEEF          5.36061    53.6061
CHK           2          20
FISH          2          20
HAM          10         100
MCH          10         100
MTL          10         100
SPG           9.30605    93.0605
TUR           2          20
```

The expression in a *data-spec* may also use operators like `sum` that define their own dummy indices. Thus a table of total production and sales by period for `steelT.mod` could be specified by

```
table SteelTotal OUT:  [TIME],
   {t in 1..T} (sum {p in PROD} Make[p,t] ~ Made,
                sum {p in PROD} Sell[p,t] ~ Sold);
```

As a two-dimensional example, a table of the amounts sold and the fractions of demand met could be specified by

```
table SteelSales OUT: [PROD, TIME], Sell,
   {p in PROD, t in 1..T} Sell[p,t]/market[p,t] ~ FracDemand;
```

The resulting external table would have key columns `PROD` and `TIME`, and data columns `Sell` and `FracDemand`.

Writing rows inferred from a key specification

An alternative form of `table` declaration specifies that one table row is to be written for each member of an explicitly specified AMPL set. For the declaration to work in this way, the *key-spec* must be written as

> *set-spec* `->` [*key-col-spec*, *key-col-spec*, ...]

In contrast to the arrow `<-` that points from a key-column list to an AMPL set, indicating values to be read into the set, this form uses an arrow `->` that points from an AMPL set to a key column list, indicating information to be written from the set into the key columns.

An explicit expression for the row index set is given by the *set-spec*, which can be the name of an AMPL set, or any AMPL set-expression enclosed in braces `{ }`. The *key-col-spec*s give the names of the corresponding key columns in the database. Dummy indices, if needed, can appear either with the *set-spec* or the *key-col-spec*s, as we will show.

The simplest case of this form involves writing database columns for model components indexed over the same one-dimensional set, as in this example for `diet.mod`:

```
table Foods OUT: FOOD -> [FoodName], f_min, Buy, f_max;
```

When `write table Foods` is executed, a table row is created for each member of the AMPL set `FOOD`. In that row, the set member is written to the key column `FoodName`, and the values of `f_min`, `Buy`, and `f_max` subscripted by the set member are written to

the like-named data columns. (For the data used in our diet example, the resulting table
would be the same as for the FoodName table given previously in this section.) If the
key column has the same name, FOOD, as the AMPL set, the appropriate table declaration becomes

```
table Foods OUT: FOOD -> [FOOD], f_min, Buy, f_max;
```

In this special case only, the *key-spec* can also be written in the abbreviated form
[FOOD] OUT.

The use of ~ with AMPL names and suffixed names is governed by the considerations
previously described, so that the example of diet slack and dual values would be written

```
table Nutrs OUT: NUTR -> [Nutrient],
    Diet.lslack ~ lb_slack, Diet.ldual ~ lb_dual,
    Diet.uslack ~ ub_slack, Diet.udual ~ ub_dual;
```

and write table Nutrs would give the same table as previously shown.

More general expressions for the values in data columns require the use of dummy
indices. Since the rows to be written are determined from the *key-spec*, however, the
dummies are also defined there (rather than in the *data-spec*s as in the alternative form
above). To specify a column containing the amount of a food bought as a percentage of
the maximum allowed, for example, it is necessary to write 100*Buy[j]/f_max[j],
which in turn requires that dummy index j be defined. The definition may appear either
in a *set-spec* of the form { *index-list* in *set-expr* }:

```
table Purchases OUT: {j in FOOD} -> [FoodName],
    Buy[j] ~ servings, 100*Buy[j]/f_max[j] ~ percent;
```

or in a *key-col-spec* of the form *index ~ key-col-name*:

```
table Purchases OUT: FOOD -> [j ~ FoodName],
    Buy[j] ~ servings, 100*Buy[j]/f_max[j] ~ percent;
```

These two forms are equivalent. Either way, as each row is written, the index j takes the
key column value, which is used in interpreting the expressions that give the values for
the data columns. For our example, the resulting table, having key column FoodName
and data columns servings and percent, is the same as previously shown. Similarly, the previous example of the table SteelTotal could be written as either

```
table SteelTotal OUT: {t in 1..T} -> [TIME],
    sum {p in PROD} Make[p,t] ~ Made,
    sum {p in PROD} Sell[p,t] ~ Sold;
```

or

```
table SteelTotal OUT: {1..T} -> [t ~ TIME],
    sum {p in PROD} Make[p,t] ~ Made,
    sum {p in PROD} Sell[p,t] ~ Sold;
```

The result will have a key column TIME containing the integers 1 through T, and data
columns Made and Sold containing the values of the two summations. (Notice that

since `1..T` is a set-expression, rather than the name of a set, it must be included in braces to be used as a *set-spec*.)

Tables corresponding to higher-dimensional sets are handled analogously, with the number of *key-col-spec*s listed in brackets being equal to the dimension of the *set-spec*. Thus a table containing the results from `steelT.mod` could be defined as

```
table SteelProd OUT:
    {PROD, 1..T} -> [PROD, TIME], Make, Sell, Inv;
```

and a subsequent `write table SteelProd` would produce a table of the form

```
PROD    TIME    Make    Sell    Inv
bands    1      5990    6000      0
bands    2      6000    6000      0
bands    3      1400    1400      0
bands    4      2000    2000      0
coils    1      1407     307   1100
coils    2      1400    2500      0
coils    3      3500    3500      0
coils    4      4200    4200      0
```

This result is not quite the same as the table produced by the previous `SteelProd` example, because the rows to be written here correspond explicitly to the members of the set `{PROD, 1..T}`, rather than being inferred from the indexing sets of `Make`, `Sell`, and `Inv`. In particular, the values of `Inv["bands",0]` and `Inv["coils",0]` do not appear in this table.

The options for dummy indices in higher dimensions are the same as in one dimension. Thus our example `SteelSales` could be written either using dummy indices defined in the *set-spec*:

```
table SteelSales OUT:
    {p in PROD, t in 1..T} -> [PROD, TIME],
    Sell[p,t] ~ sold, Sell[p,t]/market[p,t] ~ met;
```

or with dummy indices added to the *key-col-spec*s:

```
table SteelSales OUT:
    {PROD,1..T} -> [p ~ PROD, t ~ TIME],
    Sell[p,t] ~ sold, Sell[p,t]/market[p,t] ~ met;
```

If dummy indices happen to appear in both the *set-spec* and the *key-col-spec*s, ones in the *key-col-spec*s take precedence.

10.5 Reading and writing the same table

To read data from a relational table and then write results to the same table, you can use a pair of `table` declarations that reference the same file and table names. You may also be able to combine these declarations into one that specifies some columns to be read and others to be written. This section gives examples and instructions for both of these possibilities.

Reading and writing using two `table` declarations

A single external table can be read by one `table` declaration and later written by another. The two `table` declarations follow the rules for reading and writing given above.

In this situation, however, one usually wants `write table` to add or rewrite selected columns, rather than overwriting the entire table. This preference can be communicated to the AMPL table handler by including input as well as output columns in the `table` declaration that is to be used for writing. Columns intended for input to AMPL can be distinguished from those intended for output to the external table by specifying a read/write status column by column (rather than for the table as a whole).

As an example, an external table for `diet.mod` might consist of columns `cost`, `f_min` and `f_max` containing input for the model, and a column `Buy` containing the results. If this is maintained as a Microsoft Access table named `Diet` within a file `diet.mdb`, the `table` declaration for reading data into AMPL could be

```
table FoodInput IN "ODBC" "diet1.mdb" "Diet":
   FOOD <- [FoodName], cost, f_min, f_max;
```

The corresponding declaration for writing the results would have a different AMPL *table-name* but would refer to the same Access table and file:

```
table FoodOutput "ODBC" "diet1.mdb" "Diet":
   [FoodName], cost IN, f_min IN, Buy OUT, f_max IN;
```

When `read table FoodInput` is executed, only the three columns listed in the `table FoodInput` declaration are read; if there is an existing column named Buy, it is ignored. Later, when the problem has been solved and `write table FoodOutput` is executed, only the one column that has read/write status `OUT` in the `table FoodOut-put` declaration is written to the Access table, while the table's other columns are left unmodified.

Although details may vary with the database software used, the general convention is that overwriting of an entire existing table or file is intended only when *all* data columns in the `table` declaration have read/write status `OUT`. Selective rewriting or addition of columns is intended otherwise. Thus if our AMPL table for output had been declared

```
table FoodOutput "ODBC" "diet1.mdb" "Diet":
            [FoodName], Buy OUT;
```

then all of the data columns in Access table `Diet` would have been deleted by `write table FoodOutput`, but the alternative

```
table FoodOutput "ODBC" "diet1.mdb" "Diet":
            [FoodName], Buy;
```

would have only overwritten the column Buy, as in the example we originally gave, since there is a data column (namely Buy itself) that does not have read/write status OUT. (The default, when no status is given, is INOUT.)

Reading and writing using the same `table` declaration

In many cases, all of the information for both reading and writing an external table can be specified in the same `table` declaration. The *key-spec* may use the arrow <- to read contents of the key columns into an AMPL set, -> to write members of an AMPL set into the key columns, or <-> to do both. A *data-spec* may specify read/write status IN for a column that will only be read into AMPL, OUT for a column that will only be written out from AMPL, or INOUT for a column that will be both read and written.

A `read table` *table-name* command reads only the key or data columns that are specified in the declaration of *table-name* as being IN or INOUT. A `write table` *table-name* command analogously writes to only the columns that are specified as OUT or INOUT.

As an example, the declarations defining FoodInput and FoodOutput above could be replaced by

```
table Foods "ODBC" "diet1.mdb" "Diet":
    FOOD <- [FoodName], cost IN, f_min IN, Buy OUT, f_max IN;
```

A `read table Foods` would then read only from key column FoodName and data columns cost, f_min and f_max. A later `write table Foods` would write only to the column Buy.

10.6 Indexed collections of tables and columns

In some circumstances, it is convenient to declare an indexed collection of tables, or to define an indexed collection of data columns within a table. This section explains how indexing of these kinds can be specified within the `table` declaration.

To illustrate indexed collections of tables, we present a script (Chapter 13) that automatically solves a series of scenarios stored separately. To illustrate indexed collections of columns, we show how a two-dimensional spreadsheet table can be read.

All of our examples of these features make use of AMPL's character-string expressions to generate names for series of files, tables, or columns. For more on string expressions, see Sections 13.7 and A.4.2.

Indexed collections of tables

AMPL table declarations can be indexed in much the same way as sets, parameters, and other model components. An optional {*indexing-expr*} follows the *table-name*:

```
table table-name {indexing-expr}opt  string-listopt  : ...
```

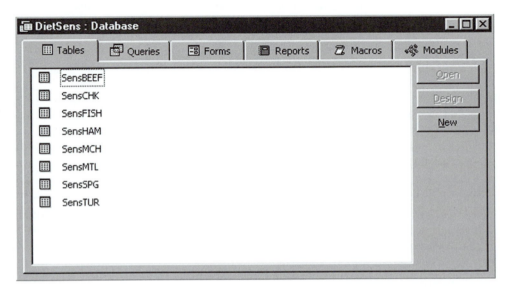

Figure 10-5: Access database with tables of sensitivity analysis.

One table is defined for each member of the set specified by the *indexing-expr*. Individual tables in this collection are denoted in the usual way, by appending a bracketed subscript or subscripts to the *table-name*.

As an example, the following declaration defines a collection of AMPL tables indexed over the set of foods in diet.mod, each table corresponding to a different database table in the Access file DietSens.mdb:

```
table DietSens {j in FOOD}
    OUT "ODBC" "DietSens.mdb" ("Sens" & j):
        [Food], f_min, Buy, f_max;
```

Following the rules for the standard ODBC table handler, the Access table names are given by the third item in the *string-list*, the string expression ("Sens" & j). Thus the AMPL table DietSens["BEEF"] is associated with the Access table SensBEEF, the AMPL table DietSens["CHK"] is associated with the Access table SensCHK, and so forth. The following AMPL script uses these tables to record the optimal diet when there is a two-for-the-price-of-one sale on each of the foods:

```
for {j in FOOD} {
    let cost[j] := cost[j] / 2;
    solve;
    write table DietSens[j];
    let cost[j] := cost[j] * 2;
}
```

Food	BuyBEEF	BuyCHK	BuyFISH	BuyHAM	BuyMCH	BuyMTL	BuyS
BEEF	10	8.5681492	6.5677749	5.3606138	5.3606138	5.3606138	5.778
CHK	2	5.0738881	2	2	2	2	
FISH	2	2	10	2	2	2	
HAM	7.0164474	10	10	10	10	10	
MCH	10	10	10	10	10	10	8.887
MTL	10	10	10	10	10	10	
SPG	6.6557018	2	2.7655584	9.3060529	9.3060529	9.3060529	
TUR	2	2	2	2	2	2	

Record: 14 ◀ | 6 ▶ ▶I ▶* of 8

Figure 10-6: Alternate Access table for sensitivity analysis.

For the data in `diet2a.dat`, the set FOOD has eight members, so eight tables are written in the Access database, as seen in Figure 10-5. If instead the `table` declaration were to give a string expression for the second string in the *string-list*, which specifies the Access filename:

```
table DietSens {j in FOOD}
    OUT "ODBC" ("DietSens" & j & ".mdb"):
        [Food], f_min, Buy, f_max;
```

then AMPL would write eight different Access database files, named `DietSensBEEF.mdb`, `DietSensCHK.mdb`, and so forth, each containing a single table named (by default) `DietSens`. (These files must have been created before the `write table` commands are executed.)

A string expression can be used in a similar way to make every member of an indexed collection of AMPL tables correspond to the same Access table, but with a different *data-col-name* for the optimal amounts:

```
table DietSens {j in FOOD} "ODBC" "DietSens.mdb":
    [Food], Buy ~ ("Buy" & j);
```

Then running the script shown above will result in the Access table of Figure 10-6. The AMPL tables in this case were deliberately left with the default read/write status, INOUT. Had the read/write status been specified as OUT, then each `write table` would have overwritten the columns created by the previous one.

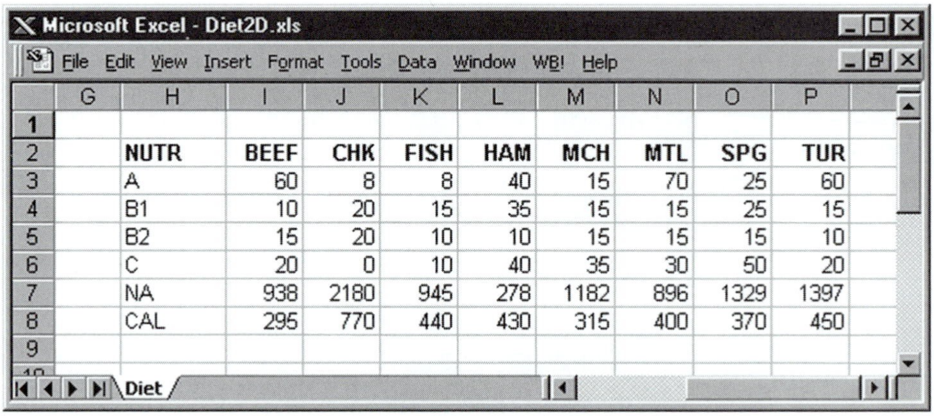

NUTR	BEEF	CHK	FISH	HAM	MCH	MTL	SPG	TUR
A	60	8	8	40	15	70	25	60
B1	10	20	15	35	15	15	25	15
B2	15	20	10	10	15	15	15	10
C	20	0	10	40	35	30	50	20
NA	938	2180	945	278	1182	896	1329	1397
CAL	295	770	440	430	315	400	370	450

Figure 10-7: Two-dimensional AMPL table in Excel.

Indexed collections of data columns

Because there is a natural correspondence between data columns of a relational table and indexed collections of entities in an AMPL model, each *data-spec* in a `table` declaration normally refers to a different AMPL parameter, variable, or expression. Occasionally the values for one AMPL entity are split among multiple data columns, however. Such a case can be handled by defining a collection of data columns, one for each member of a specified indexing set.

The most common use of this feature is to read or write two-dimensional tables. For example, the data for the parameter

```
param amt {NUTR,FOOD} >= 0;
```

from `diet.mod` might be represented in an Excel spreadsheet as a table with nutrients labeling the rows and foods the columns (Figure 10-7). To read this table using AMPL's external database features, we must regard it as having one key column, under the heading `NUTR`, and data columns headed by the names of individual foods. Thus we require a `table` declaration whose *key-spec* is one-dimensional and whose *data-spec*s are indexed over the AMPL set `FOOD`:

```
table dietAmts IN "ODBC" "Diet2D.xls":
   [i ~ NUTR], {j in FOOD} <amt[i,j] ~ (j)>;
```

The *key-spec* `[i ~ NUTR]` associates the first table column with the set `NUTR`. The *data-spec* `{j in FOOD} <...>` causes AMPL to generate an individual *data-spec* for each member of set `FOOD`. Specifically, for each `j in FOOD`, AMPL generates the *data-spec* `amt[i,j] ~ (j)`, where `(j)` is the AMPL string expression for the heading of the external table column for food `j`, and `amt[i,j]` denotes the parameter to which the values in that column are to be written. (According to the convention used here and in other AMPL declarations and commands, the parentheses around (j) cause it to be interpreted as an expression for a string; without the parentheses it would denote a column name consisting of the single character j.)

A similar approach works for writing two-dimensional tables to spreadsheets. As an example, after steelT.mod is solved, the results could be written to a spreadsheet using the following table declaration:

```
table Results1 OUT "ODBC" "steel1out.xls":
    {p in PROD} -> [Product],
        Inv[p,0] ~ Inv0,
        {t in 1..T} < Make[p,t] ~ ('Make' & t),
                      Sell[p,t] ~ ('Sell' & t),
                      Inv[p,t]  ~ ('Inv' & t) >;
```

or, equivalently, using display-style indexing:

```
table Results2 OUT "ODBC" "steel2out.xls":
    [Product],
        {p in PROD} Inv[p,0] ~ Inv0,
        {t in 1..T} < {p in PROD} (Make[p,t] ~ ('Make' & t),
                                   Sell[p,t] ~ ('Sell' & t),
                                   Inv[p,t]  ~ ('Inv' & t) ) >;
```

The key column labels the rows with product names. The data columns include one for the initial inventories, and then three representing production, sales, and inventories, respectively, for each period, as in Figure 10-8. Conceptually, there is a symmetry between the row and column indexing of a two-dimensional table. But because the tables in these examples are being treated as relational tables, the table declaration must treat the row indexing and the column indexing in different ways. As a result, the expressions describing row indexing are substantially different from those describing column indexing.

As these examples suggest, the general form for specifying an indexed collection of table columns is

> {*indexing-expr*} < *data-spec*, *data-spec*, *data-spec*, ... >

where each *data-spec* has any of the forms previously given. For each member of the set specified by the *indexing-expr*, AMPL generates one copy of each *data-spec* within the angle brackets <...>. The *indexing-expr* also defines one or more dummy indices that run over the index set; these indices are used in expressions within the *data-spec*s, and also appear in string expressions that give the names of columns in the external database.

10.7 Standard and built-in table handlers

To work with external database files, AMPL relies on table handlers. These are add-ons, usually in the form of shared or dynamic link libraries, that can be loaded as needed.

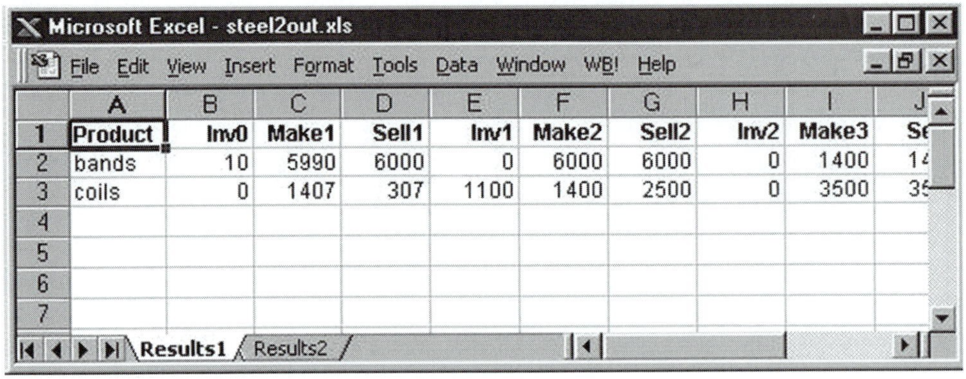

Figure 10-8: Another two-dimensional Excel table.

AMPL is distributed with a "standard" table handler that runs under Microsoft Windows and communicates via the Open Database Connectivity (ODBC) application programming interface; it recognizes relational tables in the formats used by Access, Excel, and any other application for which an ODBC driver exists on your computer. Additional handlers may be supplied by vendors of AMPL or of database software.

In addition to any supplied handlers, minimal ASCII and binary relational table file handlers are built into AMPL for testing. Vendors may include other built-in handlers. If you are not sure which handlers are currently seen by your copy of AMPL, the features described in A.13 can get you a list of active handlers and brief instructions for using them.

As the introductory examples of this chapter have shown, AMPL communicates with handlers through the *string-list* in the `table` declaration. The form and interpretation of the *string-list* are specific to each handler. The remainder of this section describes the *string-list*s that are recognized by AMPL's standard ODBC handler. Following a general introduction, specific instructions are provided for the two applications, Access and Excel, that are used in many of the examples in preceding sections. A final subsection describes the string-lists recognized by the built-in binary and ASCII table handlers.

Using the standard ODBC table handler

In the context of a declaration that begins `table` *table-name* ...., the general form of the *string-list* for the standard ODBC table handler is

> `"ODBC"` *"connection-spec"* *"external-table-spec"* $_{opt}$ `"verbose"` $_{opt}$

The first string tells AMPL that data transfers using this table declaration should employ the standard ODBC handler. Subsequent strings then provide directions to that handler.

The second string identifies the external database file that is to be read or written upon execution of `read table` *table-name* or `write table` *table-name* commands. There are several possibilities, depending on the form of the *connection-spec* and the configuration of ODBC on your computer.

If the *connection-spec* is a filename of the form *name.ext*, where *ext* is a 3-letter extension associated with an installed ODBC driver, then the named file is the database file. This form can be seen in a number of our examples, where filenames of the forms *name.*mdb and *name.*xls refer to Access and Excel files, respectively.

Other forms of *connection-spec* are more specific to ODBC, and are explained in online documentation. Information about your computer's configuration of ODBC drivers, data source names, file data sources, and related entities can be examined and changed through the Windows ODBC control panel.

The third string normally gives the name of the relational table, within the specified file, that is to be read or written upon execution of `read table` or `write table` commands. If the third string is omitted, the name of the relational table is taken to be the same as the *table-name* of the containing `table` declaration. For writing, if the indicated table does not exist, it is created; if the table exists but all of the `table` declaration's *data-spec*s have read/write status `OUT`, then it is overwritten. Otherwise, writing causes the existing table to be modified; each column written either overwrites an existing column of the same name, or becomes a new column appended to the table.

Alternatively, if the third string has the special form

    "SQL=*sql-query*"

the table declaration applies to the relational table that is (temporarily) created by a statement in the Structured Query Language, commonly abbreviated SQL. Specifically, a relational table is first constructed by executing the SQL statement given by *sql-query*, with respect to the database file given by the second string in the `table` declaration's *string-list*. Then the usual interpretations of the `table` declaration are applied to the constructed table. All columns specified in the declaration should have read/write status `IN`, since it would make no sense to write to a temporary table. Normally the *sql-query* is a `SELECT` statement, which is SQL's primary device for operating on tables to create new ones.

As an example, if you wanted to read as data for `diet.mod` only those foods having a cost of $2.49 or less, you could use an SQL query to extract the relevant records from the `Foods` table of your database:

```
table cheapFoods IN "ODBC" "diet.mdb"
   "SQL=SELECT * FROM Foods WHERE cost <= 2.49":
   FOOD <- [FOOD], cost, f_min, f_max;
```

Then to read the relevant data for parameter `amt`, which is indexed over nutrients and foods, you would want to read only those records that pertained to a food having a cost of

$2.49 or less. Here is one way that an SQL query could be used to extract the desired records:

```
option selectAmts "SELECT NUTR, Amts.FOOD, amt "
    "FROM Amts, Foods "
        "WHERE Amts.FOOD = Foods.FOOD and cost <= 2.49";

table cheapAmts IN "ODBC" "diet.mdb" ("SQL=" & $selectAmts):
    [NUTR, FOOD], amt;
```

Here we have used an AMPL option to store the string containing the SQL query. Then the table declaration's third string can be given by the relatively short string expression `"SQL="` & `$selectAmts`.

The string `verbose` after the first three strings requests diagnostic messages — such as the DSN= string that ODBC reports using — whenever the containing table declaration is used by a `read table` or `write table` command.

Using the standard ODBC table handler with Access and Excel

To set up a relational table correspondence for reading or writing Microsoft Access files, specify the *ext* in the second string of the *string-list* as mdb:

> `"ODBC"` `"`*filename*`.mdb"` `"`*external-table-spec*`"` *opt*

The file named by the second string must exist, though for writing it may be a database that does not yet contain any tables.

To set up a relational table correspondence for reading or writing Microsoft Excel spreadsheets, specify the *ext* in the second string of the *string-list* as xls:

> `"ODBC"` `"`*filename*`.xls"` `"`*external-table-spec*`"` *opt*

In this case, the second string identifies the external Excel workbook file that is to be read or written. For writing, the file specified by the second string is created if it does not exist already.

The *external-table-spec* specified by the third string identifies a spreadsheet range, within the specified file, that is to be read or written; if this string is absent, it is taken to be the *table-name* given at the start of the `table` declaration. For reading, the specified range must exist in the Excel file. For writing, if the range does not exist, it is created, at the upper left of a new worksheet having the same name. If the range exists but all of the table declaration's *data-spec*s have read/write status OUT, it is overwritten. Otherwise, writing causes the existing range to be modified. Each column written either overwrites an existing column of the same name, or becomes a new column appended to the table; each row written either overwrites entries in an existing row having the same key column entries, or becomes a new row appended to the table.

When writing causes an existing range to be extended, rows or columns are added at the bottom or right of the range, respectively. The cells of added rows or columns must be empty; otherwise, the attempt to write the table fails and the `write table` command elicits an error message. After a table is successfully written, the corresponding range is created or adjusted to contain exactly the cells of that table.

Built-in table handlers for text and binary files

For debugging and demonstration purposes, AMPL has built-in handlers for two very simple relational table formats. These formats store one table per file and convey equivalent information. One produces ASCII files that can be examined in any text editor, while the other creates binary files that are much faster to read and write.

For these handlers, the `table` declaration's *string-list* contains at most one string, identifying the external file that contains the relational table. If the string has the form

> "*filename*.tab"

the file is taken to be an ASCII text file; if it has the form

> "*filename*.bit"

it is taken to be a binary file. If no *string-list* is given, a text file *table-name*.tab is assumed.

For reading, the indicated file must exist. For writing, if the file does not exist, it is created. If the file exists but all of the `table` declaration's *data-spec*s have read/write status OUT, it is overwritten. Otherwise, writing causes the existing file to be modified; each column written either replaces an existing column of the same name, or becomes a new column added to the table.

The format for the text files can be examined by writing one and viewing the results in a text editor. For example, the following AMPL session,

```
ampl:  model diet.mod;
ampl:  data diet2a.dat;
ampl:  solve;
MINOS 5.5: optimal solution found.
13 iterations, objective 118.0594032
ampl:  table ResultList OUT "DietOpt.tab":
ampl?     [FOOD], Buy, Buy.rc, {j in FOOD} Buy[j]/f_max[j];
ampl:  write table ResultList;
```

produces a file DietOpt.tab with the following content:

```
ampl.tab 1 3
FOOD Buy Buy.rc 'Buy[j]/f_max[j]'
BEEF 5.360613810741701 8.881784197001252e-16 0.5360613810741701
CHK 2 1.1888405797101402 0.2
FISH 2 1.1444075021312856 0.2
HAM 10 -0.30265132139812223 1
MCH 10 -0.5511508951406658 1
MTL 10 -1.3289002557544745 1
SPG 9.306052855924973 -8.881784197001252e-16 0.9306052855924973
TUR 1.9999999999999998 2.7316197783461176 0.1999999999999998
```

In the first line, `ampl.tab` identifies this as an AMPL relational table text file, and is followed by the numbers of key and non-key columns, respectively. The second line gives the names of the table's columns, which may be any strings. (Use of the ~ operator to specify valid column-names is not necessary in this case.) Each subsequent line gives the values in one table row; numbers are written in full precision, with no special formatting or alignment.

11

Modeling Commands

AMPL provides a variety of commands like `model`, `solve`, and `display` that tell the AMPL modeling system what to do with models and data. Although these commands may use AMPL expressions, they are not part of the modeling language itself. They are intended to be used in an environment where you give a command, wait for the system to display a response, then decide what command to give next. Commands might be typed directly, or given implicitly by menus in a graphical user interface like the ones available on the AMPL web site. Commands also appear in scripts, the subject of Chapter 13.

This chapter begins by describing the general principles of the command environment. Section 11.2 then presents the commands that you are likely to use most for setting up and solving optimization problems.

After solving a problem and looking at the results, the next step is often to make a change and solve again. The remainder of this chapter explains the variety of changes that can be made without restarting the AMPL session from the beginning. Section 11.3 describes commands for re-reading data and modifying specific data values. Section 11.4 describes facilities for completely deleting or redefining model components, and for temporarily dropping constraints, fixing variables, or relaxing integrality of variables. (Convenient commands for examining model information can be found in Chapter 12, especially in Section 12.6.)

11.1 General principles of commands and options

To begin an interactive AMPL session, you must start the AMPL program, for example by typing the command *ampl* in response to a prompt or by selecting it from a menu or clicking on an icon. The startup procedure necessarily varies somewhat from one operating system to another; for details, you should refer to the system-specific instructions that come with your AMPL software.

Commands

If you are using a text-based interface, after starting AMPL, the first thing you should see is AMPL's prompt:

```
ampl:
```

Whenever you see this prompt, AMPL is ready to read and interpret what you type. As with most command interpreters, AMPL waits until you press the ''enter'' or ''return'' key, then processes everything you typed on the line.

An AMPL command ends with a semicolon. If you enter one or more complete commands on a line, AMPL processes them, prints any appropriate messages in response, and issues the `ampl:` prompt again. If you end a line in the middle of a command, you are prompted to continue it on the next line; you can tell that AMPL is prompting you to continue a command, because the prompt ends with a question mark rather than a colon:

```
ampl: display {i in ORIG, j in DEST}
ampl? sum {p in PROD} Trans[i,j,p];
```

You can type any number of characters on a line (up to whatever limit your operating system might impose), and can continue a command on any number of lines.

Several commands use filenames for reading or writing information. A filename can be any sequence of printing characters (except for semicolon ; and quotes " or ') or any sequence of any characters enclosed in matching quotes. The rules for correct filenames are determined by the operating system, however, not by AMPL. For the examples in this book we have used filenames like `diet.mod` that are acceptable to almost any operating system.

To conclude an AMPL session, type `end` or `quit`.

If you are running AMPL from a graphical interface, the details of your interaction will be different, but the command interface is likely to be accessible, and in fact is being used behind the scenes as well, so it's well worth understanding how to use it effectively.

Options

The behavior of AMPL commands depends not only on what you type directly, but on a variety of options for choosing alternative solvers, controlling the display of results, and the like.

Each option has a name, and a value that may be a number or a character string. For example, the options `prompt1` and `prompt2` are strings that specify the prompts. The option `display_width` has a numeric value, which says how many characters wide the output produced by the `display` command may be.

The `option` command displays and sets option values. If `option` is followed by a list of option names, AMPL replies with the current values:

```
ampl: option prompt1, display_width;
option prompt1 'ampl: ';
option display_width 79;
ampl:
```

A * in an option name is a "wild card" that matches any sequence of characters:

```
ampl: option prom*;
option prompt1 'ampl: ';
option prompt2 'ampl? ';
ampl:
```

The command `option *`, or just `option` alone, lists all current options and values.

When `option` is followed by a name and a value, it resets the named option to the specified value. In the following example we change the prompt and the display width, and then verify that the latter has been changed:

```
ampl: option prompt1 "A> ", display_width 60;
A> option display_width;
option display_width 60;
A>
```

You can specify any string value by surrounding the string in matching quotes '...' or "..." as above; the quotes may be omitted if the string looks like a name or number. Two consecutive quotes (' ' or " ") denote an empty string, which is a meaningful value for some options. At the other extreme, if you want to spread a long string over several lines, place the backslash character \ at the end of each intermediate line.

When AMPL starts, it sets many options to initial, or default, values. The `prompt1` option is initialized to `'ampl: '`, for instance, so prompts appear in the standard way. The `display_width` option has a default value of 79. Other options, especially ones that pertain to particular solvers, are initially unset:

```
ampl: option cplex_options;
option cplex_options ''; #not defined
```

To return all options to their default values, use the command `reset options`.

AMPL maintains no master list of valid options, but rather accepts any new option that you define. Thus if you mis-type an option name, you will most likely define a new option by mistake, as the following example demonstrates:

```
ampl: option display_wdith 60;
ampl: option display_w*;
option display_wdith 60;
option display_width 79;
```

The `option` statement also doesn't check to see if you have assigned a meaningful value to an option. You will be informed of a value error only when an option is used by some subsequent command. In these respects, AMPL options are much like operating system or shell "environment variables." In fact you can use the settings of environment variables to override AMPL's option defaults; see your system-specific documentation for details.

11.2 Setting up and solving models and data

To apply a solver to an instance of a model, the examples in this book use `model`, `data`, and `solve` commands:

```
ampl: model diet.mod; data diet.dat; solve;
MINOS 5.5: optimal solution found.
6 iterations, objective 88.2
```

The `model` command names a file that contains model declarations (Chapters 5 through 8), and the `data` command names a file that contains data values for model components (Chapter 9). The `solve` command causes a description of the optimization problem to be sent to a solver, and the results to be retrieved for examination. This section takes a closer look at the main AMPL features for setting up and solving models. Features for subsequently changing and re-solving models are covered in Section 11.4.

Entering models and data

AMPL maintains a "current" model, which is the one that will be sent to the solver if you type `solve`. At the beginning of an interactive session, the current model is empty. A `model` command reads declarations from a file and adds them to the current model; a `data` command reads data statements from a file to supply values for components already in the current model. Thus you may use several `model` or `data` commands to build up the description of an optimization problem, reading different parts of the model and data from different files.

You can also type parts of a model and its data directly at an AMPL prompt. Model declarations such as `param`, `var` and `subject to` act as commands that add components to the current model. The data statements of Chapter 9 also act as commands, which supply data values for already defined components such as sets and parameters. Because model and data statements look much alike, however, you need to tell AMPL which you will be typing. AMPL always starts out in "model mode"; the statement `data` (without a filename) switches the interpreter to "data mode", and the statement `model` (without a filename) switches it back. Any command (like `option`, `solve` or `subject to`) that does not begin like a data statement also has the effect of switching data mode back to model mode.

If a model declares more than one objective function, AMPL by default passes all of them to the solver. Most solvers deal only with one objective function and usually select the first by default. The `objective` command lets you select a single objective function to pass to the solver; it consists of the keyword `objective` followed by a name from a `minimize` or `maximize` declaration:

```
objective Total_Number;
```

If a model has an indexed collection of objectives, you must supply a subscript to indicate which one is to be chosen:

```
objective Total_Cost["A&P"];
```

The uses of multiple objectives are illustrated by two examples in Section 8.3.

Solving a model

The `solve` command sets in motion a series of activities. First, it causes AMPL to generate a specific optimization problem from the model and data that you have supplied. If you have neglected to provide some needed data, an error message is printed; you will also get error messages if your data values violate any restrictions imposed by qualification phrases in `var` or `param` declarations or by `check` statements. AMPL waits to verify data restrictions until you type `solve`, because a restriction may depend in a complicated way on many different data values. Arithmetic errors like dividing by zero are also caught at this stage.

After the optimization problem is generated, AMPL enters a ''presolve'' phase that tries to make the problem easier for the solver. Sometimes presolve so greatly reduces the size of a problem that it become substantially easier to solve. Normally the work of presolve goes on behind the scenes, however, and you need not be concerned about it. In rare cases, presolve can substantially affect the optimal values of the variables — when there is more than one optimal solution — or can interfere with other preprocessing routines that are built into your solver software. Also presolve sometimes detects that no feasible solution is possible, and so does not bother sending your program to the solver. For example, if you drastically reduce the availability of one resource in `steel4.mod`, then AMPL produces an error message:

```
ampl: model steel4.mod;
ampl: data steel4.dat;
ampl: let avail['reheat'] := 10;
ampl: solve;
presolve: constraint Time['reheat'] cannot hold:
        body <= 10 cannot be >= 11.25; difference = -1.25
```

For these cases you should consult the detailed description of presolve in Section 14.1.

The generated optimization problem, as possibly modified by presolve, is finally sent by AMPL to the solver of your choice. Every version of AMPL is distributed with some default solver that will be used automatically if you give no other instructions; type `option solver` to see its name:

```
ampl: option solver;
option solver minos;
```

If you have more than one solver, you can switch among them by changing the `solver` option:

```
ampl: model steelT.mod; data steelT.dat;
ampl: solve;
MINOS 5.5: optimal solution found.
15 iterations, objective 515033
```

```
ampl: reset;

ampl: model steelT.mod;
ampl: data steelT.dat;

ampl: option solver cplex;

ampl: solve;
CPLEX 8.0.0: optimal solution; objective 515033
16 dual simplex iterations (0 in phase I)

ampl: reset;

ampl: model steelT.mod;
ampl: data steelT.dat;

ampl: option solver snopt;

ampl: solve;
SNOPT 6.1-1: Optimal solution found.
15 iterations, objective 515033
```

In this example we reset the problem between solves, so that the solvers are invoked with the same initial conditions and their performance can be compared. Without reset, information about the solution found by one solver would be passed along to the next one, possibly giving the latter a substantial advantage. Passing information from one solve to the next is most useful when a series of similar LPs are to be sent to the same solver; we discuss this case in more detail in Section 14.2.

Almost any solver can handle a linear program, although those specifically designed for linear programming generally give the best performance. Other kinds of optimization problems, such as nonlinear (Chapter 18) and integer (Chapter 20), can be handled only by solvers designed for them. A message such as "ignoring integrality" or "can't handle nonlinearities" is an indication that you have not chosen a solver appropriate for your model.

If your optimization problems are not too difficult, you should be able to use AMPL without referring to instructions for a specific solver: set the solver option appropriately, type solve, and wait for the results.

If your solver takes a very long time to return with a solution, or returns to AMPL without any "optimal solution" message, then it's time to read further. Each solver is a sophisticated collection of algorithms and algorithmic strategies, from which many combinations of choices can be made. For most problems the solver makes good choices automatically, but you can also pass along your own choices through AMPL options. The details may vary with each solver, so for more information you must look to the solver-specific instructions that accompany your AMPL software.

If your problem takes a long time to optimize, you will want some evidence of the solver's progress to appear on your screen. Directives for this purpose are also described in the solver-specific instructions.

11.3 Modifying data

Many modeling projects involve solving a series of problem instances, each defined by somewhat different data. We describe here AMPL's facilities for resetting parameter values while leaving the model as is. They include commands for resetting data mode input and for resampling random parameters, as well as the `let` command for directly assigning new values.

Resetting

To delete the current data for several model components, without changing the current model itself, use the `reset data` command, as in:

```
reset data MINREQ, MAXREQ, amt, n_min, n_max;
```

You may then use `data` commands to read in new values for these sets and parameters. To delete all data, type `reset data`.

The `update data` command works similarly, but does not actually delete any data until new values are assigned. Thus if you type

```
update data MINREQ, MAXREQ, amt, n_min, n_max;
```

but you only read in new values for MINREQ, amt and n_min, the previous values for MAXREQ and n_max will remain. If instead you used `reset data`, MAXREQ and n_max would be without values, and you would get an error message when you next tried to solve.

Resampling

The `reset data` command also acts to resample the randomly computed parameters described in Section 7.6. Continuing with the variant of `steel4.mod` introduced in that section, if the definition of parameter `avail` is changed so that its value is given by a random function:

```
param avail_mean {STAGE} >= 0;
param avail_variance {STAGE} >= 0;

param avail {s in STAGE} =
    Normal(avail_mean[s], avail_variance[s]);
```

with corresponding data:

```
param avail_mean := reheat 35 roll 40 ;
param avail_variance := reheat 5 roll 2 ;
```

then AMPL will take new samples from the `Normal` distribution after each `reset data`. Different samples result in different solutions, and hence in different optimal objective values:

```
ampl: model steel4r.mod;
ampl: data steel4r.dat;
ampl: solve;
MINOS 5.5: optimal solution found.
3 iterations, objective 187632.2489

ampl: display avail;
reheat  32.3504
  roll  43.038 ;

ampl: reset data avail;
ampl: solve;
MINOS 5.5: optimal solution found.
4 iterations, objective 158882.901

ampl: display avail;
reheat  32.0306
  roll  32.6855 ;
```

Only reset data has this effect; if you issue a reset command then AMPL's random number generator is reset, and the values of avail repeat from the beginning. (Section 7.6 explains how to reset the generator's ''seed'' to get a different sequence of random numbers.)

The let command

The let command also permits you to change particular data values while leaving the model the same, but it is more convenient for small or easy-to-describe changes than reset data or update data. You can use it, for example, to solve the diet model of Figure 5-1, trying out a series of upper bounds f_max["CHK"] on the purchases of food CHK:

```
ampl: model dietu.mod;
ampl: data dietu.dat;

ampl: solve;
MINOS 5.5: optimal solution found.
5 iterations, objective 74.27382022

ampl: let f_max["CHK"] := 11;
ampl: solve;
MINOS 5.5: optimal solution found.
1 iterations, objective 73.43818182

ampl: let f_max["CHK"] := 12;
ampl: solve;
MINOS 5.5: optimal solution found.
0 iterations, objective 73.43818182
```

Relaxing the bound to 11 reduces the cost somewhat, but further relaxation apparently has no benefit.

An indexing expression may be given after the keyword `let`, in which case a change is made for each member of the specified indexing set. You could use this feature to change all upper bounds to 8:

```
let {j in FOOD} f_max[j] := 8;
```

or to increase all upper bounds by 10 percent:

```
let {j in FOOD} f_max[j] := 1.1 * f_max[j];
```

In general this command consists of the keyword `let`, an indexing expression if needed, and an assignment. Any set or parameter whose declaration does not define it using an = phrase may be specified to the left of the assignment's `:=` operator, while to the right may appear any appropriate expression that can currently be evaluated.

Although AMPL does not impose any restrictions on what you can change using `let`, you should take care in changing any set or parameter that affects the indexing of other data.

For example, after solving the multiperiod production problem of Figures 4-4 and 4-5, it might be tempting to change the number of weeks T from 4 (as given in the original data) to 3:

```
ampl: let T := 3;
ampl: solve;
Error executing "solve" command:
error processing param avail:
        invalid subscript avail[4] discarded.
error processing param market:
        2 invalid subscripts discarded:
        market['bands',4]
        market['coils',4]
error processing param revenue:
        2 invalid subscripts discarded:
        revenue['bands',4]
        revenue['coils',4]
error processing var Sell['coils',1]:
        invalid subscript market['bands',4]
```

The problem here is that AMPL still has current data for 4th-week parameters such as `avail[4]`, which has become invalid with the change of T to 3. If you want to properly reduce the number of weeks in the linear program while using the same data, you must declare two parameters:

```
param Tdata integer > 0;
param T integer <= Tdata;
```

Use `1..Tdata` for indexing in the `param` declarations, while retaining `1..T` for the variables, objective and constraints; then you can use `let` to change T as you like.

You can also use the `let` command to change the current values of variables. This is sometimes a convenient feature for exploring an alternative solution. For example, here

ceil(x)	ceiling of x (next higher integer)
floor(x)	floor of x (next lower integer)
precision(x, n)	x rounded to n significant digits
round(x, n)	x rounded to n digits past the decimal point
round(x)	x rounded to the nearest integer
trunc(x, n)	x truncated to n digits past the decimal point
trunc(x)	x truncated to an integer (fractional part dropped)

Table 11-1: Rounding functions.

is what happens when we solve for the optimal diet as above, then round the optimal solution down to the next lowest integer number of packages:

```
ampl: model dietu.mod; data dietu.dat; solve;
MINOS 5.5: optimal solution found.
5 iterations, objective 74.27382022

ampl: let {j in FOOD} Buy[j] := floor(Buy[j]);
ampl: display Total_Cost, n_min, Diet_Min.slack;;
Total_Cost = 70.8

:       n_min Diet_Min.slack      :=
A         700            231
B1          0            580
B2          0            475
C         700            -40
CAL     16000           -640
;
```

Because we have used let to change the values of the variables, the objective and the slacks are automatically computed from the new, rounded values. The cost has dropped by about $3.50, but the solution is now short of the requirements for C by nearly 6 percent and for CAL by 4 percent.

AMPL provides a variety of rounding functions that can be used in this way. They are summarized in Table 11-1.

11.4 Modifying models

Several commands are provided to help you make limited changes to the current model, without modifying the model file or issuing a full reset. This section describes commands that completely remove or redefine model components, and that temporarily drop constraints, fix variables, or relax integrality restrictions on variables.

Removing or redefining model components

The delete command removes a previously declared model component, provided that no other components use it in their declarations. The form of the command is simply delete followed by a comma-separated list of names of model components:

```
ampl: model dietobj.mod;
ampl: data dietobj.dat;
ampl: delete Total_Number, Diet_Min;
```

Normally you cannot delete a set, parameter, or variable, because it is declared for use later in the model; but you can delete any objective or constraint. You can also specify a component "name" of the form check *n* to delete the *n*th check statement in the current model.

The purge command has the same form, but with the keyword purge in place of delete. It removes not only the listed components, but also all components that *depend* on them either directly (by referring to them) or indirectly (by referring to their dependents). Thus for example in diet.mod we have

```
param f_min {FOOD} >= 0;
param f_max {j in FOOD} >= f_min[j];
var Buy {j in FOOD} >= f_min[j], <= f_max[j];
minimize Total_Cost:  sum {j in FOOD} cost[j] * Buy[j];
```

The command purge f_min deletes parameter f_min and the components whose declarations refer to f_min, including parameter f_max and variable Buy. It also deletes objective Total_Cost, which depends indirectly on f_min through its reference to Buy.

If you're not sure which components depend on some given component, you can use the xref command to find out:

```
ampl: xref f_min;
# 4 entities depend on f_min:
f_max
Buy
Total_Cost
Diet
```

Like delete and purge, the xref command can be applied to any list of model components.

Once a component has been removed by delete or purge, any previously hidden meaning of the component's name becomes visible again. After a constraint named prod is deleted, for instance, AMPL again recognizes prod as an iterated multiplication operator (Table 7-1).

If there is no previously hidden meaning, the name of a component removed by delete or purge becomes again unused, and may subsequently be declared as the name of any new component of any type. If you only want to make some relatively limited modifications to a declaration, however, then you will probably find redeclare to be more convenient. You can change any component's declaration by writing the keyword `redeclare` followed by the complete revised declaration that you would like to substitute. Looking again at `diet.mod`, for example,

> ampl: **redeclare param f_min {FOOD} > 0 integer;**

changes only the validity conditions on `f_min`. The declarations of all components that depend on `f_min` are left unchanged, as are any values previously read for `f_min`.

A list of all component types to which `delete`, `purge`, `xref`, and `redeclare` may be applied is given in A.18.5.

Changing the model: *fix, unfix; drop, restore*

The simplest (but most drastic) way to change the model is by issuing the command `reset`, which expunges all of the current model and data. Following `reset`, you can issue new `model` and `data` commands to set up a different optimization problem; the effect is like typing `quit` and then restarting AMPL, except that options are not reset to their default values. If your operating system or your graphical environment for AMPL allows you to edit files while keeping AMPL active, `reset` is valuable for debugging and experimentation; you may make changes to the model or data files, type `reset`, then read in the modified files. (If you need to escape from AMPL to run a text editor, you can use the `shell` command described in Section A.21.1.)

The `drop` command instructs AMPL to ignore certain constraints or objectives of the current model. As an example, the constraints of Figure 5-1 initially include

```
subject to Diet_Max {i in MAXREQ}:
    sum {j in FOOD} amt[i,j] * Buy[j] <= n_max[i];
```

A `drop` command can specify a particular one of these constraints to ignore:

```
drop Diet_Max["CAL"];
```

or it may specify all constraints or objectives indexed by some set:

```
drop {i in MAXNOT} Diet_Max[i];
```

where MAXNOT has previously been defined as some subset of MAXREQ. The entire collection of constraints can be ignored by

```
drop {i in MAXREQ} Diet_Max[i];
```

or more simply:

```
drop Diet_Max;
```

In general, this command consists of the keyword `drop`, an optional indexing expression, and a constraint name that may be subscripted. Successive `drop` commands have a cumulative effect.

The `restore` command reverses the effect of `drop`. It has the same syntax, except for the keyword `restore`.

The `fix` command fixes specified variables at their current values, as if there were a constraint that the variables must equal these values; the `unfix` command reverses the effect. These commands have the same syntax as `drop` and `restore`, except that they name variables rather than constraints. For example, here we initialize all variables of our diet problem to their lower bounds, fix all variables representing foods that have more than 1200 mg of sodium per package, and optimize over the remaining variables:

```
ampl: let {j in FOOD} Buy[j] := f_min[j];
ampl: fix {j in FOOD: amt["NA",j] > 1200} Buy[j];
ampl: solve;
MINOS 5.5: optimal solution found.
7 iterations, objective 86.92
Objective = Total_Cost['A&P']

ampl: display {j in FOOD} (Buy[j].lb,Buy[j],amt["NA",j]);
:       Buy[j].lb     Buy[j]  amt['NA',j]    :=
BEEF       2            2          938
CHK        2            2         2180
FISH       2           10          945
HAM        2            2          278
MCH        2         9.42857      1182
MTL        2           10          896
SPG        2            2         1329
TUR        2            2         1397
;
```

Rather than setting and fixing the variables in separate statements, you can add an assignment phrase to the `fix` command:

```
ampl: fix {j in FOOD: amt["NA",j] > 1200} Buy[j] := f_min[j];
```

The `unfix` command works in the same way, to reverse the effect of fix and optionally also reset the value of a variable.

Relaxing integrality

Changing option `relax_integrality` from its default of 0 to any nonzero value:

```
option relax_integrality 1;
```

tells AMPL to ignore all restrictions of variables to integer values. Variables declared `integer` get whatever bounds you specified for them, while variables declared `binary` are given a lower bound of zero and an upper bound of one. To restore the integrality restrictions, set the `relax_integrality` option back to 0.

A variable's name followed by the suffix `.relax` indicates its current integrality relaxation status: 0 if integrality is enforced, nonzero otherwise. You can make use of this suffix to relax integrality on selected variables only. For example,

```
let Buy['CHK'].relax = 1
```

relaxes integrality only on the variable Buy['CHK'], while

```
    let {j in FOOD: f_min[j] > allow_frac} Buy[j].relax := 1;
```

relaxes integrality on all Buy variables for foods that have a minimum purchase of at least some cutoff parameter allow_frac.

Some of the solvers that work with AMPL provide their own directives for relaxing integrality, but these do not necessarily have the same effect as AMPL's relax_integrality option or .relax suffix. The distinction is due to the effects of AMPL's problem simplification, or presolve, stage (Section 14.1). AMPL drops integrality restrictions *before* the presolve phase, so that the solver receives a true continuous relaxation of the original integer problem. If the relaxation is performed by the solver, however, then the integrality restrictions are still in effect during AMPL's presolve phase, and AMPL may perform some additional tightening and simplification as a result.

As a simple example, suppose that diet model variable declarations are written to allow the food limits f_max to be adjusted by setting an additional parameter, scale:

```
    var Buy {j in FOOD} integer >= f_min[j], <= scale * f_max[j];
```

In our example of Figure 2-3, all of the f_max values are 10; suppose that also we set scale to 0.95. First, here are the results of solving the unrelaxed problem:

```
ampl: option relax_integrality;
option relax_integrality 0;

ampl: let scale := 0.95;
ampl: solve;
CPLEX 8.0.0: optimal integer solution; objective 122.89
6 MIP simplex iterations
0 branch-and-bound nodes
```

When no relaxation is specified in AMPL, presolve sees that all the variables have upper limits of 9.5, and since it knows that the variables must take integer values, it rounds these limits down to 9. Then these limits are sent to the solver, where they remain even if we specify a *solver* directive for integrality relaxation:

```
ampl: option cplex_options 'relax';
ampl: solve;
CPLEX 8.0.0: relax
Ignoring integrality of 8 variables.
CPLEX 8.0.0: optimal solution; objective 120.2421057
2 dual simplex iterations (0 in phase I)

ampl: display Buy;
Buy [*] :=
BEEF  8.39898
 CHK  2
FISH  2
 HAM  9
 MCH  9
 MTL  9
 SPG  8.93436
 TUR  2
;
```

If instead option `relax_integrality` is set to 1, presolve leaves the upper limits at 9.5 and sends those to the solver, with the result being a less constrained problem and hence a lower objective value:

```
ampl: option relax_integrality 1;
ampl: solve;
CPLEX 8.0.0: optimal solution; objective 119.1507545
3 dual simplex iterations (0 in phase I)

ampl: display Buy;
Buy [*] :=
BEEF   6.8798
 CHK   2
FISH   2
 HAM   9.5
 MCH   9.5
 MTL   9.5
 SPG   9.1202
 TUR   2
 ;
```

Variables that were at upper bound 9 in the previous solution are now at upper bound 9.5.

The same situation can arise in much less obvious circumstances, and can lead to unexpected results. In general, the optimal value of an integer program under AMPL's `relax_integrality` option may be lower (for minimization) or higher (for maximization) than the optimal value reported by the solver's relaxation directive.

12

Display Commands

AMPL offers a rich variety of commands and options to help you examine and report the results of optimization. Section 12.1 introduces `display`, the most convenient command for arranging set members and numerical values into lists and tables; Sections 12.2 and 12.3 provide a detailed account of `display` options that give you more control over how the lists and tables are arranged and how numbers appear in them. Section 12.4 describes `print` and `printf`, two related commands that are useful for preparing data to be sent to other programs, and for formatting simple reports.

Although our examples are based on the display of sets, parameters and variables — and expressions involving them — you can use the same commands to inspect dual values, slacks, reduced costs, and other quantities associated with an optimal solution; the rules for doing so are explained in Section 12.5.

AMPL also provides ways to access modeling and solving information. Section 12.6 describes features that can be useful when, for example, you want to view a parameter's declaration at the command-line, display a particular constraint from a problem instance, list the values and bounds of all variables regardless of their names, or record timings of AMPL and solver activities.

Finally, Section 12.7 addresses general facilities for manipulating output of AMPL commands. These include features for redirection of command output, logging of output, and suppression of error messages.

12.1 Browsing through results: the `display` command

The easiest way to examine data and result values is to type `display` and a description of what you want to look at. The `display` command automatically formats the values in an intuitive and familiar arrangement; as much as possible, it uses the same list and table formats as the data statements described in Chapter 9. Our examples use parameters and variables from models defined in other chapters.

As we will describe in more detail in Section 12.7, it is possible to capture the output of `display` commands in a file, by adding >*filename* to the end of a `display` command; this redirection mechanism applies as well to most other commands that produce output.

Displaying sets

The contents of sets are shown by typing `display` and a list of set names. This example is taken from the model of Figure 6-2a:

```
ampl: display ORIG, DEST, LINKS;
set ORIG := GARY CLEV PITT;
set DEST := FRA DET LAN WIN STL FRE LAF;
set LINKS :=
(GARY,DET)    (GARY,LAF)    (CLEV,LAN)    (CLEV,LAF)    (PITT,STL)
(GARY,LAN)    (CLEV,FRA)    (CLEV,WIN)    (PITT,FRA)    (PITT,FRE)
(GARY,STL)    (CLEV,DET)    (CLEV,STL)    (PITT,WIN) ;
```

If you specify the name of an indexed collection of sets, each set in the collection is shown (from Figure 6-3):

```
ampl: display PROD, AREA;
set PROD := bands coils;
set AREA[bands] := east north;
set AREA[coils] := east west export;
```

Particular members of an indexed collection can be viewed by subscripting, as in `display AREA["bands"]`.

The argument of `display` need not be a declared set; it can be any of the expressions described in Chapter 5 or 6 that evaluate to sets. For example, you can show the union of all the sets `AREA[p]`:

```
ampl: display union {p in PROD} AREA[p];
set  union {p in PROD}  AREA[p] := east north west export;
```

or the set of all transportation links on which the shipping cost is greater than 500:

```
ampl: display {(i,j) in LINKS: cost[i,j] * Trans[i,j] > 500};
set {(i,j) in LINKS: cost[i,j]*Trans[i,j] > 500}   :=
(GARY,STL)    (CLEV,DET)    (CLEV,WIN)    (PITT,FRA)    (PITT,FRE)
(GARY,LAF)    (CLEV,LAN)    (CLEV,LAF)    (PITT,STL) ;
```

Because the membership of this set depends upon the current values of the variables `Trans[i,j]`, you could not refer to it in a model, but it is legal in a `display` command, where variables are treated the same as parameters.

Displaying parameters and variables

The `display` command can show the value of a scalar model component:

```
ampl: display T;
T = 4
```

or the values of individual components from an indexed collection (Figure 1-6b):

```
ampl: display avail["reheat"], avail["roll"];
avail['reheat'] = 35
avail['roll'] = 40
```

or an arbitrary expression:

```
ampl: display sin(1)^2 + cos(1)^2;
sin(1)^2 + cos(1)^2 = 1
```

The major use of display, however, is to show whole indexed collections of data. For "one-dimensional" data — parameters or variables indexed over a simple set — AMPL uses a column format (Figure 4-6b):

```
ampl: display avail;
avail [*] :=
reheat   35
  roll   40
;
```

For "two-dimensional" parameters or variables — indexed over a set of pairs or two simple sets — AMPL forms a list for small amounts of data (Figure 4-1):

```
ampl: display supply;
supply :=
CLEV bands      700
CLEV coils     1600
CLEV plate      300
GARY bands      400
GARY coils      800
GARY plate      200
PITT bands      800
PITT coils     1800
PITT plate      300
;
```

or a table for larger amounts:

```
ampl: display demand;

demand [*,*]
:    bands coils plate   :=
DET    300   750   100
FRA    300   500   100
FRE    225   850   100
LAF    250   500   250
LAN    100   400     0
STL    650   950   200
WIN     75   250    50
;
```

You can control the choice between formats by setting option `display_1col`, which is described in the next section.

A parameter or variable (or any other model entity) indexed over a set of ordered pairs is also considered to be a two-dimensional object and is displayed in a similar manner. Here is the display for a parameter indexed over the set LINKS that was displayed earlier in this section (from Figure 6-2a):

```
ampl: display cost;
cost :=
CLEV DET     9
CLEV FRA    27
CLEV LAF    17
CLEV LAN    12
CLEV STL    26
CLEV WIN     9
GARY DET    14
GARY LAF     8
GARY LAN    11
GARY STL    16
PITT FRA    24
PITT FRE    99
PITT STL    28
PITT WIN    13
;
```

This, too, can be made to appear in a table format, as the next section will show.

To display values indexed in three or more dimensions, AMPL again forms lists for small amounts of data. Multi-dimensional entities more often involve data in large quantities, however, in which case AMPL ''slices'' the values into two-dimensional tables, as in the case of this variable from Figure 4-6:

```
ampl: display Trans;
Trans [CLEV,*,*]
:   bands coils plate      :=
DET     0    750     0
FRA     0      0     0
FRE     0      0     0
LAF     0    500     0
LAN     0    400     0
STL     0     50     0
WIN     0    250     0

  [GARY,*,*]
:   bands coils plate      :=
DET     0      0      0
FRA     0      0      0
FRE   225    850    100
LAF   250      0      0
LAN     0      0      0
STL   650    900    200
WIN     0      0      0
```

```
   [PITT,*,*]
   :    bands coils plate    :=
   DET    300      0   100
   FRA    300    500   100
   FRE      0      0     0
   LAF      0      0   250
   LAN    100      0     0
   STL      0      0     0
   WIN     75      0    50
   ;
```

At the head of the first table, the template [CLEV, *, *] indicates that the slice is through CLEV in the first component, so the entry in row LAF and column coils says that Trans["CLEV","LAF","coils"] is 500. Since the first index of Trans is always CLEV, GARY or PITT in this case, there are three slice tables in all. But AMPL does not always slice through the first component; it picks the slices so that the display will contain the fewest possible tables.

A display of two or more components of the same dimensionality is always presented in a list format, whether the components are one-dimensional (Figure 4-4):

```
ampl: display inv0, prodcost, invcost;
:       inv0 prodcost invcost    :=
bands    10      10       2.5
coils     0      11       3
;
```

or two-dimensional (Figure 4-6):

```
ampl: display rate, make_cost, Make;
:          rate make_cost  Make    :=
CLEV bands  190     190       0
CLEV coils  130     170    1950
CLEV plate  160     185       0
GARY bands  200     180    1125
GARY coils  140     170    1750
GARY plate  160     180     300
PITT bands  230     190     775
PITT coils  160     180     500
PITT plate  170     185     500
;
```

or any higher dimension. The indices appear in a list to the left, with the last one changing most rapidly.

As you can see from these examples, display normally arranges row and column labels in alphabetical or numerical order, regardless of the order in which they might have been given in your data file. When the labels come from an ordered set, however, the original ordering is honored (Figure 5-3):

```
ampl: display avail;
avail [*] :=
27sep  40
04oct  40
11oct  32
18oct  40
;
```

For this reason, it can be worthwhile to declare certain sets of your model to be ordered, even if their ordering plays no explicit role in your formulation.

Displaying indexed expressions

The display command can show the value of any arithmetic expression that is valid in an AMPL model. Single-valued expressions pose no difficulty, as in the case of these three profit components from Figure 4-4:

```
ampl: display sum {p in PROD,t in 1..T} revenue[p,t]*Sell[p,t],
ampl?             sum {p in PROD,t in 1..T} prodcost[p]*Make[p,t],
ampl?             sum {p in PROD,t in 1..T} invcost[p]*Inv[p,t];
sum{p in PROD, t in 1 .. T} revenue[p,t]*Sell[p,t] = 787810
sum{p in PROD, t in 1 .. T} prodcost[p]*Make[p,t] = 269477
sum{p in PROD, t in 1 .. T} invcost[p]*Inv[p,t] = 3300
```

Suppose, however, that you want to see all the individual values of revenue[p,t] * Sell[p,t]. Since you can type display revenue, Sell to display the separate values of revenue[p,t] and Sell[p,t], you might want to ask for the products of these values by typing:

```
ampl: display revenue * Sell;
syntax error
context:  display revenue  >>> *  <<< Sell;
```

AMPL does not recognize this kind of array arithmetic. To display an indexed collection of expressions, you must specify the indexing explicitly:

```
ampl: display {p in PROD, t in 1..T} revenue[p,t]*Sell[p,t];
revenue[p,t]*Sell[p,t] [*,*] (tr)
:   bands    coils     :=
1   150000     9210
2   156000    87500
3    37800   129500
4    54000   163800
;
```

To apply the same indexing to two or more expressions, enclose a list of them in parentheses after the indexing expression:

```
ampl: display {p in PROD, t in 1..T}
ampl?     (revenue[p,t]*Sell[p,t], prodcost[p]*Make[p,t]);
:       revenue[p,t]*Sell[p,t] prodcost[p]*Make[p,t]     :=
bands 1            150000                 59900
bands 2            156000                 60000
bands 3             37800                 14000
bands 4             54000                 20000
coils 1              9210                 15477
coils 2             87500                 15400
coils 3            129500                 38500
coils 4            163800                 46200
;
```

An indexing expression followed by an expression or parenthesized list of expressions is treated as a single display item, which specifies some indexed collection of values. A display command may contain one of these items as above, or a comma-separated list of them.

The presentation of the values for indexed expressions follows the same rules as for individual parameters and variables. In fact, you can regard a command like

```
display revenue, Sell
```

as shorthand for

```
ampl: display {p in PROD, t in 1..T} (revenue[p,t],Sell[p,t]);
:       revenue[p,t] Sell[p,t]     :=
bands 1       25        6000
bands 2       26        6000
bands 3       27        1400
bands 4       27        2000
coils 1       30         307
coils 2       35        2500
coils 3       37        3500
coils 4       39        4200
;
```

If you rearrange the indexing expression so that t in 1..T comes first, however, the rows of the list are instead sorted first on the members of 1..T:

```
ampl: display {t in 1..T, p in PROD} (revenue[p,t],Sell[p,t]);
:       revenue[p,t] Sell[p,t]     :=
1 bands       25        6000
1 coils       30         307
2 bands       26        6000
2 coils       35        2500
3 bands       27        1400
3 coils       37        3500
4 bands       27        2000
4 coils       39        4200
;
```

This change in the default presentation can only be achieved by placing an explicit indexing expression after display.

In addition to indexing individual display items, you can specify a set over which the whole `display` command is indexed — that is, you can ask that the command be executed once for each member of an indexing set. This feature is particularly useful for rearranging slices of multidimensional tables. When, earlier in this section, we displayed the three-dimensional variable `Trans` indexed over `{ORIG, DEST, PROD}`, AMPL chose to slice the values through members of `ORIG` to produce a series of two-dimensional tables.

What if you want to display slices through `PROD`? Rearranging the indexing expression, as in our previous example, will not reliably have the desired effect; the `display` command always picks the smallest indexing set, and where there is more than one that is smallest, it does not necessarily choose the first. Instead, you can say explicitly that you want a separate display for each p in `PROD`:

```
ampl: display {p in PROD}:
ampl?    {i in ORIG, j in DEST} Trans[i,j,p];
Trans[i,j,'bands'] [*,*] (tr)
:    CLEV   GARY   PITT     :=
DET   0      0     300
FRA   0      0     300
FRE   0     225      0
LAF   0     250      0
LAN   0      0     100
STL   0     650      0
WIN   0      0      75
;

Trans[i,j,'coils'] [*,*] (tr)
:    CLEV   GARY   PITT     :=
DET  750     0      0
FRA   0      0     500
FRE   0     850      0
LAF  500     0      0
LAN  400     0      0
STL   50    900      0
WIN  250     0      0
;

Trans[i,j,'plate'] [*,*] (tr)
:    CLEV   GARY   PITT     :=
DET   0      0     100
FRA   0      0     100
FRE   0     100      0
LAF   0      0     250
LAN   0      0      0
STL   0     200      0
WIN   0      0      50
;
```

As this example shows, if a `display` command specifies an indexing expression right at the beginning, followed by a colon, the indexing set applies to the whole command. For

display_1col	maximum elements for a table to be displayed in list format (20)
display_transpose	transpose tables if rows – columns < display_transpose (0)
display_width	maximum line width (79)
gutter_width	separation between table columns (3)
omit_zero_cols	if not 0, omit all-zero columns from displays (0)
omit_zero_rows	if not 0, omit all-zero rows from displays (0)

Table 12-1: Formatting options for display (with default values).

each member of the set, the expressions following the colon are evaluated and displayed separately.

12.2 Formatting options for display

The display command uses a few simple rules for choosing a good arrangement of data. By changing several options, you can control overall arrangement, handling of zero values, and line width. These options are summarized in Table 12-1, with default values shown in parentheses.

Arrangement of lists and tables

The display of a one-dimensional parameter or variable can produce a very long list, as in this example from the scheduling model of Figure 16-5:

```
ampl: display required;
required [*] :=
Fri1   100
Fri2    78
Fri3    52
Mon1   100
Mon2    78
Mon3    52
Sat1   100
Sat2    78
Thu1   100
Thu2    78
Thu3    52
Tue1   100
Tue2    78
Tue3    52
Wed1   100
Wed2    78
Wed3    52
 ;
```

The option `display_1col` can be used to request a more compact format:

```
ampl: option display_1col 0;
ampl: display required;
required [*] :=
Fri1 100    Mon1 100    Sat1 100    Thu2  78    Tue2  78    Wed2  78
Fri2  78    Mon2  78    Sat2  78    Thu3  52    Tue3  52    Wed3  52
Fri3  52    Mon3  52    Thu1 100    Tue1 100    Wed1 100
;
```

The one-column list format is used when the number of values to be displayed is less than or equal to `display_1col`, and the compact format is used otherwise. The default for `display_1col` is 20; set it to zero to force the compact format, or to a very large number to force the list format.

Multi-dimensional displays are affected by option `display_1col` in an analogous way. The one-column list format is used when the number of values is less than or equal to `display_1col`, while the appropriate compact format — in this case, a table — is used otherwise. We showed an example of the difference in the previous section, where the display for `supply` appeared as a list because it had only 9 values, while the display for `demand` appeared as a table because its 21 values exceed the default setting of 20 for option `display_1col`.

Since a parameter or variable indexed over a set of ordered pairs is also considered to be two-dimensional, the value of `display_1col` affects its display as well. Here is the table format for the parameter `cost` indexed over the set `LINKS` (from Figure 6-2a) that was displayed in the preceding section:

```
ampl: option display_1col 0;
ampl: display cost;

cost [*,*] (tr)
:   CLEV GARY PITT     :=
DET    9   14     .
FRA   27    .    24
FRE    .    .    99
LAF   17    8     .
LAN   12   11     .
STL   26   16    28
WIN    9    .    13
;
```

A dot (`.`) entry indicates a nonexistent combination in the index set. Thus in the GARY column of the table, there is a dot in the FRA row because the pair (GARY, FRA) is not a member of LINKS; no `cost["GARY","FRA"]` is defined for this problem. On the other hand, LINKS does contain the pair (GARY, LAF), and `cost["GARY","LAF"]` is shown as 8 in the table.

In choosing an orientation for tables, the `display` command by default favors rows over columns; that is, if the number of columns would exceed the number of rows, the table is transposed. Thus the table for `demand` in the previous section has rows labeled by the first coordinate and columns by the second, because it is indexed over DEST with

7 members and then PROD with 3 members. By contrast, the table for cost has columns labeled by the first coordinate and rows by the second, because it is indexed over ORIG with 3 members and then DEST with 7 members. A transposed table is indicated by a (tr) in its first line.

The transposition status of a table can be reversed by changing the display_transpose option. Positive values tend to force transposition:

```
ampl: option display_transpose 5;
ampl: display demand;
demand [*,*] (tr)
:         DET    FRA    FRE    LAF    LAN    STL    WIN        :=
bands     300    300    225    250    100    650     75
coils     750    500    850    500    400    950    250
plate     100    100    100    250      0    200     50
;
```

while negative values tend to suppress it:

```
ampl: option display_transpose -5;
ampl: display cost;
cost [*,*]
:       DET   FRA   FRE   LAF   LAN   STL   WIN      :=
CLEV      9    27     .    17    12    26     9
GARY     14     .     .     8    11    16     .
PITT      .    24    99     .     .    28    13
;
```

The rule is as follows: a table is transposed only when the number of rows minus the number of columns would be less than display_transpose. At its default value of zero, display_transpose gives the previously described default behavior.

Control of line width

The option display_width gives the maximum number of characters on a line generated by display (as seen in the model of Figure 16-4):

```
ampl: option display_width 50, display_1col 0;
ampl: display required;

required [*] :=
Fri1 100    Mon3   52    Thu3   52    Wed2   78
Fri2  78    Sat1  100    Tue1  100    Wed3   52
Fri3  52    Sat2   78    Tue2   78
Mon1 100    Thu1  100    Tue3   52
Mon2  78    Thu2   78    Wed1  100
;
```

When a table would be wider than display_width, it is cut vertically into two or more tables. The row names in each table are the same, but the columns are different:

```
ampl: option display_width 50; display cost;
cost [*,*]
:           C118 C138 C140 C246 C250 C251 D237 D239        :=
Coullard       6    9    8    7   11   10    4    5
Daskin        11    8    7    6    9   10    1    5
Hazen          9   10   11    1    5    6    2    7
Hopp          11    9    8   10    6    5    1    7
Iravani        3    2    8    9   10   11    1    5
Linetsky      11    9   10    5    3    4    6    7
Mehrotra       6   11   10    9    8    7    1    2
Nelson        11    5    4    6    7    8    1    9
Smilowitz     11    9   10    8    6    5    7    3
Tamhane        5    6    9    8    4    3    7   10
White         11    9    8    4    6    5    3   10

:           D241 M233 M239        :=
Coullard       3    2    1
Daskin         4    2    3
Hazen          8    3    4
Hopp           4    2    3
Iravani        4    6    7
Linetsky       8    1    2
Mehrotra       5    4    3
Nelson        10    2    3
Smilowitz      4    1    2
Tamhane       11    2    1
White          7    2    1
;
```

If a table's column headings are much wider than the values, display introduces abbreviations to keep all columns together (Figure 4-4):

```
ampl: option display_width 40;
ampl: display {p in PROD, t in 1..T} (revenue[p,t]*Sell[p,t],
ampl?    prodcost[p]*Make[p,t], invcost[p]*Inv[p,t]);
# $1 = revenue[p,t]*Sell[p,t]
# $2 = prodcost[p]*Make[p,t]
# $3 = invcost[p]*Inv[p,t]
:           $1      $2      $3       :=
bands 1   150000   59900       0
bands 2   156000   60000       0
bands 3    37800   14000       0
bands 4    54000   20000       0
coils 1     9210   15477    3300
coils 2    87500   15400       0
coils 3   129500   38500       0
coils 4   163800   46200       0
;
```

On the other hand, where the headings are narrower than the values, you may be able to squeeze more on a line by reducing the option gutter_width — the number of spaces between columns — from its default value of 3 to 2 or 1.

Suppression of zeros

In some kinds of linear programs that have many more variables than constraints, most of the variables have an optimal value of zero. For instance in the assignment problem of Figure 3-2, the optimal values of all the variables form this table, in which there is a single 1 in each row and each column:

```
ampl: display Trans;

Trans [*,*]
:          C118 C138 C140 C246 C250 C251 D237 D239 D241 M233 M239  :=
Coullard    1    0    0    0    0    0    0    0    0    0    0
Daskin      0    0    0    0    0    0    0    0    1    0    0
Hazen       0    0    0    1    0    0    0    0    0    0    0
Hopp        0    0    0    0    0    0    1    0    0    0    0
Iravani     0    1    0    0    0    0    0    0    0    0    0
Linetsky    0    0    0    0    1    0    0    0    0    0    0
Mehrotra    0    0    0    0    0    0    0    1    0    0    0
Nelson      0    0    1    0    0    0    0    0    0    0    0
Smilowitz   0    0    0    0    0    0    0    0    0    1    0
Tamhane     0    0    0    0    0    1    0    0    0    0    0
White       0    0    0    0    0    0    0    0    0    0    1
;
```

By setting `omit_zero_rows` to 1, all the zero values are suppressed, and the list comes down to the entries of interest:

```
ampl: option omit_zero_rows 1;

ampl: display Trans;

Trans :=
Coullard  C118   1
Daskin    D241   1
Hazen     C246   1
Hopp      D237   1
Iravani   C138   1
Linetsky  C250   1
Mehrotra  D239   1
Nelson    C140   1
Smilowitz M233   1
Tamhane   C251   1
White     M239   1
;
```

If the number of nonzero entries is less than the value of `display_1col`, the data is printed as a list, as it is here. If the number of nonzeros is greater than `display_1col`, a table format would be used, and the `omit_zero_rows` option would only suppress table rows that contain all zero entries.

For example, the display of the three-dimensional variable `Trans` from earlier in this chapter would be condensed to the following:

```
ampl: display Trans;
Trans [CLEV,*,*]
:     bands coils plate      :=
DET      0    750     0
LAF      0    500     0
LAN      0    400     0
STL      0     50     0
WIN      0    250     0

  [GARY,*,*]
:     bands coils plate      :=
FRE    225    850   100
LAF    250      0     0
STL    650    900   200

  [PITT,*,*]
:     bands coils plate      :=
DET    300      0   100
FRA    300    500   100
LAF      0      0   250
LAN    100      0     0
WIN     75      0    50
;
```

A corresponding option `omit_zero_cols` suppresses all-zero columns when set to 1, and would eliminate two columns from `Trans[CLEV,*,*]`.

12.3 Numeric options for `display`

The numbers in a table or list produced by `display` are the result of a transformation from the computer's internal numeric representation to a string of digits and symbols. AMPL's options for adjusting this transformation are shown in Table 12-2. In this section we first consider options that affect only the appearance of numbers, and then options that affect underlying solution values as well.

`display_eps`	smallest magnitude displayed differently from zero (0)
`display_precision`	digits of precision to which displayed numbers are rounded; full precision if 0 (6)
`display_round`	digits left or (if negative) right of decimal place to which displayed numbers are rounded, overriding `display_precision` ("")
`solution_precision`	digits of precision to which solution values are rounded; full precision if 0 (0)
`solution_round`	digits left or (if negative) right of decimal place to which solution values are rounded, overriding `solution_precision` ("")

Table 12-2: Numeric options for `display` (with default values).

Appearance of numeric values

In all of our examples so far, the display command shows each numerical value to the same number of significant digits:

```
ampl: display {p in PROD, t in 1..T} Make[p,t]/rate[p];
Make[p,t]/rate[p] [*,*] (tr)
:     bands     coils     :=
1    29.95     10.05
2    30        10
3    20        12
4    32.1429    7.85714
;

ampl: display {p in PROD, t in 1..T} prodcost[p]*Make[p,t];
prodcost[p]*Make[p,t] [*,*] (tr)
:     bands     coils     :=
1    59900     15477
2    60000     15400
3    40000     18480
4    64285.7   12100
;
```

(see Figures 6-3 and 6-4). The default is to use six significant digits, whether the result comes out as 7.85714 or 64285.7. Some numbers seem to have fewer digits, but only because trailing zeros have been dropped; 29.95 represents the number that is exactly 29.9500 to six digits, for example, and 59900 represents 59900.0.

By changing the option display_precision to a value other than six, you can vary the number of significant digits reported:

```
ampl: option display_precision 3;
ampl: display Make['bands',4] / rate['bands'],
ampl?       prodcost['bands'] * Make['bands',4];
Make['bands',4]/rate['bands'] = 32.1
prodcost['bands']*Make['bands',4] = 64300

ampl: option display_precision 9;
ampl: display Make['bands',4] / rate['bands'],
ampl?       prodcost['bands'] * Make['bands',4];
Make['bands',4]/rate['bands'] = 32.1428571
prodcost['bands']*Make['bands',4] = 64285.7143

ampl: option display_precision 0;
ampl: display Make['bands',4] / rate['bands'],
ampl?       prodcost['bands'] * Make['bands',4];
Make['bands',4]/rate['bands'] = 32.14285714285713
prodcost['bands']*Make['bands',4] = 64285.71428571427
```

In the last example, a display_precision of 0 is interpreted specially; it tells display to represent numbers as exactly as possible, using however many digits are necessary. (To be precise, the displayed number is the shortest decimal representation that, when correctly rounded to the computer's representation, gives the value exactly as stored in the computer.)

Displays to a given precision provide the same degree of useful information about each number, but they can look ragged due to the varying numbers of digits after the decimal point. To specify rounding to a fixed number of decimal places, regardless of the resulting precision, you may set the option `display_round`. A nonnegative value specifies the number of digits to appear after the decimal point:

```
ampl: option display_round 2;
ampl: display {p in PROD, t in 1..T} Make[p,t]/rate[p];
Make[p,t]/rate[p] [*,*] (tr)
:    bands    coils    :=
1    29.95    10.05
2    30.00    10.00
3    20.00    12.00
4    32.14     7.86
;
```

A negative value indicates rounding before the decimal point. For example, when `display_round` is –2, all numbers are rounded to hundreds:

```
ampl: option display_round -2;
ampl: display {p in PROD, t in 1..T} prodcost[p]*Make[p,t];
prodcost[p]*Make[p,t] [*,*] (tr)
:    bands    coils    :=
1    59900    15500
2    60000    15400
3    40000    18500
4    64300    12100
;
```

Any integer value of `display_round` overrides the effect of `display_precision`. To turn off `display_round`, set it to some non-integer such as the empty string `' '`.

Depending on the solver you employ, you may find that some of the solution values that ought to be zero do not always quite come out that way. For example, here is one solver's report of the objective function terms `cost[i,j] * Trans[i,j]` for the assignment problem of Section 3.3:

```
ampl: option omit_zero_rows 1;
ampl: display {i in ORIG, j in DEST} cost[i,j] * Trans[i,j];
cost[i,j]*Trans[i,j] :=
Coullard   C118    6
Coullard   D241    2.05994e-17
Daskin     D237    1
Hazen      C246    1
Hopp       D237    6.86647e-18
Hopp       D241    4
...    9 lines omitted
White      C246    2.74659e-17
White      C251    -3.43323e-17
White      M239    1
;
```

Minuscule values like 6.86647e–18 and –3.43323e–17 have no significance in the context of this problem; they would be zeros in an exact solution, but come out slightly nonzero as an artifact of the way that the solver's algorithm interacts with the computer's representation of numbers.

To avoid viewing these numbers in meaningless precision, you can pick a reasonable setting for display_round — in this case 0, since there are no digits of interest after the decimal point:

```
ampl: option display_round 0;
ampl: display {i in ORIG, j in DEST} cost[i,j] * Trans[i,j];
cost[i,j]*Trans[i,j]  :=
Coullard   C118    6
Coullard   D241    0
Daskin     D237    1
Hazen      C246    1
Hopp       D237    0
Hopp       D241    4
Iravani    C118    0
Iravani    C138    2
Linetsky   C250    3
Mehrotra   D239    2
Nelson     C138    0
Nelson     C140    4
Smilowitz  M233    1
Tamhane    C118   -0
Tamhane    C251    3
White      C246    0
White      C251   -0
White      M239    1
;
```

The small numbers are now represented only as 0 if positive or –0 if negative. If you want to suppress their appearance entirely, however, you must set a separate option, display_eps:

```
ampl: option display_eps 1e-10;
ampl: display {i in ORIG, j in DEST} cost[i,j] * Trans[i,j];
cost[i,j]*Trans[i,j]  :=
Coullard   C118    6
Daskin     D237    1
Hazen      C246    1
Hopp       D241    4
Iravani    C138    2
Linetsky   C250    3
Mehrotra   D239    2
Nelson     C140    4
Smilowitz  M233    1
Tamhane    C251    3
White      M239    1
;
```

Any value whose magnitude is less than the value of `display_eps` is treated as an exact zero in all output of `display`.

Rounding of solution values

The options `display_precision`, `display_round` and `display_eps` affect only the appearance of numbers, not their actual values. You can see this if you try to display the set of all pairs of i in ORIG and j in DEST that have a positive value in the preceding table, by comparing `cost[i,j]*Trans[i,j]` to 0:

```
ampl: display {i in ORIG, j in DEST: cost[i,j]*Trans[i,j] > 0};
set {i in ORIG, j in DEST: cost[i,j]*Trans[i,j] > 0}  :=
(Coullard,C118)      (Iravani,C118)       (Smilowitz,M233)
(Coullard,D241)      (Iravani,C138)       (Tamhane,C251)
(Daskin,D237)        (Linetsky,C250)      (White,C246)
(Hazen,C246)         (Mehrotra,D239)      (White,M239)
(Hopp,D237)          (Nelson,C138)
(Hopp,D241)          (Nelson,C140);
```

Even though a value like 2.05994e–17 is treated as a zero for purposes of `display`, it tests greater than zero. You could fix this problem by changing > 0 above to, say, > 0.1. As an alternative, you can set the option `solution_round` so that AMPL rounds the solution values to a reasonable precision when they are received from the solver:

```
ampl: option solution_round 10;
ampl: solve;
MINOS 5.5: optimal solution found.
40 iterations, objective 28
ampl: display {i in ORIG, j in DEST: cost[i,j]*Trans[i,j] > 0};
set {i in ORIG, j in DEST: cost[i,j]*Trans[i,j] > 0}  :=
(Coullard,C118)      (Iravani,C138)       (Smilowitz,M233)
(Daskin,D237)        (Linetsky,C250)      (Tamhane,C251)
(Hazen,C246)         (Mehrotra,D239)      (White,M239)
(Hopp,D241)          (Nelson,C140);
```

The options `solution_precision` and `solution_round` work in the same way as `display_precision` and `display_round`, except that they are applied only to solution values upon return from a solver, and they permanently change the returned values rather than only their appearance.

Rounded values can make a difference even when they are not near zero. As an example, we first use several `display` options to get a compact listing of the fractional solution to the scheduling model of Figure 16-4:

```
ampl: model sched.mod;
ampl: data sched.dat;

ampl: solve;
MINOS 5.5: optimal solution found.
19 iterations, objective 265.6
```

```
ampl: option display_width 60;
ampl: option display_1col 5;

ampl: option display_eps 1e-10;
ampl: option omit_zero_rows 1;
ampl: display Work;
Work [*] :=
  10 28.8     30 14.4     71 35.6     106 23.2    123 35.6
  18  7.6     35  6.8     73 28       109 14.4
  24  6.8     66 35.6     87 14.4     113 14.4
;
```

Each value Work[j] represents the number of workers assigned to schedule j. We can get a quick practical schedule by rounding the fractional values up to the next highest integer; using the ceil function to perform the rounding, we see that the total number of workers needed should be:

```
ampl: display sum {j in SCHEDS} ceil(Work[j]);
sum{j in SCHEDS} ceil(Work[j]) = 273
```

If we copy the numbers from the preceding table and round them up by hand, however, we find that they only sum to 271. The source of the difficulty can be seen by displaying the numbers to full precision:

```
ampl: option display_eps 0;
ampl: option display_precision 0;

ampl: display Work;
Work [*] :=
  10 28.799999999999997       73 28.000000000000018
  18  7.599999999999998       87 14.399999999999995
  24  6.7999999999999          95 -5.876671973951407e-15
  30 14.40000000000001        106 23.200000000000006
  35  6.799999999999995       108  4.685288280240683e-16
  55 -4.939614313857677e-15   109 14.4
  66 35.6                     113 14.4
  71 35.599999999999994       123 35.59999999999999
;
```

Half the problem is due to the minuscule positive value of Work[108], which was rounded up to 1. The other half is due to Work[73]; although it is 28 in an exact solution, it comes back from the solver with a slightly larger value of 28.000000000000018, so it gets rounded up to 29.

The easiest way to ensure that our arithmetic works correctly in this case is again to set solution_round before solve:

```
ampl: option solution_round 10;
ampl: solve;
MINOS 5.5: optimal solution found.
19 iterations, objective 265.6

ampl: display sum {j in SCHEDS} ceil(Work[j]);
sum{j in SCHEDS} ceil(Work[j]) = 271
```

We picked a value of 10 for `solution_round` because we observed that the slight inaccuracies in the solver's values occurred well past the 10th decimal place.

The effect of `solution_round` or `solution_precision` applies to all values returned by the solver. To modify only certain values, use the assignment (`let`) command described in Section 11.3 together with the rounding functions in Table 11-1.

12.4 Other output commands: `print` and `printf`

Two additional AMPL commands have much the same syntax as `display`, but do not automatically format their output. The `print` command does no formatting at all, while the `printf` command requires an explicit description of the desired formatting.

The `print` command

A `print` command produces a single line of output:

```
ampl: print sum {p in PROD, t in 1..T} revenue[p,t]*Sell[p,t],
ampl?        sum {p in PROD, t in 1..T} prodcost[p]*Make[p,t],
ampl?        sum {p in PROD, t in 1..T} invcost[p]*Inv[p,t];
787810 269477 3300

ampl: print {t in 1..T, p in PROD} Make[p,t];
5990 1407 6000 1400 1400 3500 2000 4200
```

or, if followed by an indexing expression and a colon, a line of output for each member of the index set:

```
ampl: print {t in 1..T}: {p in PROD} Make[p,t];
5990 1407
6000 1400
1400 3500
2000 4200
```

Printed entries are normally separated by a space, but option `print_separator` can be used to change this. For instance, you might set `print_separator` to a tab for data to be imported by a spreadsheet; to do this, type `option print_separator "→"`, where → stands for the result of pressing the tab key.

The keyword `print` (with optional indexing expression and colon) is followed by a print item or comma-separated list of print items. A print item can be a value, or an indexing expression followed by a value or parenthesized list of values. Thus a print item is much like a display item, except that only individual values may appear; although you can say `display rate`, you must explicitly specify `print {p in PROD} rate[p]`. Also a set may not be an argument to `print`, although its members may be:

```
ampl: print PROD;
syntax error
context:  print  >>> PROD;  <<<

ampl: print {p in PROD} (p, rate[p]);
bands 200 coils 140
```

Unlike `display`, however, `print` allows indexing to be nested within an indexed item:

```
ampl: print {p in PROD} (p, rate[p], {t in 1..T} Make[p,t]);
bands 200 5990 6000 1400 2000 coils 140 1407 1400 3500 4200
```

The representation of numbers in the output of `print` is governed by the `print_precision` and `print_round` options, which work exactly like the `display_precision` and `display_round` options for the `display` command. Initially `print_precision` is 0 and `print_round` is an empty string, so that by default `print` uses as many digits as necessary to represent each value as precisely as possible. For the above examples, `print_round` has been set to 0, so that the numbers are rounded to integers.

Working interactively, you may find `print` useful for viewing a few values on your screen in a more compact format than `display` produces. With output redirected to a file, `print` is useful for writing unformatted results in a form convenient for spreadsheets and other data analysis tools. As with `display`, just add *>filename* to the end of the `print` command.

The `printf` command

The syntax of `printf` is exactly the same as that of `print`, except that the first print item is a character string that provides formatting instructions for the remaining items:

```
ampl: printf "Total revenue is $%6.2f.\n",
ampl?    sum {p in PROD, t in 1..T} revenue[p,t]*Sell[p,t];
Total revenue is $787810.00.
```

The format string contains two types of objects: ordinary characters, which are copied to the output, and conversion specifications, which govern the appearance of successive remaining print items. Each conversion specification begins with the character % and ends with a conversion character. For example, %6.2f specifies conversion to a decimal representation at least six characters wide with two digits after the decimal point. The complete rules are much the same as for the `printf` function in the C programming language; a summary appears in Section A.16 of the Appendix.

The output from `printf` is not automatically broken into lines. A line break must be indicated explicitly by the combination \n, representing a ''newline'' character, in the format string. To produce a series of lines, use the indexed version of `printf`:

```
ampl: printf {t in 1..T}: "%3i%12.2f%12.2f\n", t,
ampl?    sum {p in PROD} revenue[p,t]*Sell[p,t],
ampl?    sum {p in PROD} prodcost[p]*Make[p,t];
  1   159210.00    75377.00
  2   243500.00    75400.00
  3   167300.00    52500.00
  4   217800.00    66200.00
```

This `printf` is executed once for each member of the indexing set preceding the colon; for each `t in 1..T` the format is applied again, and the `\n` character generates another line break.

The `printf` command is mainly useful, in conjunction with redirection of output to a file, for printing short summary reports in a readable format. Because the number of conversion specifications in the format string must match the number of values being printed, `printf` cannot conveniently produce tables in which the number of items on a line may vary from run to run, such as a table of all `Make[p,t]` values.

12.5 Related solution values

Sets, parameters and variables are the most obvious things to look at in interpreting the solution of a linear program, but AMPL also provides ways of examining objectives, bounds, slacks, dual prices and reduced costs associated with the optimal solution.

As we have shown in numerous examples already, AMPL distinguishes the various values associated with a model component by use of ''qualified'' names that consist of a variable or constraint identifier, a dot (.), and a predefined ''suffix'' string. For instance, the upper bounds for the variable `Make` are called `Make.ub`, and the upper bound for `Make["coils",2]` is written `Make["coils",2].ub`. (Note that the suffix comes after the subscript.) A qualified name can be used like an unqualified one, so that `display Make.ub` prints a table of upper bounds on the `Make` variables, while `display Make, Make.ub` prints a list of the optimal values and upper bounds.

Objective functions

The name of the objective function (from a `minimize` or `maximize` declaration) refers to the objective's value computed from the current values of the variables. This name can be used to represent the optimal objective value in `display`, `print`, or `printf`:

```
ampl: print 100 * Total_Profit /
ampl?    sum {p in PROD, t in 1..T} revenue[p,t] * Sell[p,t];
65.37528084182735
```

If the current model declares several objective functions, you can refer to any of them, even though only one has been optimized.

Bounds and slacks

The suffixes `.lb` and `.ub` on a variable denote its lower and upper bounds, while `.slack` denotes the difference of a variable's value from its nearer bound. Here's an example from Figure 5-1:

```
ampl: display Buy.lb, Buy, Buy.ub, Buy.slack;
:        Buy.lb      Buy     Buy.ub  Buy.slack      :=
BEEF      2         2           10      0
CHK       2        10           10      0
FISH      2         2           10      0
HAM       2         2           10      0
MCH       2         2           10      0
MTL       2         6.23596     10      3.76404
SPG       2         5.25843     10      3.25843
TUR       2         2           10      0
;
```

The reported bounds are those that were sent to the solver. Thus they include not only the bounds specified in `>=` and `<=` phrases of `var` declarations, but also certain bounds that were deduced from the constraints by AMPL's presolve phase. Other suffixes let you look at the original bounds and at additional bounds deduced by presolve; see the discussion of presolve in Section 14.1 for details.

Two equal bounds denote a fixed variable, which is normally eliminated by presolve. Thus in the planning model of Figure 4-4, the constraint `Inv[p,0] = inv0[p]` fixes the initial inventories:

```
ampl: display {p in PROD} (Inv[p,0].lb,inv0[p],Inv[p,0].ub);
:         Inv[p,0].lb inv0[p] Inv[p,0].ub      :=
bands         10          10          10
coils          0           0           0
;
```

In the production-and-transportation model of Figure 4-6, the constraint

```
sum {i in ORIG} Trans[i,j,p] = demand[j,p]
```

leads presolve to fix three variables at zero, because `demand["LAN","plate"]` is zero:

```
ampl: display {i in ORIG}
ampl?     (Trans[i,"LAN","plate"].lb,Trans[i,"LAN","plate"].ub);
:     Trans[i,'LAN','plate'].lb Trans[i,'LAN','plate'].ub     :=
CLEV               0                              0
GARY               0                              0
PITT               0                              0
;
```

As this example suggests, presolve's adjustments to the bounds may depend on the data as well as the structure of the constraints.

The concepts of bounds and slacks have an analogous interpretation for the constraints of a model. Any AMPL constraint can be put into the standard form

$$lower\ bound \le body \le upper\ bound$$

where the *body* is a sum of all terms involving variables, while the *lower bound* and *upper bound* depend only on the data. The suffixes `.lb`, `.body` and `.ub` give the current values of these three parts of the constraint. For example, in the diet model of Figure 5-1 we have the declarations

```
subject to Diet_Min {i in MINREQ}:
    sum {j in FOOD} amt[i,j] * Buy[j] >= n_Min[i];

subject to Diet_Max {i in MAXREQ}:
    sum {j in FOOD} amt[i,j] * Buy[j] <= n_Max[i];
```

and the following constraint bounds:

```
ampl: display Diet_Min.lb, Diet_Min.body, Diet_Min.ub;
:    Diet_Min.lb Diet_Min.body Diet_Min.ub    :=
A          700         1013.98      Infinity
B1           0          605         Infinity
B2           0          492.416     Infinity
C          700          700         Infinity
CAL      16000        16000         Infinity
;

ampl: display Diet_Max.lb, Diet_Max.body, Diet_Max.ub;
:    Diet_Max.lb Diet_Max.body Diet_Max.ub    :=
A      -Infinity     1013.98       20000
CAL    -Infinity    16000          24000
NA     -Infinity    43855.9        50000
;
```

Naturally, `<=` constraints have no lower bounds, and `>=` constraints have no upper bounds; AMPL uses `-Infinity` and `Infinity` in place of a number to denote these cases. Both the lower and the upper bound can be finite, if the constraint is specified with two `<=` or `>=` operators; see Section 8.4. For an `=` constraint the two bounds are the same.

The suffix `.slack` refers to the difference between the body and the nearer bound:

```
ampl: display Diet_Min.slack;
Diet_Min.slack [*] :=
  A   313.978
  B1  605
  B2  492.416
   C    0
CAL    0
;
```

For constraints that have a single `<=` or `>=` operator, the slack is always the difference between the expressions to the left and right of the operator, even if there are variables on both sides. The constraints that have a slack of zero are the ones that are truly constraining at the optimal solution.

Dual values and reduced costs

Associated with each constraint in a linear program is a quantity variously known as the dual variable, marginal value or shadow price. In the AMPL command environment, these dual values are denoted by the names of the constraints, without any qualifying suffix. Thus for example in Figure 4-6 there is a collection of constraints named `Demand`:

```
subject to Demand {j in DEST, p in PROD}:
    sum {i in ORIG} Trans[i,j,p] = demand[j,p];
```

and a table of the dual values associated with these constraints can be viewed by

```
ampl: display Demand;
Demand [*,*]
:       bands     coils  plate      :=
DET     201       190.714   199
FRA     209       204       211
FRE     266.2     273.714   285
LAF     201.2     198.714   205
LAN     202       193.714     0
STL     206.2     207.714   216
WIN     200       190.714   198
;
```

Solvers return optimal dual values to AMPL along with the optimal values of the "primal" variables. We have space here only to summarize the most common interpretation of dual values; the extensive theory of duality and applications of dual variables can be found in any textbook on linear programming.

To start with an example, consider the constraint `Demand["DET","bands"]` above. If we change the value of the parameter `demand["DET","bands"]` in this constraint, the optimal value of the objective function `Total_Cost` changes accordingly. If we were to plot the optimal value of `Total_Cost` versus all possible values of `demand["DET","bands"]`, the result would be a cost "curve" that shows how overall cost varies with demand for bands at Detroit.

Additional computation would be necessary to determine the entire cost curve, but you can learn something about it from the optimal dual values. After you solve the linear program using a particular value of `demand["DET","bands"]`, the dual price for the constraint tells you the *slope* of the cost curve, at the demand's current value. In our example, reading from the table above, we find that the slope of the curve at the current demand is 201. This means that total production and shipping cost is increasing at the rate of $201 for each extra ton of bands demanded at `DET`, or is decreasing by $201 for each reduction of one ton in the demand.

As an example of an inequality, consider the following constraint from the same model:

```
subject to Time {i in ORIG}:
    sum {p in PROD} (1/rate[i,p]) * Make[i,p] <= avail[i];
```

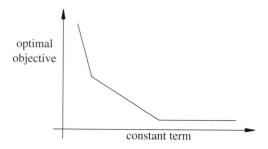

Figure 12-1: Piecewise-linear plot of objective function.

Here it is revealing to look at the dual values together with the slacks:

```
ampl: display Time, Time.slack;
:           Time    Time.slack    :=
CLEV    -1522.86       0
GARY    -3040          0
PITT        0         10.5643
;
```

Where the slack is positive, the dual value is zero. Indeed, the positive slack implies that the optimal solution does not use all of the time available at PITT; hence changing avail["PITT"] somewhat does not affect the optimum. On the other hand, where the slack is zero the dual value may be significant. In the case of GARY the value is –3040, implying that the total cost is decreasing at a rate of $3040 for each extra hour available at GARY, or is increasing at a rate of $3040 for each hour lost.

In general, if we plot the optimal objective versus a constraint's constant term, the curve will be convex piecewise-linear (Figure 12-1) for a minimization, or concave piecewise-linear (the same, but upside-down) for a maximization.

In terms of the standard form *lower bound* ≤ *body* ≤ *upper bound* introduced previously, the optimal dual values can be viewed as follows. If the slack of the constraint is positive, the dual value is zero. If the slack is zero, the body of the constraint must equal one (or both) of the bounds, and the dual value pertains to the equaled bound or bounds. Specifically, the dual value is the slope of the plot of the objective versus the bound, evaluated at the current value of the bound; equivalently it is the rate of change of the optimal objective with respect to the bound value.

A nearly identical analysis applies to the bounds on a variable. The role of the dual value is played by the variable's so-called reduced cost, which can be viewed from the AMPL command environment by use of the suffix .rc. As an example, here are the bounds and reduced costs for the variables in Figure 5-1:

```
ampl: display Buy.lb, Buy, Buy.ub, Buy.rc;
:      Buy.lb      Buy    Buy.ub     Buy.rc        :=
BEEF     2         2        10       1.73663
CHK      2        10        10      -0.853371
FISH     2         2        10       0.255281
HAM      2         2        10       0.698764
MCH      2         2        10       0.246573
MTL      2         6.23596  10       0
SPG      2         5.25843  10       0
TUR      2         2        10       0.343483
;
```

Since Buy["MTL"] has slack with both its bounds, its reduced cost is zero. Buy["HAM"] is at its lower bound, so its reduced cost indicates that total cost is increasing at about 70 cents per unit increase in the lower bound, or is decreasing at about 70 cents per unit decrease in the lower bound. On the other hand, Buy["CHK"] is at its upper bound, and its negative reduced cost indicates that total cost is decreasing at about 85 cents per unit increase in the upper bound, or is increasing at about 85 cents per unit decrease in the upper bound.

If the current value of a bound happens to lie right at a breakpoint in the relevant curve — one of the places where the slope changes abruptly in Figure 12-1 — the objective will change at one rate as the bound increases, but at a different rate as the bound decreases. In the extreme case either of these rates may be infinite, indicating that the linear program becomes infeasible if the bound is increased or decreased by any further amount. A solver reports only one optimal dual price or reduced cost to AMPL, however, which may be the rate in either direction, or some value between them.

In any case, moreover, a dual price or reduced cost can give you only one slope on the piecewise-linear curve of objective values. Hence these quantities should only be used as an initial guide to the objective's sensitivity to certain variable or constraint bounds. If the sensitivity is very important to your application, you can make a series of runs with different bound settings; see Section 11.3 for ways to quickly change a small part of the data. (There do exist algorithms for finding part or all of the piecewise-linear curve, given a linear change in one or more bounds, but they are not directly supported by the current version of AMPL.)

12.6 Other display features for models and instances

We gather in this section various utility commands and other features for displaying information about models or about problem instances generated from models.

Two commands let you review an AMPL model from the command-line: show lists the names of model components and displays the definitions of individual components, while xref lists all components that depend on a given component. The expand command displays selected objectives and constraints that AMPL has generated from a model and data, or analogous information for variables. AMPL's "generic" names for variables, constraints, or objectives permit listings or tests that apply to all variables, constraints, or objectives.

Displaying model components: the `show` command:

By itself, the `show` command lists the names of all components of the current model:

```
ampl: model multmip3.mod;
ampl: show;
parameters:  demand fcost limit maxserve minload supply vcost
sets:   DEST   ORIG   PROD
variables:   Trans   Use
constraints:   Demand   Max_Serve   Min_Ship   Multi   Supply
objective:   Total_Cost
checks: one, called check 1.
```

This display may be restricted to components of one or more types:

```
ampl: show vars;
variables:   Trans   Use
ampl: show obj, constr;
objective:   Total_Cost
constraints:   Demand   Max_Serve   Min_Ship   Multi   Supply
```

The `show` command can also display the declarations of individual components, saving you the trouble of looking them up in the model file:

```
ampl: show Total_Cost;
minimize Total_Cost: sum{i in ORIG, j in DEST, p in PROD}
vcost[i,j,p]*Trans[i,j,p] + sum{i in ORIG, j in DEST}
fcost[i,j]*Use[i,j];
ampl: show vcost, fcost, Trans;
param vcost{ORIG, DEST, PROD}  >= 0;
param fcost{ORIG, DEST}  >= 0;
var Trans{ORIG, DEST, PROD}  >= 0;
```

If an item following `show` is the name of a component in the current model, the declaration of that component is displayed. Otherwise, the item is interpreted as a component type according to its first letter or two; see Section A.19.1. (Displayed declarations may differ in inessential ways from their appearance in your model file, due to transformations that AMPL performs when the model is parsed and translated.)

Since the `check` statements in a model do not have names, AMPL numbers them in the order that they appear. Thus to see the third check statement you would type

```
ampl: show check 3;
check{p in PROD}  :
    sum{i in ORIG} supply[i,p] == sum{j in DEST} demand[j,p];
```

By itself, `show checks` indicates the number of `check` statements in the model.

Displaying model dependencies: the `xref` command

The `xref` command lists all model components that depend on a specified component, either directly (by referring to it) or indirectly (by referring to its dependents). If more than one component is given, the dependents are listed separately for each. Here is an example from `multmip3.mod`:

```
ampl: xref demand, Trans;
# 2 entities depend on demand:
check 1
Demand
# 5 entities depend on Trans:
Total_Cost
Supply
Demand
Multi
Min_Ship
```

In general, the command is simply the keyword `xref` followed by a comma-separated list of any combination of set, parameter, variable, objective and constraint names.

Displaying model instances: the `expand` command

In checking a model and its data for correctness, you may want to look at some of the specific constraints that AMPL is generating. The `expand` command displays all constraints in a given indexed collection or specific constraints that you identify:

```
ampl: model nltrans.mod;
ampl: data nltrans.dat;
ampl: expand Supply;
subject to Supply['GARY']:
        Trans['GARY','FRA'] + Trans['GARY','DET'] +
        Trans['GARY','LAN'] + Trans['GARY','WIN'] +
        Trans['GARY','STL'] + Trans['GARY','FRE'] +
        Trans['GARY','LAF'] = 1400;

subject to Supply['CLEV']:
        Trans['CLEV','FRA'] + Trans['CLEV','DET'] +
        Trans['CLEV','LAN'] + Trans['CLEV','WIN'] +
        Trans['CLEV','STL'] + Trans['CLEV','FRE'] +
        Trans['CLEV','LAF'] = 2600;

subject to Supply['PITT']:
        Trans['PITT','FRA'] + Trans['PITT','DET'] +
        Trans['PITT','LAN'] + Trans['PITT','WIN'] +
        Trans['PITT','STL'] + Trans['PITT','FRE'] +
        Trans['PITT','LAF'] = 2900;
```

(See Figures 18-4 and 18-5.) The ordering of terms in an expanded constraint does not necessarily correspond to the order of the symbolic terms in the constraint's declaration.

Objectives may be expanded in the same way:

```
ampl: expand Total_Cost;
minimize Total_Cost:
     39*Trans['GARY','FRA']/(1 - Trans['GARY','FRA']/500) + 14*
     Trans['GARY','DET']/(1 - Trans['GARY','DET']/1000) + 11*
     Trans['GARY','LAN']/(1 - Trans['GARY','LAN']/1000) + 14*
     Trans['GARY','WIN']/(1 - Trans['GARY','WIN']/1000) + 16*
     ...   15 lines omitted
     Trans['PITT','FRE']/(1 - Trans['PITT','FRE']/500) + 20*
     Trans['PITT','LAF']/(1 - Trans['PITT','LAF']/900);
```

When expand is applied to a variable, it lists all of the nonzero coefficients of that variable in the linear terms of objectives and constraints:

```
ampl: expand Trans;
Coefficients of Trans['GARY','FRA']:
        Supply['GARY']    1
        Demand['FRA']     1
        Total_Cost        0 + nonlinear

Coefficients of Trans['GARY','DET']:
        Supply['GARY']    1
        Demand['DET']     1
        Total_Cost        0 + nonlinear

Coefficients of Trans['GARY','LAN']:
        Supply['GARY']    1
        Demand['LAN']     1
        Total_Cost        0 + nonlinear

Coefficients of Trans['GARY','WIN']:
        Supply['GARY']    1
        Demand['WIN']     1
        Total_Cost        0 + nonlinear
   ...   17 terms omitted
```

When a variable also appears in nonlinear expressions within an objective or a constraint, the term + nonlinear is appended to represent those expressions.

The command expand alone produces an expansion of all variables, objectives and constraints in a model. Since a single expand command can produce a very long listing, you may want to redirect its output to a file by placing >*filename* at the end as explained in Section 12.7 below.

The formatting of numbers in the expanded output is governed by the options expand_precision and expand_round, which work like the display command's display_precision and display_round described in Section 12.3.

The output of expand reflects the ''modeler's view'' of the problem; it is based on the model and data as they were initially read and translated. But AMPL's presolve phase (Section 14.1) may make significant simplifications to the problem before it is sent to the solver. To see the expansion of the ''solver's view'' of the problem following presolve, use the keyword solexpand in place of expand.

Generic synonyms for variables, constraints and objectives

Occasionally it is useful to make a listing or a test that applies to all variables, constraints, or objectives. For this purpose, AMPL provides automatically updated parameters that hold the numbers of variables, constraints, and objectives in the currently generated problem instance:

```
_nvars      number of variables in the current problem
_ncons      number of constraints in the current problem
_nobjs      number of objectives in the current problem
```

Correspondingly indexed parameters contain the AMPL names of all the components:

```
_varname{1.._nvars}      names of variables in the current problem
_conname{1.._ncons}      names of constraints in the current problem
_objname{1.._nobjs}      names of objectives in the current problem
```

Finally, the following synonyms for the components are made available:

```
_var{1.._nvars}      synonyms for variables in the current problem
_con{1.._ncons}      synonyms for constraints in the current problem
_obj{1.._nobjs}      synonyms for objectives in the current problem
```

These synonyms let you refer to components by number, rather than by the usual indexed names. Using the variables as an example, _var[5] refers to the fifth variable in the problem, _var[5].ub to its upper bound, _var[5].rc to its reduced cost, and so forth, while _varname[5] is a string giving the variable's true AMPL name. Table A-13 lists all of the generic synonyms for sets, variables, and the like.

Generic names are useful for tabulating properties of all variables, where the variables have been defined in several different var declarations:

```
ampl: model net3.mod
ampl: data net3.dat
ampl: solve;
MINOS 5.5: optimal solution found.
3 iterations, objective 1819

ampl: display {j in 1.._nvars}
ampl?    (_varname[j],_var[j],_var[j].ub,_var[j].rc);

:             _varname[j]       _var[j] _var[j].ub      _var[j].rc   :=
1     "PD_Ship['NE']"            250       250         -0.5
2     "PD_Ship['SE']"            200       250         -1.11022e-16
3     "DW_Ship['NE','BOS']"       90        90          0
4     "DW_Ship['NE','EWR']"      100       100         -1.1
5     "DW_Ship['NE','BWI']"       60       100          0
6     "DW_Ship['SE','EWR']"       20       100          2.22045e-16
7     "DW_Ship['SE','BWI']"       60       100          2.22045e-16
8     "DW_Ship['SE','ATL']"       70        70          0
9     "DW_Ship['SE','MCO']"       50        50          0
;
```

Another use is to list all variables having some property, such as being away from the upper bound in the optimal solution:

```
ampl: display {j in 1.._nvars:
ampl?     _var[j] < _var[j].ub - 0.00001} _varname[j];
_varname[j] [*] :=
2   "PD_Ship['SE']"
5   "DW_Ship['NE','BWI']"
6   "DW_Ship['SE','EWR']"
7   "DW_Ship['SE','BWI']"
;
```

The same comments apply to constraints and objectives. More precise formatting of this information can be obtained with printf (12.4, A.16) instead of display.

As in the case of the expand command, these parameters and generic synonyms reflect the modeler's view of the problem; their values are determined from the model and data as they were initially read and translated. AMPL's presolve phase may make significant simplifications to the problem before it is sent to the solver. To work with parameters and generic synonyms that reflect the solver's view of the problem following presolve, replace _ by _s in the names given above; for example in the case of variables, use _snvars, _svarname and _svar.

Additional predefined sets and parameters represent the names and dimensions (arities) of the model components. They are summarized in A.19.4.

Resource listings

Changing option show_stats from its default of 0 to a nonzero value requests summary statistics on the size of the optimization problem that AMPL generates:

```
ampl: model steelT.mod;
ampl: data steelT.dat;
ampl: option show_stats 1;
ampl: solve;

Presolve eliminates 2 constraints and 2 variables.
Adjusted problem:
24 variables, all linear
12 constraints, all linear; 38 nonzeros
1 linear objective; 24 nonzeros.

MINOS 5.5: optimal solution found.
15 iterations, objective 515033
```

Additional lines report the numbers of integer and variables and nonlinear components where appropriate.

Changing option times from its default of 0 to a nonzero value requests a summary of the AMPL translator's time and memory requirements. Similarly, by changing option gentimes to a nonzero value, you can get a detailed summary of the resources that AMPL's genmod phase consumes in generating a model instance.

When AMPL appears to hang or takes much more time than expected, the display produced by `gentimes` can help associate the difficulty with a particular model component. Typically, some parameter, variable or constraint has been indexed over a set far larger than intended or anticipated, with the result that excessive amounts of time and memory are required.

The timings given by these commands apply only to the AMPL translator, not to the solver. A variety of predefined parameters (Table A-14) let you work with both AMPL and solver times. For example, `_solve_time` always equals total CPU seconds required for the most recent `solve` command, and `_ampl_time` equals total CPU seconds for AMPL excluding time spent in solvers and other external programs.

Many solvers also have directives for requesting breakdowns of the solve time. The specifics vary considerably, however, so information on requesting and interpreting these timings is provided in the documentation of AMPL's links to individual solvers, rather than in this book.

12.7 General facilities for manipulating output

We describe here how some or all of AMPL's output can be directed to a file, and how the volume of warning and error messages can be regulated.

Redirection of output

The examples in this book all show the outputs of commands as they would appear in an interactive session, with typed commands and printed responses alternating. You may direct all such output to a file instead, however, by adding a > and the name of the file:

```
ampl: display ORIG, DEST, PROD >multi.out;
ampl: display supply >multi.out;
```

The first command that specifies >multi.out creates a new file by that name (or overwrites any existing file of the same name). Subsequent commands add to the end of the file, until the end of the session or a matching `close` command:

```
ampl: close multi.out;
```

To open a file and append output to whatever is already there (rather than overwriting), use >> instead of >. Once a file is open, subsequent uses of > and >> have the same effect.

Output logs

The `log_file` option instructs AMPL to save subsequent commands and responses to a file. The option's value is a string that is interpreted as a filename:

```
ampl: option log_file 'multi.tmp';
```

The log file collects all AMPL statements and the output that they produce, with a few exceptions described below. Setting `log_file` to the empty string:

```
ampl: option log_file '';
```

turns off writing to the file; the empty string is the default value for this option.

When AMPL reads from an input file by means of a `model` or `data` command (or an `include` command as defined in Chapter 13), the statements from that file are not normally copied to the log file. To request that AMPL echo the contents of input files, change option `log_model` (for input in model mode) or `log_data` (for input in data mode) from the default value of 0 to a nonzero value.

When you invoke a solver, AMPL logs at least a few lines summarizing the objective value, solution status and work required. Through solver-specific directives, you can typically request additional solver output such as logs of iterations or branch-and-bound nodes. Many solvers automatically send all of their output to AMPL's log file, but this compatibility is not universal. If a solver's output does not appear in your log file, you should consult the supplementary documentation for that solver's AMPL interface; possibly that solver accepts nonstandard directives for diverting its output to files.

Limits on messages

By specifying `option eexit` n, where n is some integer, you determine how AMPL handles error messages. If n is not zero, any AMPL statement is terminated after it has produced `abs`(n) error messages; a negative value causes only the one statement to be terminated, while a positive value results in termination of the entire AMPL session. The effect of this option is most often seen in the use of `model` and `data` statements where something has gone badly awry, like using the wrong file:

```
ampl: option eexit -3;
ampl: model diet.mod;
ampl: data diet.mod;
diet.mod, line 4 (offset 32):
        expected ; ( [ : or symbol
context:  param cost  >>> { <<< FOOD} > 0;

diet.mod, line 5 (offset 56):
        expected ; ( [ : or symbol
context:  param f_min  >>> { <<< FOOD} >= 0;

diet.mod, line 6 (offset 81):
        expected ; ( [ : or symbol
context:  param f_max  >>> { <<< j in FOOD} >= f_min[j];

Bailing out after 3 warnings.
```

The default value for `eexit` is −10. Setting it to 0 causes all error messages to be displayed.

The `eexit` setting also applies to infeasibility warnings produced by AMPL's presolve phase after you type `solve`. The number of these warnings is simultaneously limited by the value of option `presolve_warnings`, which is typically set to a smaller value; the default is 5.

An AMPL `data` statement may specify values that correspond to illegal combinations of indices, due to any number of mistakes such as incorrect index sets in the model, indices in the wrong order, misuse of `(tr)`, and typing errors. Similar errors may be caused by `let` statements that change the membership of index sets. AMPL catches these errors after `solve` is typed. The number of invalid combinations displayed is limited to the value of the option `bad_subscripts`, whose default value is 3.

13

Command Scripts

You will probably find that your most intensive use of AMPL's command environment occurs during the initial development of a model, when the results are unfamiliar and changes are frequent. When the formulation eventually settles down, you may find yourself typing the same series of commands over and over to solve for different collections of data. To accelerate this process, you can arrange to have AMPL read often-used sequences of commands from files or to repeat command sequences automatically, determining how to proceed and when to stop on the basis of intermediate results.

A *script* is a sequence of commands, captured in a file, to be used and re-used. Scripts can contain any AMPL commands, and may include programming language constructs like `for`, `repeat`, and `if` to repeat statements and perform them conditionally. In effect, these and related commands let you write small programs in the AMPL command language. Another collection of commands permit stepping through a script for observation or debugging. This chapter introduces AMPL command scripts, using formatted printing and sensitivity analysis as examples.

AMPL command scripts are able to work directly with the sets of character strings that are central to the definition of models. A `for` statement can specify commands to be executed once for each member of some set, for example. To support scripts that work with strings, AMPL provides a variety of string functions and operators, whose use is described in the last section of this chapter.

13.1 Running scripts: `include` and `commands`

AMPL provides several commands that cause input to be taken from a file. The command

    include *filename*

is replaced by the contents of the named file. An `include` can even appear in the middle of some other statement, and does not require a terminating semicolon.

The `model` and `data` commands that appear in most of our examples are special cases of `include` that put the command interpreter into model or data mode before reading the specified file. By contrast, `include` leaves the mode unchanged. To keep things simple, the examples in this book always assume that `model` reads a file of model declarations, and that `data` reads a file of data values. You may use any of `model`, `data` and `include` to read any file, however; the only difference is the mode that is set when reading starts. Working with a small model, for example, you might find it convenient to store in one file all the model declarations, a `data` command, and all the data statements; either a `model` or an `include` command could read this file to set up both model and data in one operation.

As an illustration, if the file `dietu.run` contains

```
model dietu.mod;
data dietu.dat;
solve;
option display_1col 5;
option display_round 1;
display Buy;
```

then including it will load the model and data, run the problem, and display the optimal values of the variables:

```
ampl: include dietu.run;
MINOS 5.5: optimal solution found.
6 iterations, objective 74.27382022

Buy [*] :=
BEEF  2.0    FISH  2.0    MCH  2.0    SPG  5.3
 CHK 10.0    HAM  2.0    MTL  6.2    TUR  2.0
;
```

When an included file itself contains an `include`, `model` or `data` command, reading of the first file is suspended while the contents of the contained file are included. In this example, the command `include dietu.run` causes the subsequent inclusion of the files `dietu.mod` and `dietu.dat`.

One particularly useful kind of `include` file contains a list of `option` commands that you want to run before any other commands, to modify the default options. You can arrange to include such a file automatically at startup; you can even have AMPL write such a file automatically at the end of a session, so that your option settings will be restored the next time around. Details of this arrangement depend on your operating system; see Sections A.14.1 and A.23.

The statement

```
commands filename ;
```

is very similar to `include`, but is a true statement that needs a terminating semicolon and can only appear in a context where a statement is legal.

To illustrate `commands`, consider how we might perform a simple sensitivity analysis on the multi-period production problem of Section 4.2. Only 32 hours of production time are available in week 3, compared to 40 hours in the other weeks. Suppose that we want to see how much extra profit could be gained for each extra hour in week 3. We can accomplish this by repeatedly solving, displaying the solution values, and increasing avail[3]:

```
ampl: model steelT.mod;
ampl: data steelT.dat;

ampl: solve;
MINOS 5.5: optimal solution found.
15 iterations, objective 515033
ampl: display Total_Profit >steelT.sens;
ampl: option display_1col 0;
ampl: option omit_zero_rows 0;
ampl: display Make >steelT.sens;
ampl: display Sell >steelT.sens;
ampl: option display_1col 20;
ampl: option omit_zero_rows 1;
ampl: display Inv >steelT.sens;

ampl: let avail[3] := avail[3] + 5;
ampl: solve;
MINOS 5.5: optimal solution found.
1 iterations, objective 532033
ampl: display Total_Profit >steelT.sens;
ampl: option display_1col 0;
ampl: option omit_zero_rows 0;
ampl: display Make >steelT.sens;
ampl: display Sell >steelT.sens;
ampl: option display_1col 20;
ampl: option omit_zero_rows 1;
ampl: display Inv >steelT.sens;

ampl: let avail[3] := avail[3] + 5;
ampl: solve;
MINOS 5.5: optimal solution found.
1 iterations, objective 549033
ampl:
```

To continue trying values of avail[3] in steps of 5 up to say 62, we must complete another four solve cycles in the same way. We can avoid having to type the same commands over and over by creating a new file containing the commands to be repeated:

```
solve;
display Total_Profit >steelT.sens;
option display_1col 0;
option omit_zero_rows 0;
display Make >steelT.sens;
display Sell >steelT.sens;
option display_1col 20;
option omit_zero_rows 1;
display Inv >steelT.sens;
let avail[3] := avail[3] + 5;
```

If we call this file `steelT.sa1`, we can execute all the commands in it by typing the single line `commands steelT.sa1`:

```
ampl: model steelT.mod;
ampl: data steelT.dat;
ampl: commands steelT.sa1;
MINOS 5.5: optimal solution found.
15 iterations, objective 515033
ampl: commands steelT.sa1;
MINOS 5.5: optimal solution found.
1 iterations, objective 532033
ampl: commands steelT.sa1;
MINOS 5.5: optimal solution found.
1 iterations, objective 549033
ampl: commands steelT.sa1;
MINOS 5.5: optimal solution found.
2 iterations, objective 565193
ampl:
```

(All output from the `display` command is redirected to the file `steelT.sens`, although we could just as well have made it appear on the screen.)

In this and many other cases, you can substitute `include` for `commands`. In general it is best to use `commands` within command scripts, however, to avoid unexpected interactions with `repeat`, `for` and `if` statements.

13.2 Iterating over a set: the `for` statement

The examples above still require that some command be typed repeatedly. AMPL provides looping commands that can do this work automatically, with various options to determine how long the looping should continue.

We begin with the `for` statement, which executes a statement or collection of statements once for each member of some set. To execute our multi-period production problem sensitivity analysis script four times, for example, we can use a single `for` statement followed by the command that we want to repeat:

```
ampl: model steelT.mod;
ampl: data steelT.dat;
ampl: for {1..4} commands steelT.sa1;
MINOS 5.5: optimal solution found.
15 iterations, objective 515033
MINOS 5.5: optimal solution found.
1 iterations, objective 532033
MINOS 5.5: optimal solution found.
1 iterations, objective 549033
MINOS 5.5: optimal solution found.
2 iterations, objective 565193
ampl:
```

The expression between `for` and the command can be any AMPL indexing expression.

As an alternative to taking the commands from a separate file, we can write them as the body of a `for` statement, enclosed in braces:

```
model steelT.mod;
data steelT.dat;

for {1..4} {
   solve;
   display Total_Profit >steelT.sens;
   option display_1col 0;
   option omit_zero_rows 0;
   display Make >steelT.sens;
   display Sell >steelT.sens;
   option display_1col 20;
   option omit_zero_rows 1;
   display Inv >steelT.sens;
   let avail[3] := avail[3] + 5;
}
```

If this script is stored in `steelT.sa2`, then the whole iterated sensitivity analysis is carried out by typing

```
ampl: commands steelT.sa2
```

This approach tends to be clearer and easier to work with, particularly as we make the loop more sophisticated. As a first example, consider how we would go about compiling a table of the objective and the dual value on constraint `Time[3]`, for successive values of `avail[3]`. A script for this purpose is shown in Figure 13-1. After the model and data are read, the script provides additional declarations for the table of values:

```
set AVAIL3;
param avail3_obj {AVAIL3};
param avail3_dual {AVAIL3};
```

The set `AVAIL3` will contain all the different values for `avail[3]` that we want to try; for each such value a, `avail3_obj[a]` and `avail3_dual[a]` will be the associated objective and dual values. Once these are set up, we assign the set value to `AVAIL3`:

```
let AVAIL3 := avail[3] .. avail[3] + 15 by 5;
```

and then use a `for` loop to iterate over this set:

```
for {a in AVAIL3} {
   let avail[3] := a;
   solve;
   let avail3_obj[a] := Total_Profit;
   let avail3_dual[a] := Time[3].dual;
}
```

We see here that a `for` loop can be over an arbitrary set, and that the index running over the set (a in this case) can be used in statements within the loop. After the loop is com-

```
model steelT.mod;
data steelT.dat;
option solver_msg 0;

set AVAIL3;
param avail3_obj {AVAIL3};
param avail3_dual {AVAIL3};
let AVAIL3 := avail[3] .. avail[3] + 15 by 5;

for {a in AVAIL3} {
    let avail[3] := a;
    solve;
    let avail3_obj[a]  := Total_Profit;
    let avail3_dual[a] := Time[3].dual;
}
display avail3_obj, avail3_dual;
```

Figure 13-1: Parameter sensitivity script (steelT.sa3).

plete, the desired table is produced by displaying avail3_obj and avail3_dual, as shown at the end of the script in Figure 13-1. If this script is stored in steelT.sa3, then the desired results are produced with a single command:

```
ampl: commands steelT.sa3;
:   avail3_obj avail3_dual :=
32     515033       3400
37     532033       3400
42     549033       3400
47     565193       2980
;
```

In this example we have suppressed the messages from the solver, by including the command option solver_msg 0 in the script.

AMPL's for loops are also convenient for generating formatted tables. Suppose that after solving the multi-period production problem, we want to display sales both in tons and as a percentage of the market limit. We could use a display command to produce a table like this:

```
ampl: display {t in 1..T, p in PROD}
ampl?    (Sell[p,t], 100*Sell[p,t]/market[p,t]);
:       Sell[p,t] 100*Sell[p,t]/market[p,t]     :=
1 bands    6000            100
1 coils     307              7.675
2 bands    6000            100
2 coils    2500            100
3 bands    1400             35
3 coils    3500            100
4 bands    2000             30.7692
4 coils    4200            100
;
```

By writing a script that uses the `printf` command (A.16), we can create a much more effective table:

```
ampl: commands steelT.tab1;
SALES           bands              coils
week 1      6000   100.0%       307     7.7%
week 2      6000   100.0%      2500   100.0%
week 3      1399    35.0%      3500   100.0%
week 4      1999    30.8%      4200   100.0%
```

The script to write this table can be as short as two `printf` commands:

```
printf "\n%s%14s%17s\n", "SALES", "bands", "coils";
printf {t in 1..T}: "week %d%9d%7.1f%%%9d%7.1f%%\n", t,
    Sell["bands",t], 100*Sell["bands",t]/market["bands",t],
    Sell["coils",t], 100*Sell["coils",t]/market["coils",t];
```

This approach is undesirably restrictive, however, because it assumes that there will always be two products and that they will always be named `coils` and `bands`. In fact the `printf` statement cannot write a table in which both the number of rows and the number of columns depend on the data, because the number of entries in its format string is always fixed.

A more general script for generating the table is shown in Figure 13-2. Each pass through the "outer" loop over {1..T} generates one row of the table. Within each pass, an "inner" loop over PROD generates the row's product entries. There are more `printf` statements than in the previous example, but they are shorter and simpler. We use several statements to write the contents of each line; `printf` does not begin a new line except where a newline (\n) appears in its format string.

Loops can be nested to any depth, and may be iterated over any set that can be represented by an AMPL set expression. There is one pass through the loop for every member of the set, and if the set is ordered — any set of numbers like 1..T, or a set declared `ordered` or `circular` — the order of the passes is determined by the ordering of the set. If the set is unordered (like PROD) then AMPL chooses the order of the passes, but

---

```
printf "\nSALES";
printf {p in PROD}: "%14s    ", p;
printf "\n";
for {t in 1..T} {
   printf "week %d", t;
   for {p in PROD} {
      printf "%9d", Sell[p,t];
      printf "%7.1f%%", 100 * Sell[p,t]/market[p,t];
   }
   printf "\n";
}
```

Figure 13-2: Generating a formatted sales table with nested loops (`steelT.tab1`).

the choice is the same every time; the Figure 13-2 script relies on this consistency to ensure that all of the entries in one column of the table refer to the same product.

13.3 Iterating subject to a condition: the `repeat` statement

A second kind of looping construct, the `repeat` statement, continues iterating as long as some logical condition is satisfied.

Returning to the sensitivity analysis example, we wish to take advantage of a property of the dual value on the constraint `Time[3]`: the additional profit that can be realized from each extra hour added to `avail[3]` is at most `Time[3].dual`. When `avail[3]` is made sufficiently large, so that there is more third-week capacity than can be used, the associated dual value falls to zero and further increases in `avail[3]` have no effect on the optimal solution.

We can specify that looping should stop once the dual value falls to zero, by writing a `repeat` statement that has one of the following forms:

```
repeat while Time[3].dual > 0 {...};
repeat until Time[3].dual = 0 {...};
repeat {...} while Time[3].dual > 0;
repeat {...} until Time[3].dual = 0;
```

The loop body, here indicated by `{...}`, must be enclosed in braces. Passes through the loop continue as long as the `while` condition is true, or as long as the `until` condition is false. A condition that appears before the loop body is tested before every pass; if a `while` condition is false or an `until` condition is true before the first pass, then the loop body is never executed. A condition that appears after the loop body is tested after every pass, so that the loop body is executed at least once in this case. If there is no `while` or `until` condition, the loop repeats indefinitely and must be terminated by other means, like the `break` statement described below.

A complete script using `repeat` is shown in Figure 13-3. For this particular application we choose the `until` phrase that is placed after the loop body, as we do not want `Time[3].dual` to be tested until after a `solve` has been executed in the first pass. Two other features of this script are worth noting, as they are relevant to many scripts of this kind.

At the beginning of the script, we don't know how many passes the `repeat` statement will make through the loop. Thus we cannot determine AVAIL3 in advance as we did in Figure 13-1. Instead, we declare it initially to be the empty set:

```
set AVAIL3 default {};
param avail3_obj {AVAIL3};
param avail3_dual {AVAIL3};
```

and add each new value of `avail[3]` to it after solving:

```
model steelT.mod;
data steelT.dat;

option solution_precision 10;
option solver_msg 0;

set AVAIL3 default {};
param avail3_obj {AVAIL3};
param avail3_dual {AVAIL3};
param avail3_step := 5;

repeat {
   solve;
   let AVAIL3 := AVAIL3 union {avail[3]};
   let avail3_obj[avail[3]] := Total_Profit;
   let avail3_dual[avail[3]] := Time[3].dual;
   let avail[3] := avail[3] + avail3_step;
} until Time[3].dual = 0;

display avail3_obj, avail3_dual;
```

Figure 13-3: Script for recording sensitivity (steelT.sa4).

```
let AVAIL3 := AVAIL3 union {avail[3]};
let avail3_obj[avail[3]] := Total_Profit;
let avail3_dual[avail[3]] := Time[3].dual;
```

By adding a new member to AVAIL3, we also create new components of the parameters avail3_obj and avail3_dual that are indexed over AVAIL3, and so we can proceed to assign the appropriate values to these components. Any change to a set is propagated to all declarations that use the set, in the same way that any change to a parameter is propagated.

Because numbers in the computer are represented with a limited number of bits of precision, a solver may return values that differ very slightly from the solution that would be computed using exact arithmetic. Ordinarily you don't see this, because the display command rounds values to six significant digits by default. For example:

```
ampl: model steelT.mod; data steelT.dat; solve;
ampl: display Make;
Make [*,*] (tr)
:   bands   coils     :=
1   5990    1407
2   6000    1400
3   1400    3500
4   2000    4200
;
```

Compare what is shown when rounding is dropped, by setting display_precision to 0:

```
ampl: option display_precision 0;
ampl: display Make;
Make [*,*] (tr)
:          bands              coils           :=
1    5989.999999999999    1407.0000000000002
2    6000                 1399.9999999999998
3    1399.9999999999995   3500
4    1999.9999999999993   4200
;
```

These seemingly tiny differences can have undesirable effects whenever a script makes a comparison that uses values returned by the solver. The rounded table would lead you to believe that `Make["coils",2] >= 1400` is true, for example, whereas from the second table you can see that really it is false.

You can avoid this kind of surprise by writing arithmetic tests more carefully; instead of `until Time[3].dual = 0`, for instance, you might say `until Time[3].dual <= 0.0000001`. Alternatively, you can round all solution values that are returned by the solver, so that numbers that are supposed to be equal really do come out equal. The statement

```
option solution_precision 10;
```

toward the beginning of Figure 13-3 has this effect; it states that solution values are to be rounded to 10 significant digits. This and related rounding options are discussed and illustrated in Section 12.3.

Note finally that the script declares set AVAIL3 as `default {}` rather than `= {}`. The former allows AVAIL3 to be changed by `let` commands as the script proceeds, whereas the latter permanently defines AVAIL3 to be the empty set.

13.4 Testing a condition: the `if-then-else` statement

In Section 7.3 we described the conditional (`if-then-else`) expression, which produces an arithmetic or set value that can be used in any expression context. The `if-then-else` *statement* uses the same syntactic form to conditionally control the execution of statements or groups of statements.

In the simplest case, the `if` statement evaluates a condition and takes a specified action if the condition is true:

```
if Make["coils",2] < 1500 then printf "under 1500\n";
```

The action may also be a series of commands grouped by braces as in the `for` and `repeat` commands:

```
if Make["coils",2] < 1500 then {
   printf "Fewer than 1500 coils in week 2.\n";
   let market["coils",2] := market["coils",2] * 1.1;
}
```

An optional `else` specifies an alternative action that also may be a single command:

```
if Make["coils",2] < 1500 then {
    printf "Fewer than 1500 coils in week 2.\n";
    let market["coils",2] := market["coils",2] * 1.1;
}
else
    printf "At least 1500 coils in week 2.\n";
```

or a group of commands:

```
if Make["coils",2] < 1500 then
    printf "under 1500\n";
else {
    printf "at least 1500\n";
    let market["coils",2] := market["coils",2] * 0.9;
}
```

AMPL executes these commands by first evaluating the logical expression following `if`. If the expression is true, the command or commands following `then` are executed. If the expression is false, the command or commands following `else`, if any, are executed.

The `if` command is most useful for regulating the flow of control in scripts. In Figure 13-2, we could suppress any occurrences of `100%` by placing the statement that prints `Sell[p,t]/market[p,t]` inside an `if`:

```
if Sell[p,t] < market[p,t] then
    printf "%7.1f%%", 100 * Sell[p,t]/market[p,t];
else
    printf "     --- ";
```

In the script of Figure 13-3, we can use an `if` command inside the `repeat` loop to test whether the dual value has changed since the previous pass through the loop, as shown in the script of Figure 13-4. This loop creates a table that has exactly one entry for each different dual value discovered.

The statement following `then` or `else` can itself be an `if` statement. In the formatting example (Figure 13-2), we could handle both `0%` and `100%` specially by writing

```
if Sell[p,t] < market[p,t] then
    if Sell[p,t] = 0 then
        printf "        ";
    else
        printf "%7.1f%%", 100 * Sell[p,t]/market[p,t];
else
    printf "    --- ";
```

or equivalently, but perhaps more clearly,

```
if Sell[p,t] = 0 then
    printf "        ";
else if Sell[p,t] < market[p,t] then
    printf "%7.1f%%", 100 * Sell[p,t]/market[p,t];
else
    printf "    --- ";
```

```
model steelT.mod; data steelT.dat;
option solution_precision 10; option solver_msg 0;

set AVAIL3 default {};
param avail3_obj {AVAIL3};
param avail3_dual {AVAIL3};

let avail[3] := 1;
param avail3_step := 1;
param previous_dual default Infinity;

repeat while previous_dual > 0 {
   solve;
   if Time[3].dual < previous_dual then {
      let AVAIL3 := AVAIL3 union {avail[3]};
      let avail3_obj[avail[3]] := Total_Profit;
      let avail3_dual[avail[3]] := Time[3].dual;
      let previous_dual := Time[3].dual;
      }
   let avail[3] := avail[3] + avail3_step;
   }

display avail3_obj, avail3_dual;
```

Figure 13-4: Testing conditions within a loop (`steelT.sa5`).

In all cases, an `else` is paired with the closest preceding available `if`.

13.5 Terminating a loop: `break` and `continue`

Two other statements work with looping statements to make some scripts easier to write. The `continue` statement terminates the current pass through a `for` or `repeat` loop; all further statements in the current pass are skipped, and execution continues with the test that controls the start of the next pass (if any). The `break` statement completely terminates a `for` or `repeat` loop, sending control immediately to the statement following the end of the loop.

As an example of both these commands, Figure 13-5 exhibits another way of writing the loop from Figure 13-4, so that a table entry is made only when there is a change in the dual value associated with `avail[3]`. After solving, we test to see if the new dual value is equal to the previous one:

```
if Time[3].dual = previous_dual then continue;
```

If it is, there is nothing to be done for this value of `avail[3]`, and the `continue` statement jumps to the end of the current pass; execution resumes with the next pass, starting at the beginning of the loop.

After adding an entry to the table, we test to see if the dual value has fallen to zero:

```
model steelT.mod;
data steelT.dat;

option solution_precision 10;
option solver_msg 0;

set AVAIL3 default {};
param avail3_obj {AVAIL3};
param avail3_dual {AVAIL3};

let avail[3] := 0;
param previous_dual default Infinity;

repeat {
   let avail[3] := avail[3] + 1;
   solve;
   if Time[3].dual = previous_dual then continue;

   let AVAIL3 := AVAIL3 union {avail[3]};
   let avail3_obj[avail[3]] := Total_Profit;
   let avail3_dual[avail[3]] := Time[3].dual;

   if Time[3].dual = 0 then break;

   let previous_dual := Time[3].dual;
}

display avail3_obj, avail3_dual;
```

Figure 13-5: Using break and continue in a loop (steelT.sa7).

```
   if Time[3].dual = 0 then break;
```

If it has, the loop is done and the break statement jumps out; execution passes to the display command that follows the loop in the script. Since the repeat statement in this example has no while or until condition, it relies on the break statement for termination.

When a break or continue lies within a nested loop, it applies only to the inner-most loop. This convention generally has the desired effect. As an example, consider how we could expand Figure 13-5 to perform a separate sensitivity analysis on each avail[t]. The repeat loop would be nested in a for {t in 1..T} loop, but the continue and break statements would apply to the inner repeat as before.

There do exist situations in which the logic of a script requires breaking out of multiple loops. In the script of Figure 13-5, for instance, we can imagine that instead of stopping when Time[3].dual is zero,

```
   if Time[3].dual = 0 then break;
```

we want to stop when Time[t].dual falls below 2700 for any t. It might seem that one way to implement this criterion is:

```
for {t in 1..T}
   if Time[t].dual < 2700 then break;
```

This statement does not have the desired effect, however, because break applies only to the inner for loop that contains it, rather than to the outer repeat loop as we desire. In such situations, we can give a name to a loop, and break or continue can specify by name the loop to which it should apply. Using this feature, the outer loop in our example could be named sens_loop:

```
repeat sens_loop {
```

and the test for termination inside it could refer to its name:

```
for {t in 1..T}
    if Time[t].dual < 2700 then break sens_loop;
```

The loop name appears right after repeat or for, and after break or continue.

13.6 Stepping through a script

If you think that a script might not be doing what you want it to, you can tell AMPL to step through it one command at a time. This facility can be used to provide an elementary form of ''symbolic debugger'' for scripts.

To step through a script that does not execute any other scripts, reset the option single_step to 1 from its default value of 0. For example, here is how you might begin stepping through the script in Figure 13-5:

```
ampl: option single_step 1;
ampl: commands steelT.sa7;
steelT.sa7:2(18)    data ...
<2>ampl:
```

The expression steelT.sa7:2(18) gives the filename, line number and character number where AMPL has stopped in its processing of the script. It is followed by the beginning of the next command (data) to be executed. On the next line you are returned to the ampl: prompt. The <2> in front indicates the level of input nesting; ''2'' means that execution is within the scope of a commands statement that was in turn issued in the original input stream.

At this point you may use the step command to execute individual commands of the script. Type step by itself to execute one command,

```
<2>ampl: step
steelT.sa7:4(36)    option ...
<2>ampl: step
steelT.sa7:5(66)    option ...
<2>ampl: step
steelT.sa7:11(167)   let ...
<2>ampl:
```

If `step` is followed by a number, that number of commands will be executed. Every command is counted except those having to do with model declarations, such as `model` and `param` in this example.

Each `step` returns you to an AMPL prompt. You may continue stepping until the script ends, but at some point you will want to display some values to see if the script is working. This sequence captures one place where the dual value changes:

```
<2>ampl: display avail[3], Time[3].dual, previous_dual;
avail[3] = 22
Time[3].dual = 3620
previous_dual = 3620

<2>ampl: step
steelT.sa7:17(317)     continue ...
<2>ampl: step
steelT.sa7:15(237)     let ...
<2>ampl: step
steelT.sa7:16(270)     solve ...
<2>ampl: step
steelT.sa7:17(280)     if ...
<2>ampl: step
steelT.sa7:19(331)     let ...
<2>ampl: display avail[3], Time[3].dual, previous_dual;
avail[3] = 23
Time[3].dual = 3500
previous_dual = 3620
<2>ampl:
```

Any series of AMPL commands may be typed while single-stepping. After each command, the `<2>ampl` prompt returns to remind you that you are still in this mode and may use `step` to continue executing the script.

To help you step through lengthy compound commands (`for`, `repeat`, or `if`) AMPL provides several alternatives to `step`. The `next` command steps past a compound command rather than into it. If we had started at the beginning, each `next` would cause the next statement to be executed; in the case of the `repeat`, the entire command would be executed, stopping again only at the `display` command on line 28:

```
ampl: option single_step 1;
ampl: commands steelT.sa7;
steelT.sa7:2(18)     data ...
<2>ampl: next
steelT.sa7:4(36)     option ...
<2>ampl: next
steelT.sa7:5(66)     option ...
<2>ampl: next
steelT.sa7:11(167)    let ...
<2>ampl: next
steelT.sa7:14(225)    repeat ...
<2>ampl: next
steelT.sa7:28(539)    display ...
<2>ampl:
```

Type next *n* to step past *n* commands in this way.

The commands skip and skip *n* work like step and step *n*, except that they skip the next 1 or *n* commands in the script rather than executing them.

13.7 Manipulating character strings

The ability to work with arbitrary sets of character strings is one of the key advantages of AMPL scripting. We describe here the string concatenation operator and several functions for building up string-valued expressions that can be used anywhere that set members can appear in AMPL statements. Further details are provided in Section A.4.2, and Table A-4 summarizes all of the string functions.

We also show how string expressions may be used to specify character strings that serve purposes other than being set members. This feature allows an AMPL script to, for example, write a different file or set different option values in each pass through a loop, according to information derived from the contents of the loop indexing sets.

String functions and operators

The concatenation operator & takes two strings as operands, and returns a string consisting of the left operand followed by the right operand. For example, given the sets NUTR and FOOD defined by diet.mod and diet2.dat (Figures 2-1 and 2-3), you could use concatenation to define a set NUTR_FOOD whose members represent nutrient-food pairs:

```
ampl: model diet.mod;
ampl: data diet2.dat;
ampl: display NUTR, FOOD;
set NUTR := A B1 B2 C NA CAL;
set FOOD := BEEF CHK FISH HAM MCH MTL SPG TUR;
ampl: set NUTR_FOOD := setof {i in NUTR,j in FOOD} i & "_" & j;
ampl: display NUTR_FOOD;
set NUTR_FOOD :=
A_BEEF      B1_BEEF     B2_BEEF     C_BEEF      NA_BEEF     CAL_BEEF
A_CHK       B1_CHK      B2_CHK      C_CHK       NA_CHK      CAL_CHK
A_FISH      B1_FISH     B2_FISH     C_FISH      NA_FISH     CAL_FISH
A_HAM       B1_HAM      B2_HAM      C_HAM       NA_HAM      CAL_HAM
A_MCH       B1_MCH      B2_MCH      C_MCH       NA_MCH      CAL_MCH
A_MTL       B1_MTL      B2_MTL      C_MTL       NA_MTL      CAL_MTL
A_SPG       B1_SPG      B2_SPG      C_SPG       NA_SPG      CAL_SPG
A_TUR       B1_TUR      B2_TUR      C_TUR       NA_TUR      CAL_TUR;
```

This is not a set that you would normally want to define, but it might be useful if you have to read data in which strings like "B2_BEEF" appear.

Numbers that appear as arguments to & are automatically converted to strings. As an example, for a multi-week model you can create a set of generically-named periods WEEK1, WEEK2, and so forth, by declaring:

```
param T integer > 1;
set WEEKS ordered = setof {t in 1..T} "WEEK" & t;
```

Numeric operands to & are always converted to full precision (or equivalently, to %.0g format) as defined in Section A.16. The conversion thus produces the expected results for concatenation of numerical constants and of indices that run over sets of integers or constants, as in our examples. Full precision conversion of computed fractional values may sometimes yield surprising results, however. The following variation on the preceding example would seem to create a set of members WEEK0.1, WEEK0.2, and so forth:

```
param T integer > 1;
set WEEKS ordered = setof {t in 1..T} "WEEK" & 0.1*t;
```

But the actual set comes out differently:

```
ampl: let T := 4;
ampl: display WEEKS;
set WEEKS :=
WEEK0.1                         WEEK0.30000000000000004
WEEK0.2                         WEEK0.4;
```

Because 0.1 cannot be stored exactly in a binary representation, the value of $0.1*3$ is slightly different from 0.3 in "full" precision. There is no easy way to predict this behavior, but it can be prevented by specifying an explicit conversion using sprintf. The sprintf function does format conversions in the same way as printf (Section A.16), except that the resulting formatted string is not sent to an output stream, but instead becomes the function's return value. For our example, "WEEK" & 0.1*t could be replaced by sprintf("WEEK%3.1f",0.1*t).

The length string function takes a string as argument and returns the number of characters in it. The match function takes two string arguments, and returns the first position where the second appears as a substring in the first, or zero if the second never appears as a substring in the first. For example:

```
ampl: display {j in FOOD} (length(j), match(j,"H"));
:      length(j) match(j, 'H')      :=
BEEF      4              0
CHK       3              2
FISH      4              4
HAM       3              1
MCH       3              3
MTL       3              0
SPG       3              0
TUR       3              0
;
```

The substr function takes a string and one or two integers as arguments. It returns a substring of the first argument that begins at the position given by the second argument; it has the length given by the third argument, or extends to the end of the string if no third argument is given. An empty string is returned if the second argument is greater than the length of the first argument, or if the third argument is less than 1.

As an example combining several of these functions, suppose that you want to use the model from diet.mod but to supply the nutrition amount data in a table like this:

```
param: NUTR_FOOD: amt_nutr :=
        A_BEEF       60
        B1_BEEF      10
        CAL_BEEF     295
        CAL_CHK      770
        ...
```

Then in addition to the declarations for the parameter amt used in the model,

```
set NUTR;
set FOOD;
param amt {NUTR,FOOD} >= 0;
```

you would declare a set and a parameter to hold the data from the ''nonstandard'' table:

```
set NUTR_FOOD;
param amt_nutr {NUTR_FOOD} >= 0;
```

To use the model, you need to write an assignment of some kind to get the data from set NUTR_FOOD and parameter amt_nutr into sets NUTR and FOOD and parameter amt. One solution is to extract the sets first, and then convert the parameters:

```
set NUTR = setof {ij in NUTR_FOOD}
                        substr(ij,1,match(ij,"_")-1);
set FOOD = setof {ij in NUTR_FOOD}
                        substr(ij,match(ij,"_")+1);
param amt {i in NUTR, j in FOOD} = amt_nutr[i & "_" & j];
```

As an alternative, you can extract the sets and parameters together with a script such as the following:

```
param iNUTR symbolic;
param jFOOD symbolic;
param upos > 0;
let NUTR := {};
let FOOD := {};

for {ij in NUTR_FOOD} {
   let upos := match(ij,"_");
   let iNUTR := substr(ij,1,upos-1);
   let jFOOD := substr(ij,upos+1);
   let NUTR := NUTR union {iNUTR};
   let FOOD := FOOD union {jFOOD};
   let amt[iNUTR,jFOOD] := amt_nutr[ij];
}
```

Under either alternative, errors such as a missing ''_'' in a member of NUTR_FOOD are eventually signaled by error messages.

AMPL provides two other functions, `sub` and `gsub`, that look for the second argument in the first, like `match`, but that then substitute a third argument for either the first occurrence (`sub`) or all occurrences (`gsub`) found. The second argument of all three of these functions is actually a *regular expression*; if it contains certain special characters, it is interpreted as a pattern that may match many sub-strings. The pattern `"^B[0-9]+_"`, for example, matches any sub-string consisting of a B followed by one or more digits and then an underscore, and occurring at the beginning of a string. Details of these features are given in Section A.4.2.

String expressions in AMPL commands

String-valued expressions may appear in place of literal strings in several contexts: in filenames that are part of commands, including `model`, `data`, and `commands`, and in filenames following `>` or `>>` to specify redirection of output; in values assigned to AMPL options by an `option` command; and in the string-list and the database row and column names specified in a `table` statement. In all such cases, the string expression must be identified by enclosing it in parentheses.

Here is an example involving filenames. This script uses a string expression to specify files for a `data` statement and for the redirection of output from a `display` statement:

```
model diet.mod;
set CASES = 1 .. 3;
for {j in CASES} {
   reset data;
   data ("diet" & j & ".dat");
   solve;
   display Buy >("diet" & j & ".out");
}
```

The result is to solve `diet.mod` with a series of different data files `diet1.dat`, `diet2.dat`, and `diet3.dat`, and to save the solution to files `diet1.out`, `diet2.out`, and `diet3.out`. The value of the index `j` is converted automatically from a number to a string as previously explained.

The following script uses a string expression to specify the value of the option `cplex_options`, which contains directions for the CPLEX solver:

```
model sched.mod;
data sched.dat;
option solver cplex;
set DIR1 = {"primal","dual"};
set DIR2 = {"primalopt","dualopt"};
for {i in DIR1, j in DIR2} {
   option cplex_options (i & " " & j);
   solve;
}
```

The loop in this script solves the same problem four times, each using a different pairing of the directives `primal` and `dual` with the directives `primalopt` and `dualopt`.

Examples of the use of string expressions in the `table` statement, to work with multiple database files, tables, or columns, are presented in Section 10.6.

14

Interactions with Solvers

This chapter describes in more detail a variety of mechanisms used by AMPL to control and adjust the problems sent to solvers, and to extract and interpret information returned by them. One of the most important is the presolve phase, which performs simplifications and transformations that can often reduce the size of the problem actually seen by the solver; this is the topic of Section 14.1. Suffixes on model components permit a variety of useful information to be returned by or exchanged with advanced solvers, as described in Sections 14.2 and 14.3. Named problems enable AMPL scripts to manage multiple problem instances within a single model and carry out iterative procedures that alternate between very different models, as we show in Sections 14.4 and 14.5.

14.1 Presolve

AMPL's presolve phase attempts to simplify a problem instance after it has been generated but before it is sent to a solver. It runs automatically when a `solve` command is given or in response to other commands that generate an instance, as explained in Section A.18.1. Any simplifications that presolve makes are reversed after a solution is returned, so that you can view the solution in terms of the original problem. Thus presolve normally proceeds silently behind the scenes. Its effects are only reported when you change option `show_stats` from its default value of 0 to 1:

```
ampl: model steelT.mod; data steelT.dat;
ampl: option show_stats 1;
ampl: solve;
Presolve eliminates 2 constraints and 2 variables.
Adjusted problem:
24 variables, all linear
12 constraints, all linear; 38 nonzeros
1 linear objective; 24 nonzeros.

MINOS 5.5: optimal solution found.
15 iterations, objective 515033
```

You can determine which variables and constraints presolve eliminated by testing, as explained in Section 14.2, to see which have a `status` of `pre`:

```
ampl: print {j in 1.._nvars:
ampl?    _var[j].status = "pre"}: _varname[j];
Inv['bands',0]
Inv['coils',0]

ampl: print {i in 1.._ncons:
ampl?    _con[i].status = "pre"}: _conname[i];
Init_Inv['bands']
Init_Inv['coils']
```

You can then use `show` and `display` to examine the eliminated components.

In this section we introduce the operations of the presolve phase and the options for controlling it from AMPL. We then explain what presolve does when it detects that no feasible solution is possible. We will not try to explain the whole presolve algorithm, however; one of the references at the end of this chapter contains a complete description.

Activities of the presolve phase

To generate a problem instance, AMPL first assigns each variable whatever bounds are specified in its `var` declaration, or the special bounds `-Infinity` and `Infinity` when no lower or upper bounds are given. The presolve phase then tries to use these bounds together with the linear constraints to deduce tighter bounds that are still satisfied by all of the problem's feasible solutions. Concurrently, presolve tries to use the tighter bounds to detect variables that can be fixed and constraints that can be dropped.

Presolve initially looks for constraints that have only one variable. Equalities of this kind fix a variable, which may then be dropped from the problem. Inequalities specify a bound for a variable, which may be folded into the existing bounds. In the example of `steelT.mod` (Figure 4-4) shown above, presolve eliminates the two constraints generated from the declaration

```
subject to Initial {p in PROD}:  Inv[p,0] = inv0[p];
```

along with the two variables fixed by these constraints.

Presolve continues by seeking constraints that can be proved redundant by the current bounds. The constraints eliminated from `dietu.mod` (Figure 5-1) provide an example:

```
ampl: model dietu.mod; data dietu.dat;
ampl: option show_stats 1;
ampl: solve;

Presolve eliminates 3 constraints.
Adjusted problem:
8 variables, all linear
5 constraints, all linear; 39 nonzeros
1 linear objective; 8 nonzeros.

MINOS 5.5: optimal solution found.
5 iterations, objective 74.27382022
```

```
ampl: print {i in 1.._ncons:
ampl?     _con[i].status = "pre"}: _conname[i];
Diet_Min['B1']
Diet_Min['B2']
Diet_Max['A']
```

On further investigation, the constraint `Diet_Min['B1']` is seen to be redundant because it is generated from

```
subject to Diet_Min {i in MINREQ}:
   sum {j in FOOD} amt[i,j] * Buy[j] >= n_min[i];
```

with `n_min['B1']` equal to zero in the data. Clearly this is satisfied by any combination of the variables, since they all have nonnegative lower bounds. A less trivial case is given by `Diet_Max['A']`, which is generated from

```
subject to Diet_Max {i in MAXREQ}:
   sum {j in FOOD} amt[i,j] * Buy[j] <= n_max[i];
```

By setting each variable to its upper bound on the left-hand side of this constraint, we get an upper bound on the total amount of the nutrient that any solution can possibly supply. In particular, for nutrient A:

```
ampl: display sum {j in FOOD} amt['A',j] * f_max[j];
sum{j in FOOD} amt['A',j]*f_max[j] = 2860
```

Since the data file gives `n_max['A']` as 20000, this is another constraint that cannot possibly be violated by the variables.

Following these tests, the first part of presolve is completed. The remainder consists of a series of passes through the problem, each attempting to deduce still tighter variable bounds from the current bounds and the linear constraints. We present here only one example of the outcome, for the problem instance generated from `multi.mod` and `multi.dat` (Figures 4-1 and 4-2):

```
ampl: model multi.mod;
ampl: data multi.dat;
ampl: option show_stats 1;
ampl: solve;

Presolve eliminates 7 constraints and 3 variables.
Adjusted problem:
60 variables, all linear
44 constraints, all linear; 165 nonzeros
1 linear objective; 60 nonzeros.

MINOS 5.5: optimal solution found.
41 iterations, objective 199500

ampl: print {j in 1.._nvars:
ampl?     _var.status[j] = "pre"}: _varname[j];
Trans['GARY','LAN','plate']
Trans['CLEV','LAN','plate']
Trans['PITT','LAN','plate']
```

```
ampl: print {i in 1.._ncons:
ampl?     _con[i].status = "pre"}: _conname[i];
Demand['LAN','plate']
Multi['GARY','LAN']
Multi['GARY','WIN']
Multi['CLEV','LAN']
Multi['CLEV','WIN']
Multi['PITT','LAN']
Multi['PITT','WIN']
```

We can see where some of the simplifications come from by expanding the eliminated demand constraint:

```
ampl: expand Demand['LAN','plate'];
subject to Demand['LAN','plate']:
    Trans['GARY','LAN','plate'] + Trans['CLEV','LAN','plate'] +
    Trans['PITT','LAN','plate'] = 0;
```

Because demand['LAN','plate'] is zero in the data, this constraint forces the sum of three nonnegative variables to be zero, as a result of which all three must have an upper limit of zero in any solution. Since they already have a lower limit of zero, they may be fixed and the constraint may be dropped. The other eliminated constraints all look like this:

```
ampl: expand Multi['GARY','LAN'];
subject to Multi['GARY','LAN']:
    Trans['GARY','LAN','bands'] + Trans['GARY','LAN','coils'] +
    Trans['GARY','LAN','plate'] <= 625;
```

They can be dropped because the sum of the upper bounds of the variables on the left is less than 625. These variables were not originally given upper bounds in the problem, however. Instead, the second part of presolve deduced their bounds. For this simple problem, it is not hard to see how the deduced bounds arise: the amount of any product shipped along any one link cannot exceed the demand for that product at the destination of the link. In the case of the destinations LAN and WIN, the total demand for the three products is less than the limit of 625 on total shipments from any origin, making the total-shipment constraints redundant.

Controlling the effects of presolve

For more complex problems, presolve's eliminations of variables and constraints may not be so easy to explain, but they can represent a substantial fraction of the problem. The time and memory needed to solve a problem instance may be reduced considerably as a result. In rare cases, presolve can also substantially affect the optimal values of the variables — when there is more than one optimal solution — or interfere with other pre-processing routines that are built into your solver software. To turn off presolve entirely, set option presolve to 0; to turn off the second part only, set it to 1. A higher value for this option indicates the maximum number of passes made in part two of presolve; the default is 10.

Following presolve, AMPL saves two sets of lower and upper bounds on the variables: ones that reflect the tightening of the bounds implied by constraints that presolve eliminated, and ones that reflect further tightening deduced from constraints that presolve could not eliminate. The problem has the same solution with either set of bounds, but the overall solution time may be lower with one or the other, depending on the optimization method in use and the specifics of the problem.

For continuous variables, normally AMPL passes to solvers the first set of bounds, but you can instruct it to pass the second set by changing option var_bounds to 2 from its default value of 1. When active-set methods (like the simplex method) are applied, the second set tends to give rise to more degenerate variables, and hence more degenerate iterations that may impede progress toward a solution.

For integer variables, AMPL rounds any fractional lower bounds up to the next higher integer and any fractional upper bounds down to the next lower integer. Due to inaccuracies of finite-precision computation, however, a bound may be calculated to have a value that is just slightly different from an integer. A lower bound that should be 7, for example, might be calculated as 7.00000000001, in which case you would not want the bound to be rounded up to 8! To deal with this possibility, AMPL subtracts the value of option presolve_inteps from each lower bound, and adds it to each upper bound, before rounding. If increasing this setting to the value of option presolve_intepsmax would make a difference to the rounded bounds of any of the variables, AMPL issues a warning. The default values of presolve_inteps and presolve_intepsmax are 1.0e–6 and 1.0e–5, respectively.

You can examine the first set of presolve bounds by using the suffixes .lb1 and .ub1, and the second set by .lb2 and .ub2. The original bounds, which are sent to the solver only if presolve is turned off, are given as .lb0 and .ub0. The suffixes .lb and .ub give the bound values currently to be passed to the solver, based on the current values of options presolve and var_bounds.

Detecting infeasibility in presolve

If presolve determines that any variable's lower bound is greater than its upper bound, then there can be no solution satisfying all the bounds and other constraints, and an error message is printed. For example, here's what would happen to steel3.mod (Figure 1-5a) if we changed market["bands"] to 500 when we meant 5000:

```
ampl: model steel3.mod;
ampl: data steel3.dat;
ampl: let market["bands"] := 500;
ampl: solve;
inconsistent bounds for var Make['bands']:
        lower bound = 1000 > upper bound = 500;
        difference = 500
```

This is a simple case, because the upper bound on variable Make["bands"] has clearly been reduced below the lower bound. Presolve's more sophisticated tests can also find infeasibilities that are not due to any one variable. As an example, consider the constraint in this model:

```
subject to Time: sum {p in PROD} 1/rate[p]*Make[p] <= avail;
```

If we reduce the value of `avail` to 13 hours, presolve deduces that this constraint can't possibly be satisfied:

```
ampl: let market["bands"] := 5000;
ampl: let avail := 13;
ampl: solve;
presolve: constraint Time cannot hold:
        body <= 13 cannot be >= 13.2589; difference = -0.258929
```

The "body" of constraint `Time` is `sum {p in PROD} 1/rate[p]*Make[p]`, the part that contains the variables (see Section 12.5). Thus, given the value of `avail` that we have set, the constraint places an upper bound of 13 on the value of the body expression. On the other hand, if we set each variable in the body expression equal to its lower bound, we get a lower bound on the value of the body in any feasible solution:

```
ampl: display sum {p in PROD} 1/rate[p]*Make[p].lb2;
sum{p in PROD} 1/rate[p]*(Make[p].lb2) = 13.2589
```

The statement from presolve that `body <= 13 cannot be >= 13.2589` is thus reporting that the upper bound on the body is in conflict with the lower bound, implying that no solution can satisfy all of the problem's bounds and constraints.

Presolve reports the difference between its two bounds for constraint `Time` as –0.258929 (to six digits). Thus in this case we can guess that 13.258929 is, approximately, the smallest value of `avail` that allows for a feasible solution, which we can verify by experiment:

```
ampl: let avail := 13.258929;
ampl: solve;
MINOS 5.5: optimal solution found.
0 iterations, objective 61750.00214
```

If we make `avail` just slightly lower, however, we again get the infeasibility message:

```
ampl: let avail := 13.258928;
ampl: solve;
presolve: constraint Time cannot hold:
          body <= 13.2589 cannot be >= 13.2589;
          difference = -5.71429e-07
Setting $presolve_eps >= 6.86e-07 might help.
```

Although the lower bound here is the same as the upper bound to six digits, it is greater than the upper bound in full precision, as the negative value of the difference indicates.

Typing `solve` a second time in this situation tells AMPL to override presolve and send the seemingly inconsistent deduced bounds to the solver:

```
ampl: solve;
MINOS 5.5: optimal solution found.
0 iterations, objective 61749.99714

ampl: option display_precision 10;

ampl: display commit, Make;
:        commit       Make          :=
bands    1000      999.9998857
coils     500      500
plate     750      750
;
```

MINOS declares that it has found an optimal solution, though with Make["bands"] being slightly less than its lower bound commit["bands"]! Here MINOS is applying an internal tolerance that allows small infeasibilities to be ignored; the AMPL/MINOS documentation explains how this tolerance works and how it can be changed. Each solver applies feasibility tolerances in its own way, so it's not surprising that a different solver gives different results:

```
ampl: option solver cplex;
ampl: option send_statuses 0;

ampl: solve;
CPLEX 8.0.0: Bound infeasibility column 'x1'.
infeasible problem.
1 dual simplex iterations (0 in phase I)
```

Here CPLEX has applied its own presolve routine and has detected the same infeasibility that AMPL did. (You may see a few additional lines about a "suffix" named dunbdd; this pertains to a direction of unboundedness that you can retrieve via AMPL's solver-defined suffix feature described in Section 14.3.)

Situations like this come about when the implied lower and upper bounds on some variable or constraint body are equal, at least for all practical purposes. Due to imprecision in the computations, the lower bound may come out slightly greater than the upper bound, causing AMPL's presolve to report an infeasible problem. To circumvent this difficulty, you can reset the option presolve_eps from its default value of 0 to some small positive value. Differences between the lower and upper bound are ignored when they are less than this value. If increasing the current presolve_eps value to a value no greater than presolve_epsmax would change presolve's handling of the problem, then presolve displays a message to this effect, such as

```
Setting $presolve_eps >= 6.86e-07 might help.
```

in the example above. The default value of option presolve_eps is zero and presolve_epsmax is 1.0e–5.

A related situation occurs when imprecision in the computations causes the implied lower bound on some variable or constraint body to come out slightly lower than the implied upper bound. Here no infeasibility is detected, but the presence of bounds that are nearly equal may make the solver's work much harder than necessary. Thus when-

ever the upper bound minus the lower bound on a variable or constraint body is positive
but less than the value of option `presolve_fixeps`, the variable or constraint body is
fixed at the average of the two bounds. If increasing the value of `presolve_fixeps`
to at most the value of `presolve_fixepsmax` would change the results of presolve, a
message to this effect is displayed.

The number of separate messages displayed by presolve is limited to the value of
`presolve_warnings`, which is 5 by default. Increasing option `show_stats` to 2
may elicit some additional information about the presolve run, including the number of
passes that made a difference to the results and the values to which `presolve_eps` and
`presolve_inteps` would have to be increased or decreased to make a difference.

14.2 Retrieving results from solvers

In addition to the solution and related numerical values, it can be useful to have certain symbolic information about the results of `solve` commands. For example, in a
script of AMPL commands, you may want to test whether the most recent `solve` encountered an unbounded or infeasible problem. Or, after you have solved a linear program by
the simplex method, you may want to use the optimal basis partition to provide a good
start for solving a related problem. The AMPL-solver interface permits solvers to return
these and related kinds of status information that you can examine and use.

Solve results

A solver finishes its work because it has identified an optimal solution or has encountered some other terminating condition. In addition to the values of the variables, the
solver may set two built-in AMPL parameters and an AMPL option that provide information about the outcome of the optimization process:

```
ampl: model diet.mod;
ampl: data diet2.dat;

ampl: display solve_result_num, solve_result;
solve_result_num = -1
solve_result = '?'

ampl: solve;
MINOS 5.5: infeasible problem.
9 iterations

ampl: display solve_result_num, solve_result;
solve_result_num = 200
solve_result = infeasible
```

```
ampl: option solve_result_table;
option solve_result_table '\
0        solved\
100      solved?\
200      infeasible\
300      unbounded\
400      limit\
500      failure\
';
```

At the beginning of an AMPL session, `solve_result_num` is -1 and `solve_result` is `'?'`. Each `solve` command resets these parameters, however, so that they describe the solver's status at the end of its run, `solve_result_num` by a number and `solve_result` by a character string. The `solve_result_table` option lists the possible combinations, which may be interpreted as follows:

`solve_result` values

number	string	interpretation
0- 99	solved	optimal solution found
100-199	solved?	optimal solution indicated, but error likely
200-299	infeasible	constraints cannot be satisfied
300-399	unbounded	objective can be improved without limit
400-499	limit	stopped by a limit that you set (such as on iterations)
500-599	failure	stopped by an error condition in the solver

Normally this status information is used in scripts, where it can be tested to distinguish among cases that must be handled in different ways. As an example, Figure 14-1 depicts an AMPL script for the diet model that reads the name of a nutrient (from the standard input, using the filename – as explained in Section 9.5), a starting upper limit for that nutrient in the diet, and a step size for reducing the limit. The loop continues running until the limit is reduced to a point where the problem is infeasible, at which point it prints an appropriate message and a table of solutions previously found. A representative run looks like this:

```
ampl: commands diet.run;
<1>ampl? NA
<1>ampl? 60000
<1>ampl? 3000
--- infeasible at 48000 ---

:         N_obj        N_dual         :=
51000    115.625    -0.0021977
54000    109.42     -0.00178981
57000    104.05     -0.00178981
60000    101.013     7.03757e-19
;
```

Here the limit on sodium (NA in the data) is reduced from 60000 in steps of 3000, until the problem becomes infeasible at a limit of 48000.

The key statement of `diet.run` that tests for infeasibility is

```
model diet.mod;
data diet2.dat;

param N symbolic in NUTR;
param nstart > 0;
param nstep > 0;
read N, nstart, nstep <- ;    # read data interactively

set N_MAX default {};
param N_obj {N_MAX};
param N_dual {N_MAX};
option solver_msg 0;

for {i in nstart .. 0 by -nstep} {
   let n_max[N] := i;
   solve;
   if solve_result = "infeasible" then {
      printf "--- infeasible at %d ---\n\n", i;
      break;
   }
   let N_MAX := N_MAX union {i};
   let N_obj[i] := Total_Cost;
   let N_dual[i] := Diet[N].dual;
}
display N_obj, N_dual;
```

Figure 14-1: Sensitivity analysis with infeasibility test (`diet.run`).

```
if solve_result = "infeasible" then {
   printf "--- infeasible at %d ---\n\n", i;
   break;
}
```

The `if` condition could equivalently be written `200 <= solve_result_num < 300`. Normally you will want to avoid this latter alternative, since it makes the script more cryptic. It can occasionally be useful, however, in making fine distinctions among different solver termination conditions. For example, here are some of the values that the CPLEX solver sets for different optimality conditions:

solve_result_num	message at termination
0	optimal solution
1	primal has unbounded optimal face
2	optimal integer solution
3	optimal integer solution within `mipgap` or `absmipgap`

The value of `solve_result` is `"solved"` in all of these cases, but you can test `solve_result_num` if you need to distinguish among them. The interpretations of `solve_result_num` are entirely solver-specific; you'll have to look at a particular solver's documentation to see which values it returns and what they mean.

AMPL's solver interfaces are set up to display a few lines like

```
MINOS 5.5: infeasible problem.
9 iterations
```

to summarize a solver run that has finished. If you are running a script that executes `solve` frequently, these messages can add up to a lot of output; you can suppress their appearance by setting the option `solver_msg` to 0. A built-in symbolic parameter, `solve_message`, still always contains the most recent solver return message, even when display of the message has been turned off. You can display this parameter to verify its value:

```
ampl: display solve_message;
solve_message = 'MINOS 5.5: infeasible problem.\
9 iterations'
```

Because `solve_message` is a symbolic parameter, its value is a character string. It is most useful in scripts, where you can use character-string functions (Section 13.7) to test the message for indications of optimality and other outcomes.

As an example, the test in `diet.run` could also have been written

```
if match(solve_message, "infeasible") > 0 then {
```

Since return messages vary from one solver to the next, however, for most situations a test of `solve_result` will be simpler and less solver-dependent.

Solve results can be returned as described above only if AMPL's invocation of the solver has been successful. Invocation can fail because the operating system is unable to find or execute the specified solver, or because some low-level error prevents the solver from attempting or completing the optimization. Typical causes include a misspelled solver name, improper installation or licensing of the solver, insufficient resources, and termination of the solver process by an execution fault (''core dump'') or a ''break'' from the keyboard. In these cases the error message that follows `solve` is generated by the operating system rather than by the solver, and you might have to consult a system guru to track down the problem. For example, a message like `can't open at8871.nl` usually indicates that AMPL is not able to write a temporary file; it might be trying to write to a disk that is full, or to a directory (folder) for which you do not have write permission. (The directory for temporary files is specified in option `TMPDIR`.)

The built-in parameter `solve_exitcode` records the success or failure of the most recent solver invocation. Initially -1, it is reset to 0 whenever there has been a successful invocation, and to some system-dependent nonzero value otherwise:

```
ampl: reset;
ampl: display solve_exitcode;
solve_exitcode = -1

ampl: model diet.mod;
ampl: data diet2.dat;
ampl: option solver xplex;
ampl: solve;
Cannot invoke xplex: No such file or directory
```

```
ampl: display solve_exitcode;
solve_exitcode = 1024
ampl: display solve_result, solve_result_num;
solve_result = '?'
solve_result_num = -1
```

Here the failed invocation, due to the misspelled solver name xplex, is reflected in a positive solve_exitcode value. The status parameters solve_result and solve_result_num are also reset to their initial values '?' and −1.

If solve_exitcode exceeds the value in option solve_exitcode_max, then AMPL aborts any currently executing compound statements (include, commands, repeat, for, if). The default value of solve_exitcode_max is 0, so that AMPL normally aborts compound statements whenever a solver invocation fails. A script that sets solve_exitcode_max to a higher value may test the value of solve_exitcode, but in general its interpretation is not consistent across operating systems or solvers.

Solver statuses of objectives and problems

Sometimes it is convenient to be able to refer to the solve result obtained when a particular objective was most recently optimized. For this purpose, AMPL associates with each built-in solve result parameter a "status" suffix:

built-in parameter	suffix
solve_result	.result
solve_result_num	.result_num
solve_message	.message
solve_exitcode	.exitcode

Appended to an objective name, this suffix indicates the value of the corresponding built-in parameter at the most recent solve in which the objective was current.

As an example, we consider again the multiple objectives defined for the assignment model in Section 8.3:

```
minimize Total_Cost:
    sum {i in ORIG, j in DEST} cost[i,j] * Trans[i,j];

minimize Pref_of {i in ORIG}:
    sum {j in DEST} cost[i,j] * Trans[i,j];
```

After minimizing three of these objectives, we can view the solve status values for all of them:

```
ampl: model transp4.mod; data assign.dat; solve;
CPLEX 8.0.0: optimal solution; objective 28
24 dual simplex iterations (0 in phase I)
Objective = Total_Cost
```

```
ampl: objective Pref_of['Coullard'];
ampl: solve;
CPLEX 8.0.0: optimal solution; objective 1
3 simplex iterations (0 in phase I)
ampl: objective Pref_of['Hazen'];
ampl: solve;
CPLEX 8.0.0: optimal solution; objective 1
5 simplex iterations (0 in phase I)

ampl: display Total_Cost.result, Pref_of.result;
Total_Cost.result = solved

Pref_of.result [*] :=
 Coullard   solved
   Daskin   '?'
    Hazen   solved
     Hopp   '?'
  Iravani   '?'
 Linetsky   '?'
 Mehrotra   '?'
   Nelson   '?'
Smilowitz   '?'
  Tamhane   '?'
    White   '?'
;
```

For the objectives that have not yet been used, the `.result` suffix is unchanged (at its initial value of `'?'` in this case).

These same suffixes can be applied to the "problem" names whose use we describe later in this chapter. When appended to a problem name, they refer to the most recent optimization carried out when that problem was current.

Solver statuses of variables

In addition to providing for return of the overall status of the optimization process as described above, AMPL lets a solver return an individual status for each variable. This feature is intended mainly for reporting the basis status of variables after a linear program is solved either by the simplex method, or by an interior-point (barrier) method followed by a "crossover" routine. The basis status is also relevant to solutions returned by certain nonlinear solvers, notably MINOS, that employ an extension of the concept of a basic solution.

In addition to the variables declared by `var` statements in an AMPL model, solvers also define "slack" or "artificial" variables that are associated with constraints. Solver statuses for these latter variables are defined in a similar way, as explained later in this section. Both variables and constraints also have an "AMPL status" that distinguishes those in the current problem from those that have been removed from the problem by presolve or by commands such as `drop`. The interpretation of AMPL statuses and their relationship to solver statuses are discussed at the end of this section.

The major use of solver status values from an optimal basic solution is to provide a good starting point for the next optimization run. The option send_statuses, when left at its default value of 1, instructs AMPL to include statuses with the information about variables sent to the solver at each solve. You can see the effect of this feature in almost any sensitivity analysis that re-solves after making some small change to the problem.

As an example, consider what happens when the multi-period production example from Figure 6-3 is solved repeatedly after increases of five percent in the availability of labor. With the send_statuses option set to 0, the solver reports about 18 iterations of the dual simplex method each time it is run:

```
ampl: model steelT3.mod;
ampl: data steelT3.dat;
ampl: option send_statuses 0;
ampl: solve;
CPLEX 8.0.0: optimal solution; objective 514521.7143
18 dual simplex iterations (0 in phase I)
ampl: let {t in 1..T} avail[t] := 1.05 * avail[t];
ampl: solve;
CPLEX 8.0.0: optimal solution; objective 537104
19 dual simplex iterations (0 in phase I)
ampl: let {t in 1..T} avail[t] := 1.05 * avail[t];
ampl: solve;
CPLEX 8.0.0: optimal solution; objective 560800.4
19 dual simplex iterations (0 in phase I)
ampl: let {t in 1..T} avail[t] := 1.05 * avail[t];
ampl: solve;
CPLEX 8.0.0: optimal solution; objective 585116.22
17 dual simplex iterations (0 in phase I)
```

With send_statuses left at its default value of 1, however, only the first solve takes 18 iterations. Subsequent runs take a few iterations at most:

```
ampl: model steelT3.mod;
ampl: data steelT3.dat;
ampl: solve;
CPLEX 8.0.0: optimal solution; objective 514521.7143
18 dual simplex iterations (0 in phase I)
ampl: let {t in 1..T} avail[t] := 1.05 * avail[t];
ampl: solve;
CPLEX 8.0.0: optimal solution; objective 537104
1 dual simplex iterations (0 in phase I)
ampl: let {t in 1..T} avail[t] := 1.05 * avail[t];
ampl: solve;
CPLEX 8.0.0: optimal solution; objective 560800.4
0 simplex iterations (0 in phase I)
ampl: let {t in 1..T} avail[t] := 1.05 * avail[t];
ampl: solve;
CPLEX 8.0.0: optimal solution; objective 585116.22
1 dual simplex iterations (0 in phase I)
```

Each `solve` after the first automatically uses the variables' basis statuses from the previous `solve` to construct a starting point that turns out to be only a few iterations from the optimum. In the case of the third `solve`, the previous basis remains optimal; the solver thus confirms optimality immediately and reports taking 0 iterations.

The following discussion explains how you can view, interpret, and change status values of variables in the AMPL environment. You don't need to know any of this to use optimal bases as starting points as shown above, but these features can be useful in certain advanced circumstances.

AMPL refers to a variable's solver status by appending `.sstatus` to its name. Thus you can print the statuses of variables with `display`. At the beginning of a session (or after a `reset`), when no problem has yet been solved, all variables have the status `none`:

```
ampl: model diet.mod;
ampl: data diet2a.dat;

ampl: display Buy.sstatus;
Buy.sstatus [*] :=
BEEF   none
 CHK   none
FISH   none
 HAM   none
 MCH   none
 MTL   none
 SPG   none
 TUR   none
;
```

After an invocation of a simplex method solver, the same `display` lists the statuses of the variables at the optimal basic solution:

```
ampl: solve;
MINOS 5.5: optimal solution found.
13 iterations, objective 118.0594032

ampl: display Buy.sstatus;
Buy.sstatus [*] :=
BEEF   bas
 CHK   low
FISH   low
 HAM   upp
 MCH   upp
 MTL   upp
 SPG   bas
 TUR   low
;
```

Two of the variables, `Buy['BEEF']` and `Buy['SPG']`, have status `bas`, which means they are in the optimal basis. Three have status `low` and three `upp`, indicating that they are nonbasic at lower and upper bounds, respectively. A table of the recognized solver status values is stored in option `sstatus_table`:

```
ampl: option sstatus_table;
option sstatus_table '\
0        none    no status assigned\
1        bas     basic\
2        sup     superbasic\
3        low     nonbasic <= (normally =) lower bound\
4        upp     nonbasic >= (normally =) upper bound\
5        equ     nonbasic at equal lower and upper bounds\
6        btw     nonbasic between bounds\
';
```

Numbers and short strings representing status values are given in the first two columns. (The numbers are mainly for communication between AMPL and solvers, though you can access them by using the suffix .sstatus_num in place of .sstatus.) The entries in the third column are comments. For nonbasic variables as defined in many textbook simplex methods, only the low status is applicable; other nonbasic statuses are required for the more general bounded-variable simplex methods in large-scale implementations. The sup status is used by solvers like MINOS to accommodate nonlinear problems. This is AMPL's standard sstatus_table; a solver may substitute its own table, in which case its documentation will indicate the interpretation of the table entries.

You can change a variable's status with the let command. This facility is sometimes useful when you want to re-solve a problem after a small, well-defined change. In a later section of this chapter, for example, we employ a pattern-cutting model (Figure 14-2a) that contains the declarations

```
param nPAT integer >= 0;    # number of patterns
set PATTERNS = 1..nPAT;   # set of patterns
var Cut {PATTERNS} integer >= 0; # rolls cut using each pattern
```

In a related script (Figure 14-3), each pass through the main loop steps nPAT by one, causing a new variable Cut[nPAT] to be created. It has an initial solver status of "none", like all new variables, but it is guaranteed, by the way that the pattern generation procedure is constructed, to enter the basis as soon as the expanded cutting problem is re-solved. Thus we give it a status of "bas" instead:

```
let Cut[nPAT].sstatus := "bas";
```

It turns out that this change tends to reduce the number of iterations in each re-optimization of the cutting problem, at least with some simplex solvers. Setting a few statuses in this way is never guaranteed to reduce the number of iterations, however. Its success depends on the particular problem and solver, and on their interaction with a number of complicating factors:

• After the problem and statuses have been modified, the statuses conveyed to the solver at the next solve may not properly define a basic solution.

• After the problem has been modified, AMPL's presolve phase may send a different subset of variables and constraints to the solver (unless option presolve is set to zero). As a result, the statuses conveyed to the solver may not correspond to a useful starting point for the next `solve`, and may not properly define a basic solution.

• Some solvers, notably MINOS, use the current values as well as the statuses of the variables in constructing the starting point at the next `solve` (unless option `reset_initial_guesses` is set to 1).

Each solver has its own way of adjusting the statuses that it receives from AMPL, when necessary, to produce an initial basic solution that it can use. Thus some experimentation is usually necessary to determine whether any particular strategy for modifying the statuses is useful.

For models that have several `var` declarations, AMPL's generic synonyms (Section 12.6) for variables provide a convenient way of getting overall summaries of statuses. For example, using expressions like `_var`, `_varname` and `_var.sstatus` in a `display` statement, you can easily specify a table of all basic variables in `steelT3.mod` along with their optimal values:

```
ampl: display {j in 1.._nvars: _var[j].sstatus = "bas"}
ampl?      (_varname[j], _var[j]);
:              _varname[j]    _var[j]      :=
1        "Make['bands',1]"    5990
2        "Make['bands',2]"    6000
3        "Make['bands',3]"    1400
4        "Make['bands',4]"    2000
5        "Make['coils',1]"    1407
6        "Make['coils',2]"    1400
7        "Make['coils',3]"    3500
8        "Make['coils',4]"    4200
15       "Inv['coils',1]"     1100
21       "Sell['bands',3]"    1400
22       "Sell['bands',4]"    2000
23       "Sell['coils',1]"     307
;
```

An analogous listing of all the variables would be produced by the command

```
display _varname, _var;
```

Solver statuses of constraints

Implementations of the simplex method typically add one variable for each constraint that they receive from AMPL. Each added variable has a coefficient of 1 or –1 in its associated constraint, and coefficients of 0 in all other constraints. If the associated constraint is an inequality, the addition is used as a ''slack'' or ''surplus'' variable; its bounds are chosen so that it has the effect of turning the inequality into an equivalent equation. If the associated constraint is an equality to begin with, the added variable is an ''artificial'' one whose lower and upper bounds are both zero.

An efficient large-scale simplex solver gains two advantages from having these ''logical'' variables added to the ''structural'' ones that it gets from AMPL: the linear program is converted to a simpler form, in which the only inequalities are the bounds on the variables, and the solver's initialization (or "crash") routines can be designed so that they find a starting basis quickly. Given any starting basis, a first phase of the simplex method finds a basis for a feasible solution (if necessary) and a second phase finds a basis for an optimal solution; the two phases are distinguished in some solvers' messages:

```
ampl: model steelP.mod;
ampl: data steelP.dat;

ampl: solve;
CPLEX 8.0.0: optimal solution; objective 1392175
27 dual simplex iterations (0 in phase I)
```

Solvers thus commonly treat all logical variables in much the same way as the structural ones, with only very minor adjustments to handle the case in which lower and upper bounds are both zero. A basic solution is defined by the collection of basis statuses of all variables, structural and logical.

To accommodate statuses of logical variables, AMPL permits a solver to return status values corresponding to the constraints as well as the variables. The solver status of a constraint, written as the constraint name suffixed by `.sstatus`, is interpreted as the status of the logical variable associated with that constraint. For example, in our diet model, where the constraints are all inequalities:

```
subject to Diet {i in NUTR}:
    n_min[i] <= sum {j in FOOD} amt[i,j] * Buy[j] <= n_max[i];
```

the logical variables are slacks that have the same variety of statuses as the structural variables:

```
ampl: model diet.mod;
ampl: data diet2a.dat;

ampl: option show_stats 1;
ampl: solve;
8 variables, all linear
6 constraints, all linear; 47 nonzeros
1 linear objective; 8 nonzeros.
MINOS 5.5: optimal solution found.
13 iterations, objective 118.0594032

ampl: display Buy.sstatus;
Buy.sstatus [*] :=
BEEF   bas
 CHK   low
FISH   low
 HAM   upp
 MCH   upp
 MTL   upp
 SPG   bas
 TUR   low
;
```

```
ampl: display Diet.sstatus;
diet.sstatus [*] :=
   A  bas
  B1  bas
  B2  low
   C  bas
 CAL  bas
  NA  upp
 ;
```

There are a total of six basic variables, equal in number to the six constraints (one for each member of set NUTR) as is always the case at a basic solution. In our transportation model, where the constraints are equalities:

```
subject to Supply {i in ORIG}:
   sum {j in DEST} Trans[i,j] = supply[i];
subject to Demand {j in DEST}:
   sum {i in ORIG} Trans[i,j] = demand[j];
```

the logical variables are artificials that receive the status "equ" when nonbasic. Here's how the statuses for all constraints might be displayed using AMPL's generic constraint synonyms (analogous to the variable synonyms previously shown):

```
ampl: model transp.mod;
ampl: data transp.dat;
ampl: solve;
MINOS 5.5: optimal solution found.
13 iterations, objective 196200

ampl: display _conname, _con.slack, _con.sstatus;
   :            _conname         _con.slack  _con.sstatus  :=
   1      "Supply['GARY']"      -4.54747e-13     equ
   2      "Supply['CLEV']"       0               equ
   3      "Supply['PITT']"      -4.54747e-13     equ
   4      "Demand['FRA']"       -6.82121e-13     bas
   5      "Demand['DET']"        0               equ
   6      "Demand['LAN']"        0               equ
   7      "Demand['WIN']"        0               equ
   8      "Demand['STL']"        0               equ
   9      "Demand['FRE']"        0               equ
  10      "Demand['LAF']"        0               equ
 ;
```

One artificial variable, on the constraint Demand['FRA'], is in the optimal basis, though at a slack value of essentially zero like all artificial variables in any feasible solution. (In fact there must be some artificial variable in every basis for this problem, due to a linear dependency among the equations in the model.)

AMPL statuses

Only those variables, objectives and constraints that AMPL actually sends to a solver can receive solver statuses on return. So that you can distinguish these from components that are removed prior to a `solve`, a separate "AMPL status" is also maintained. You can work with AMPL statuses much like solver statuses, by using the suffix `.astatus` in place of `.sstatus`, and referring to option `astatus_table` for a summary of the recognized values:

```
ampl: option astatus_table;
option astatus_table '\
0       in      normal state (in problem)\
1       drop    removed by drop command\
2       pre     eliminated by presolve\
3       fix     fixed by fix command\
4       sub     defined variable, substituted out\
5       unused  not used in current problem\
';
```

Here's an example of the most common cases, using one of our diet models:

```
ampl: model dietu.mod;
ampl: data dietu.dat;
ampl: drop Diet_Min['CAL'];
ampl: fix Buy['SPG'] := 5;
ampl: fix Buy['CHK'] := 3;
ampl: solve;
MINOS 5.5: optimal solution found.
3 iterations, objective 54.76

ampl: display Buy.astatus;
Buy.astatus [*] :=
BEEF   in
 CHK   fix
FISH   in
 HAM   in
 MCH   in
 MTL   in
 SPG   fix
 TUR   in
;
ampl: display Diet_Min.astatus;
Diet_Min.astatus [*] :=
  A    in
 B1    pre
 B2    pre
  C    in
CAL    drop
;
```

An AMPL status of `in` indicates components that are in the problem sent to the solver, such as variable `Buy['BEEF']` and constraint `Diet_Min['A']`. Three other statuses indicate components left out of the problem:

   • Variables `Buy['CHK']` and `Buy['SPG']` have AMPL status `"fix"` because the `fix` command was used to specify their values in the solution.

- Constraint `Diet_Min['CAL']` has AMPL status `"drop"` because it was removed by the `drop` command.

- Constraints `Diet_Min['B1']` and `Diet_Min['B2']` have AMPL status `"pre"` because they were removed from the problem by simplifications performed in AMPL's presolve phase.

Not shown here are the AMPL status `"unused"` for a variable that does not appear in any objective or constraint, and `"sub"` for variables and constraints eliminated by substitution (as explained in Section 18.2). The `objective` command, and the `problem` commands to be defined later in this chapter, also have the effect of fixing or dropping model components that are not in use.

For a variable or constraint, you will normally be interested in only one of the statuses at a time: the solver status if the variable or constraint was included in the problem sent most recently to the solver, or the AMPL status otherwise. Thus AMPL provides the suffix `.status` to denote the one status of interest:

```
ampl: display Buy.status, Buy.astatus, Buy.sstatus;
:      Buy.status Buy.astatus Buy.sstatus    :=
BEEF   low         in           low
CHK    fix         fix          none
FISH   low         in           low
HAM    low         in           low
MCH    bas         in           bas
MTL    low         in           low
SPG    fix         fix          none
TUR    low         in           low
;

ampl: display Diet_Min.status, Diet_Min.astatus,
ampl?     Diet_Min.sstatus;
:    Diet_Min.status Diet_Min.astatus Diet_Min.sstatus   :=
A       bas              in               bas
B1      pre              pre              none
B2      pre              pre              none
C       low              in               low
CAL     drop             drop             none
;
```

In general, *name*`.status` is equal to *name*`.sstatus` if *name*`.astatus` is `"in"`, and is equal to *name*`.astatus` otherwise.

14.3 Exchanging information with solvers via suffixes

We have seen that to represent values associated with a model component, AMPL employs various qualifiers or *suffixes* appended to component names. A suffix consists of a period or "dot" (`.`) followed by a (usually short) identifier, so that for example the reduced cost associated with a variable `Buy[j]` is written `Buy[j].rc`, and the reduced costs of all such variables can be viewed by giving the command `display Buy.rc`. There are numerous *built-in* suffixes of this kind, summarized in the tables in A.11.

AMPL cannot anticipate all of the values that a solver might associate with model components, however. The values recognized as input or computed as output depend on the design of each solver and its algorithms. To provide for open-ended representation of such values, new suffixes may be defined for the duration of an AMPL session, either by the user for sending values to a solver, or by a solver for returning values.

This section introduces both user-defined and solver-defined suffixes, illustrated by features of the CPLEX solver. We show how user-defined suffixes can pass preferences for variable selection and branch direction to an integer programming solver. Sensitivity analysis provides an example of solver-defined suffixes that have numeric values, while infeasibility diagnosis shows how a symbolic (string-valued) suffix works. Reporting a direction of unboundedness gives an example of a solver-defined suffix in an AMPL script, where it must be declared before it can be used.

User-defined suffixes: integer programming directives

Most solvers recognize a variety of algorithmic choices or settings, each governed by a single value that applies to the entire problem being solved. Thus you can alter selected settings by setting up a single string of directives, as in this example applying the CPLEX solver to an integer program:

```
ampl: model multmip3.mod;
ampl: data multmip3.dat;

ampl: option solver cplex;
ampl: option cplex_options 'nodesel 3 varsel 1 backtrack 0.1';

ampl: solve;
CPLEX 8.0.0: nodesel 3
varsel 1
backtrack 0.1
CPLEX 8.0.0: optimal integer solution; objective 235625
1052 MIP simplex iterations
75 branch-and-bound nodes
```

A few kinds of solver settings are more complex, however, in that they require separate values to be set for individual model components. These settings are far too numerous to be accommodated in a directive string. Instead the solver interface can be set up to recognize new suffixes that the user defines specially for the solver's purposes.

As an example, for each variable in an integer program, CPLEX recognizes a separate branching priority and a separate preferred branching direction, represented by an integer in $[0, 9999]$ and in $[-1, 1]$ respectively. AMPL's CPLEX driver recognizes the suffixes `.priority` and `.direction` as giving these settings. To use these suffixes, we begin by giving a `suffix` command to define each one for the current AMPL session:

```
ampl: suffix priority IN, integer, >= 0, <= 9999;
ampl: suffix direction IN, integer, >= -1, <= 1;
```

The effect of these statements is to define expressions of the form *name*.priority and *name*.direction, where *name* denotes any variable, objective or constraint of the current model. The argument IN specifies that values corresponding to these suffixes are to be read in by the solver, and the subsequent phrases place restrictions on the values that will be accepted (much as in a param declaration).

The newly defined suffixes may be assigned values by the let command (Section 11.3) or by later declarations as described in Sections A.8, A.9, A.10, and A.18.8. For our current example we want to use these suffixes to assign CPLEX priority and direction values corresponding to the binary variables Use[i,j]. Normally such values are chosen based on knowledge of the problem and past experience with similar problems. Here is one possibility:

```
ampl: let {i in ORIG, j in DEST}
ampl?    Use[i,j].priority := sum {p in PROD} demand[j,p];
ampl: let Use["GARY","FRE"].direction := -1;
```

Variables not assigned a .priority or .direction value get a default value of zero (as do all constraints and objectives in this example), as you can check:

```
ampl: display Use.direction;

Use.direction [*,*] (tr)
:   CLEV GARY PITT     :=
DET   0     0   0
FRA   0     0   0
FRE   0    -1   0
LAF   0     0   0
LAN   0     0   0
STL   0     0   0
WIN   0     0   0
;
```

With the suffix values assigned as shown, CPLEX's search for a solution turns out to require fewer simplex iterations and fewer branch-and-bound nodes:

```
ampl: option reset_initial_guesses 1;

ampl: solve;
CPLEX 8.0.0: nodesel 3
varsel 1
backtrack 0.1
CPLEX 8.0.0: optimal integer solution; objective 235625
799 MIP simplex iterations
69 branch-and-bound nodes
```

(We have set option reset_initial_guesses to 1 so that the optimal solution from the first CPLEX run won't be sent back to the second.)

Further information about the suffixes recognized by CPLEX and how to determine the corresponding settings can be found in the CPLEX driver documentation. Other solver interfaces may recognize different suffixes for different purposes; you'll need to check separately for each solver you want to use.

Solver-defined suffixes: sensitivity analysis

When the keyword `sensitivity` is included in CPLEX's list of directives, classical sensitivity ranges are computed and are returned in three new suffixes, `.up`, `.down`, and `.current`:

```
ampl: model steelT.mod; data steelT.dat;
ampl: option solver cplex;
ampl: option cplex_options 'sensitivity';

ampl: solve;
CPLEX 8.0.0: sensitivity
CPLEX 8.0.0: optimal solution; objective 515033
16 dual simplex iterations (0 in phase I)

suffix up OUT;
suffix down OUT;
suffix current OUT;
```

The three lines at the end of the output from the `solve` command show the `suffix` commands that are executed by AMPL in response to the results from the solver. These statements are executed automatically; you do not need to type them. The argument `OUT` in each command says that these are suffixes whose values will be written out by the solver (in contrast to the previous example, where the argument `IN` indicated suffix values that the solver was to read in).

The sensitivity suffixes are interpreted as follows. For variables, suffix `.current` indicates the objective function coefficient in the current problem, while `.down` and `.up` give the smallest and largest values of the objective coefficient for which the current LP basis remains optimal:

```
ampl: display Sell.down, Sell.current, Sell.up;
:        Sell.down Sell.current   Sell.up     :=
bands 1    23.3         25        1e+20
bands 2    25.4         26        1e+20
bands 3    24.9         27           27.5
bands 4    10           27           29.1
coils 1    29.2857      30           30.8571
coils 2    33           35        1e+20
coils 3    35.2857      37        1e+20
coils 4    35.2857      39        1e+20
;
```

For constraints, the interpretation is similar except that it applies to a constraint's constant term (the so-called right-hand-side value):

```
ampl: display Time.down, Time.current, Time.up;
: Time.down Time.current   Time.up     :=
1   37.8071      40        66.3786
2   37.8071      40        47.8571
3   25           32        45
4   30           40        62.5
;
```

You can use generic synonyms (Section 12.6) to display a table of ranges for *all* variables or constraints, similar to the tables produced by the standalone version of CPLEX. (Values of $-1e+20$ in the `.down` column and $1e+20$ in the `.up` column correspond to what CPLEX calls $-infinity$ and $+infinity$ in its tables.)

Solver-defined suffixes: infeasibility diagnosis

For a linear program that has no feasible solution, you can ask CPLEX to find an *irreducible infeasible subset* (or *IIS*): a collection of constraints and variable bounds that is infeasible but that becomes feasible when any one constraint or bound is removed. If a small IIS exists and can be found, it can provide valuable clues as to the source of the infeasibility. You turn on the IIS finder by changing the `iisfind` directive from its default value of 0 to either 1 (for a relatively fast version) or 2 (for a slower version that tends to find a smaller IIS).

The following example shows how IIS finding might be applied to the infeasible diet problem from Chapter 2. After `solve` detects that there is no feasible solution, it is repeated with the directive `'iisfind 1'`:

```
ampl: model diet.mod; data diet2.dat; option solver cplex;
ampl: solve;
CPLEX 8.0.0: infeasible problem.
4 dual simplex iterations (0 in phase I)
constraint.dunbdd returned
suffix dunbdd OUT;

ampl: option cplex_options 'iisfind 1';
ampl: solve;
CPLEX 8.0.0: iisfind 1
CPLEX 8.0.0: infeasible problem.
0 simplex iterations (0 in phase I)
Returning iis of 7 variables and 2 constraints.
constraint.dunbdd returned

suffix iis symbolic OUT;

option iis_table '\
0       non     not in the iis\
1       low     at lower bound\
2       fix     fixed\
3       upp     at upper bound\
';
```

Again, AMPL shows any `suffix` statement that has been executed automatically. Our interest is in the new suffix named `.iis`, which is `symbolic`, or string-valued. An associated option `iis_table`, also set up by the solver driver and displayed automatically by `solve`, shows the strings that may be associated with `.iis` and gives brief descriptions of what they mean.

You can use `display` to look at the `.iis` values that have been returned:

```
ampl: display _varname, _var.iis, _conname, _con.iis;
:        _varname    _var.iis      _conname    _con.iis     :=
1      "Buy['BEEF']"    upp       "Diet['A']"     non
2      "Buy['CHK']"     low       "Diet['B1']"    non
3      "Buy['FISH']"    low       "Diet['B2']"    low
4      "Buy['HAM']"     upp       "Diet['C']"     non
5      "Buy['MCH']"     non       "Diet['NA']"    upp
6      "Buy['MTL']"     upp       "Diet['CAL']"   non
7      "Buy['SPG']"     low          .             .
8      "Buy['TUR']"     low          .             .
;
```

This information indicates that the IIS consists of four lower and three upper bounds on
the variables, plus the constraints providing the lower bound on B2 and the upper bound
on NA in the diet. Together these restrictions have no feasible solution, but dropping any
one of them will permit a solution to be found to the remaining ones.

If dropping the bounds is not of interest, then you may want to list only the constraints in the IIS. A print statement produces a concise listing:

```
ampl: print {i in 1.._ncons:
ampl?    _con[i].iis <> "non"}: _conname[i];
Diet['B2']
Diet['NA']
```

You could conclude in this case that, to avoid violating the bounds on amounts purchased, you might need to accept either less vitamin B2 or more sodium, or both, in the
diet. Further experimentation would be necessary to determine how much less or more,
however, and what other changes you might need to accept in order to gain feasibility.
(A linear program can have several irreducible infeasible subsets, but CPLEX's IIS-
finding algorithm detects only one IIS at a time.)

Solver-defined suffixes: direction of unboundedness

For an unbounded linear program — one that has in effect a minimum of
-Infinity or a maximum of +Infinity — a solver can return a ray of feasible solutions of the form $X + \alpha d$, where $\alpha \geq 0$. On return from CPLEX, the feasible solution X
is given by the values of the variables, while the direction of unboundedness d is given by
an additional value associated with each variable through the solver-defined suffix
.unbdd.

An application of the direction of unboundedness can be found in a model
trnloc1d.mod and script trnloc1d.run for Benders decomposition applied to a
combination of a warehouse-location and transportation problem; the model, data and
script are available from the AMPL web site. We won't try to describe the whole decomposition scheme here, but rather concentrate on the subproblem obtained by fixing the
zero-one variables Build[i], which indicate the warehouses that are to be built, to trial
values build[i]. In its dual form, this subproblem is:

```
var Supply_Price {ORIG} <= 0;
var Demand_Price {DEST};
maximize Dual_Ship_Cost:
    sum {i in ORIG} Supply_Price[i] * supply[i] * build[i] +
    sum {j in DEST} Demand_Price[j] * demand[j];
subject to Dual_Ship {i in ORIG, j in DEST}:
    Supply_Price[i] + Demand_Price[j] <= var_cost[i,j];
```

When all values build[i] are set to zero, no warehouses are built, and the primal sub-problem is infeasible. As a result, the dual formulation of the subproblem, which always has a feasible solution, must be unbounded.

As the remainder of this chapter will explain, we solve a subproblem by collecting its components into an AMPL "problem" and then directing AMPL to solve only that problem. When this approach is applied to the dual subproblem from the AMPL command-line, CPLEX returns the direction of unboundedness in the expected way:

```
ampl: model trnloc1d.mod;
ampl: data trnloc1.dat;
ampl: problem Sub: Supply_Price, Demand_Price,
ampl?    Dual_Ship_Cost, Dual_Ship;
ampl: let {i in ORIG} build[i] := 0;
ampl: option solver cplex, cplex_options 'presolve 0';
ampl: solve;
CPLEX 8.0.0: presolve 0
CPLEX 8.0.0: unbounded problem.
25 dual simplex iterations (25 in phase I)
variable.unbdd returned
6 extra simplex iterations for ray (1 in phase I)

suffix unbdd OUT;
```

The suffix message indicates that .unbdd has been created automatically. You can use this suffix to display the direction of unboundedness, which is simple in this case:

```
ampl: display Supply_Price.unbdd;
Supply_Price.unbdd [*] :=
 1 -1    4 -1    7 -1   10 -1   13 -1   16 -1   19 -1   22 -1   25 -1
 2 -1    5 -1    8 -1   11 -1   14 -1   17 -1   20 -1   23 -1
 3 -1    6 -1    9 -1   12 -1   15 -1   18 -1   21 -1   24 -1
;
ampl: display Demand_Price.unbdd;
Demand_Price.unbdd [*] :=
A3   1
A6   1
A8   1
A9   1
B2   1
B4   1
;
```

Our script for Benders decomposition (trnloc1d.run) solves the subproblem repeatedly, with differing build[i] values generated from the master problem. After each solve, the result is tested for unboundedness and an extension of the master problem is constructed accordingly. The essentials of the main loop are as follows:

```
repeat {
   solve Sub;
   if Dual_Ship_Cost <= Max_Ship_Cost + 0.00001 then break;
   if Sub.result = "unbounded" then {
      let nCUT := nCUT + 1;
      let cut_type[nCUT] := "ray";
      let {i in ORIG}
         supply_price[i,nCUT] := Supply_Price[i].unbdd;
      let {j in DEST}
         demand_price[j,nCUT] := Demand_Price[j].unbdd;
   } else {
      let nCUT := nCUT + 1;
      let cut_type[nCUT] := "point";
      let {i in ORIG} supply_price[i,nCUT] := Supply_Price[i];
      let {j in DEST} demand_price[j,nCUT] := Demand_Price[j];
   }
   solve Master;
   let {i in ORIG} build[i] := Build[i];
}
```

An attempt to use .unbdd in this context fails, however:

```
ampl: commands trnloc1d.run;
trnloc1d.run, line 39 (offset 931):
        Bad suffix .unbdd for Supply_Price
context:  let {i in ORIG} supply_price[i,nCUT] :=
                >>> Supply_Price[i].unbdd; <<<
```

The difficulty here is that AMPL scans all commands in the `repeat` loop before beginning to execute any of them. As a result it encounters the use of .unbdd before any infeasible subproblem has had a chance to cause this suffix to be defined. To make this script run as intended, it is necessary to place the statement

```
suffix unbdd OUT;
```

in the script before the `repeat` loop, so that .unbdd is already defined at the time the loop is scanned.

Defining and using suffixes

A new AMPL suffix is defined by a statement consisting of the keyword `suffix` followed by a *suffix-name* and then one or more optional qualifiers that indicate what values may be associated with the suffix and how it may be used. For example, we have seen the definition

```
suffix priority IN, integer, >= 0, <= 9999;
```

for the suffix `priority` with *in-out*, *type*, and *bound* qualifiers.

The `suffix` statement causes AMPL to recognize *suffixed expressions* of the form *component-name . suffix-name*, where *component-name* refers to any currently declared variable, constraint, or objective (or problem, as defined in the next section). The definition of a suffix remains in effect until the next `reset` command or the end of the current AMPL session. The *suffix-name* is subject to the same rules as other names in AMPL. Suffixes have a separate name space, however, so a suffix may have the same name as a parameter, variable, or other model component. The optional qualifiers of the `suffix` statement may appear in any order; their forms and effects are described below.

The optional *type* qualifier in a `suffix` statement indicates what values may be associated with the suffixed expressions, with all numeric values being the default:

suffix type	values allowed
none specified	any numeric value
integer	integer numeric values
binary	0 or 1
symbolic	character strings listed in option *suffix-name*_table

All numeric-valued suffixed expressions have an initial value of 0. Their permissible values may be further limited by one or two *bound* qualifiers of the form

```
>=  arith-expr
<=  arith-expr
```

where *arith-expr* is any arithmetic expression not involving variables.

For each `symbolic` suffix, AMPL automatically defines an associated numeric suffix, *suffix-name*_num. An AMPL option *suffix-name*_table must then be created to define a relation between the . *suffix-name* and . *suffix-name*_num values, as in the following example:

```
suffix iis symbolic OUT;
option iis_table '\
0       non     not in the iis\
1       low     at lower bound\
2       fix     fixed\
3       upp     at upper bound\
';
```

Each line of the table consists of an integer value, a string value, and an optional comment. Every string value is associated with its adjacent integer value, and with any higher integer values that are less than the integer on the next line. Assigning a string value to a . *suffix-name* expression is equivalent to assigning the associated numeric value to a . *suffix-name*_num expression. The latter expressions are initially assigned the value 0, and are subject to any type and bound qualifiers as described above. (Normally the string values of `symbolic` suffixes are used in AMPL commands and scripts, while the numeric values are used in communication with solvers.)

The optional *in-out* qualifier determines how suffix values interact with the solver:

in-out	handling of suffix values
IN	written by AMPL before invoking the solver, then read in by solver
OUT	written out by solver, then read by AMPL after the solver is finished
INOUT	both read and written, as for IN and OUT above
LOCAL	neither read nor written

INOUT is the default if no *in-out* keyword is specified.

We have seen that suffixed expressions can be assigned or reassigned values by a `let` statement:

```
let Use["GARY","FRE"].direction := -1;
```

Here just one variable is assigned a suffixed value, but often there are suffixed values for all variables in an indexed collection:

```
var Use {ORIG,DEST} binary;
let {i in ORIG, j in DEST}
   Use[i,j].priority := sum {p in PROD} demand[j,p];
```

In this case the assignment of suffix values can be combined with the variable's declaration:

```
var Use {i in ORIG, j in DEST} binary,
   suffix priority sum {p in PROD} demand[j,p];
```

In general, one or more of the phrases in a `var` declaration may consist of the keyword `suffix` followed by a previously-defined *suffix-name* and an expression for evaluating the associated suffix expressions.

14.4 Alternating between models

Chapter 13 described how ''scripts'' of AMPL commands can be set up to run as programs that perform repetitive actions. In several examples, a script solves a series of related model instances, by including a `solve` statement inside a loop. The result is a simple kind of sensitivity analysis algorithm, programmed in AMPL's command language.

Much more powerful algorithmic procedures can be constructed by using two models. An optimal solution for one model yields new data for the other, and the two are solved in alternation in such a way that some termination condition must eventually be reached. Classic methods of column and cut generation, decomposition, and Lagrangian relaxation are based on schemes of this kind, which are described in detail in references cited at the end of this chapter.

To use two models in this manner, a script must have some way of switching between them. Switching can be done with previously defined AMPL features, or more clearly and efficiently by defining separately-named problems and environments.

We illustrate these possibilities through a script for a basic form of the ''roll trim'' or ''cutting stock'' problem, using a well-known, elementary column-generation procedure. In the interest of brevity, we give only a sketchy description of the procedure here, while the references at the end of this chapter provide sources for thorough descriptions. There are several other examples of generation, decomposition, and relaxation schemes on the AMPL web site, and we will also use a few excerpts from them later, without showing the whole models.

In a roll trim problem, we wish to cut up long raw widths of some commodity such as rolls of paper into a combination of smaller widths that meet given orders with as little waste as possible. This problem can be viewed as deciding, for each raw-width roll, where the cuts should be made so as to produce one or more of the smaller widths that were ordered. Expressing such a problem in terms of decision variables is awkward, however, and leads to an integer program that is difficult to solve except for very small instances.

To derive a more manageable model, the so-called Gilmore-Gomory procedure defines a cutting *pattern* to be any one feasible way in which a raw roll can be cut up. A pattern thus consists of a certain number of rolls of each desired width, such that their total width does not exceed the raw width. If (as in Exercise 2-6) the raw width is 110", and there are demands for widths of 20", 45", 50", 55" and 75", then two rolls of 45" and one of 20" make an acceptable pattern, as do one of 50" and one of 55" (with 5" of waste). Given this view, the two simple linear programs in Figure 14-2 can be made to work together to find an efficient cutting plan.

The cutting optimization model (Figure 14-2a) finds the minimum number of raw rolls that need be cut, given a collection of known cutting patterns that may be used. This is actually a close cousin to the diet model, with the variables representing patterns cut rather than food items bought, and the constraints enforcing a lower limit on cut widths rather than nutrients provided.

The pattern generation model (Figure 14-2b) seeks to identify a new pattern that can be used in the cutting optimization, either to reduce the number of raw rolls needed, or to determine that no such new pattern exists. The variables of this model are the numbers of each desired width in the new pattern; the single constraint ensures that the total width of the pattern does not exceed the raw width. We won't try to explain the objective here, except to note that the coefficient of a variable is given by its corresponding ''dual value'' or ''dual price'' from the linear relaxation of the cutting optimization model.

We can search for a good cutting plan by solving these two problems repeatedly in alternation. First the continuous-variable relaxation of the cutting optimization problem generates some dual prices, then the pattern generation problem uses the prices to generate a new pattern, and then the procedure repeats with the collection of patterns extended by one. We stop repeating when the pattern generation problem indicates that no new pattern can lead to an improvement. We then have the best possible solution in terms of (possibly) fractional numbers of raw rolls cut. We may make one last run of the cutting optimization model with the integrality restriction restored, to get the best integral solu-

```
param roll_width > 0;         # width of raw rolls

set WIDTHS;                    # set of widths to be cut
param orders {WIDTHS} > 0;    # number of each width to be cut

param nPAT integer >= 0;      # number of patterns
set PATTERNS = 1..nPAT;       # set of patterns

param nbr {WIDTHS,PATTERNS} integer >= 0;
   check {j in PATTERNS}:
      sum {i in WIDTHS} i * nbr[i,j] <= roll_width;
                              # defn of patterns: nbr[i,j] = number
                              # of rolls of width i in pattern j

var Cut {PATTERNS} integer >= 0; # rolls cut using each pattern

minimize Number:                    # minimize total raw rolls cut
   sum {j in PATTERNS} Cut[j];

subject to Fill {i in WIDTHS}:
   sum {j in PATTERNS} nbr[i,j] * Cut[j] >= orders[i];
         # for each width, total rolls cut meets total orders
```

Figure 14-2a: Pattern-based model for cutting optimization problem (cut.mod).

```
param price {WIDTHS} default 0.0; # prices from cutting opt
var Use {WIDTHS} integer >= 0;
                          # numbers of each width in pattern
minimize Reduced_Cost:
   1 - sum {i in WIDTHS} price[i] * Use[i];

subject to Width_Limit:
   sum {i in WIDTHS} i * Use[i] <= roll_width;
```

Figure 14-2b: Knapsack model for pattern generation problem (cut.mod, continued).

tion using the patterns generated, or we may simply round the fractional numbers of rolls up to the next largest integers if that gives an acceptable result.

This is the Gilmore-Gomory procedure. In terms of our two AMPL models, its steps may be described as follows:

pick initial patterns sufficient to meet demand
repeat
      solve the (fractional) cutting optimization problem
      let price[i] equal Fill[i].dual for each pattern i
      solve the pattern generation problem
      if the optimal value is < 0 then
            add a new pattern that cuts Use[i] rolls of each width i
      else
            find a final integer solution and stop

An easy way to initialize is to generate one pattern for each width, containing as many copies of that width as will fit inside the raw roll. These patterns clearly can cover any demands, though not necessarily in an economical way.

An implementation of the Gilmore-Gomory procedure as an AMPL script is shown in Figure 14-3. The file `cut.mod` contains both the cutting optimization and pattern generation models in Figure 14-2. Since these models have no variables or constraints in common, it would be possible to write the script with simple `solve` statements using alternating objective functions:

```
repeat {
   objective Number;
   solve;
   ...
   objective Reduced_Cost;
   solve;
   ...
}
```

Under this approach, however, every `solve` would send the solver all of the variables and constraints generated by both models. Such an arrangement is inefficient and prone to error, especially for larger and more complex iterative procedures.

We could instead ensure that only the immediately relevant variables and constraints are sent to the solver, by using `fix` and `drop` commands to suppress the others. Then the outline of our loop would look like this:

```
repeat {
   unfix Cut; restore Fill; objective Number;
   fix Use; drop Width_Limit;
   solve;
   ...
   unfix Use; restore Width_Limit; objective Reduced_Cost;
   fix Cut; drop Fill;
   solve;
   ...
}
```

Before each `solve`, the previously fixed variables and dropped constraints must also be brought back, by use of `unfix` and `restore`. This approach is efficient, but it remains highly error-prone, and makes scripts difficult to read.

As an alternative, therefore, AMPL allows models to be distinguished by use of the `problem` statement seen in Figure 14-3:

```
problem Cutting_Opt: Cut, Number, Fill;
option relax_integrality 1;

problem Pattern_Gen: Use, Reduced_Cost, Width_Limit;
option relax_integrality 0;
```

The first statement defines a problem named `Cutting_Opt` that consists of the `Cut` variables, the `Fill` constraints, and the objective `Number`. This statement also makes

```
model cut.mod;
data cut.dat;
option solver cplex, solution_round 6;
option display_1col 0, display_transpose -10;

problem Cutting_Opt: Cut, Number, Fill;
option relax_integrality 1;

problem Pattern_Gen: Use, Reduced_Cost, Width_Limit;
option relax_integrality 0;

let nPAT := 0;
for {i in WIDTHS} {
   let nPAT := nPAT + 1;
   let nbr[i,nPAT] := floor (roll_width/i);
   let {i2 in WIDTHS: i2 <> i} nbr[i2,nPAT] := 0;
}

repeat {
   solve Cutting_Opt;
   let {i in WIDTHS} price[i] := Fill[i].dual;

   solve Pattern_Gen;
   if Reduced_Cost < -0.00001 then {
      let nPAT := nPAT + 1;
      let {i in WIDTHS} nbr[i,nPAT] := Use[i];
   }
   else break;
}
display nbr, Cut;

option Cutting_Opt.relax_integrality 0;
solve Cutting_Opt;
display Cut;
```

Figure 14-3: Gilmore-Gomory procedure for cutting-stock problem (cut.run).

Cutting_Opt the current problem; uses of the var, minimize, maximize, subject to, and option statements now apply to this problem only. Thus by setting option relax_integrality to 1 above, for example, we assure that the integrality condition on the Cut variables will be relaxed whenever Cutting_Opt is current. In a similar way, we define a problem Pattern_Gen that consists of the Use variables, the Width_Limit constraint, and the objective Reduced_Cost; this in turn becomes the current problem, and this time we set relax_integrality to 0 because only integer solutions to this problem are meaningful.

The for loop in Figure 14-3 creates the initial cutting patterns, after which the main repeat loop carries out the Gilmore-Gomory procedure as described previously. The statement

```
solve Cutting_Opt;
```

```
param roll_width := 110 ;
param: WIDTHS: orders :=
            20      48
            45      35
            50      24
            55      10
            75       8 ;
```

Figure 14-4: Data for cutting-stock model (cut.dat)

restores Cutting_Opt as the current problem, along with its environment, and solves the associated linear program. Then the assignment

```
let {i in WIDTHS} price[i] := Fill[i].dual;
```

transfers the optimal dual prices from Cutting_Opt to the parameters price[i] that will be used by Pattern_Gen. All sets and parameters are global in AMPL, so they can be referenced or changed whatever the current problem.

The second half of the main loop makes problem Pattern_Gen and its environment current, and sends the associated integer program to the solver. If the resulting objective is sufficiently negative, the pattern returned by the Use[i] variables is added to the data that will be used by Cutting_Opt in the next pass through the loop. Otherwise no further progress can be made, and the loop is terminated.

The script concludes with the following statements, to solve for the best integer solution using all patterns generated:

```
option Cutting_Opt.relax_integrality 0;
solve Cutting_Opt;
```

The expression Cutting_Opt.relax_integrality stands for the value of the relax_integrality option in the Cutting_Opt environment. We discuss these kinds of names and their uses at greater length in the next section.

As an example of how this works, Figure 14-4 shows data for cutting 110" raw rolls, to meet demands of 48, 35, 24, 10 and 8 for finished rolls of widths 20, 45, 50, 55 and 75, respectively. Figure 14-5 shows the output that occurs when Figure 14-3's script is run with the model and data as shown in Figures 14-2 and 14-4. The best fractional solution cuts 46.25 raw rolls in five different patterns, using 48 rolls if the fractional values are rounded up to the next integer. The final solve using integer variables shows how a collection of six of the patterns can be applied to meet demand using only 47 raw rolls.

14.5 Named problems

As our cutting-stock example has shown, the key to writing a clear and efficient script for alternating between two (or more) models lies in working with *named problems* that

```
ampl: commands cut.run;

CPLEX 8.0.0: optimal solution; objective 52.1
0 dual simplex iterations (0 in phase I)
CPLEX 8.0.0: optimal integer solution; objective -0.2
1 MIP simplex iterations
0 branch-and-bound nodes
CPLEX 8.0.0: optimal solution; objective 48.6
2 dual simplex iterations (0 in phase I)
CPLEX 8.0.0: optimal integer solution; objective -0.2
2 MIP simplex iterations
0 branch-and-bound nodes
CPLEX 8.0.0: optimal solution; objective 47
1 dual simplex iterations (0 in phase I)
CPLEX 8.0.0: optimal integer solution; objective -0.1
2 MIP simplex iterations
0 branch-and-bound nodes
CPLEX 8.0.0: optimal solution; objective 46.25
2 dual simplex iterations (0 in phase I)
CPLEX 8.0.0: optimal integer solution; objective -1e-06
8 MIP simplex iterations
0 branch-and-bound nodes

nbr [*,*]
:    1    2    3    4    5    6    7    8       :=
20   5    0    0    0    0    1    1    3
45   0    2    0    0    0    2    0    0
50   0    0    2    0    0    0    0    1
55   0    0    0    2    0    0    0    0
75   0    0    0    0    1    0    1    0
;

Cut [*] :=
1  0    2  0    3  8.25    4  5    5  0    6  17.5    7  8    8  7.5
;

CPLEX 8.0.0: optimal integer solution; objective 47
5 MIP simplex iterations
0 branch-and-bound nodes

Cut [*] :=
1  0    2  0    3  8    4  5    5  0    6  18    7  8    8  8
;
```

Figure 14-5: Output from execution of Figure 14-3 cutting-stock script.

represent different subsets of model components. In this section we describe in more detail how AMPL's problem statement is employed to define, use, and display named problems. At the end we also introduce a similar idea, *named environments,* which facilitates switching between collections of AMPL options.

Illustrations in this section are taken from the cutting-stock script and from some of the other example scripts on the AMPL web site. An explanation of the logic behind these scripts is beyond the scope of this book; some suggestions for learning more are given in the references at the end of the chapter.

Defining named problems

At any point during an AMPL session, there is a *current* problem consisting of a list of variables, objectives and constraints. The current problem is named `Initial` by default, and comprises all variables, objectives and constraints defined so far. You can define other ''named'' problems consisting of subsets of these components, however, and can make them current. When a named problem is made current, all of the model components in the problem's subset are made active, while all other variables, objectives and constraints are made inactive. More precisely, variables in the problem's subset are unfixed and the remainder are fixed at their current values. Objectives and constraints in the problem's subset are restored and the remainder are dropped. (Fixing and dropping are discussed in Section 11.4.)

You can define a problem most straightforwardly through a `problem` declaration that gives the problem's name and its list of components. Thus in Figure 14-3 we have:

```
problem Cutting_Opt: Cut, Number, Fill;
```

A new problem named `Cutting_Opt` is defined, and is specified to contain all of the `Cut` variables, the objective `Number`, and all of the `Fill` constraints from the model in Figure 14-2. At the same time, `Cutting_Opt` becomes the current problem. Any fixed `Cut` variables are unfixed, while all other declared variables are fixed at their current values. The objective `Number` is restored if it had been previously dropped, while all other declared objectives are dropped; and similarly any dropped `Fill` constraints are restored, while all other declared constraints are dropped.

For more complex models, the list of a problem's components typically includes several collections of variables and constraints, as in this example from `stoch1.run` (one of the examples from the AMPL web site):

```
problem Sub: Make, Inv, Sell,
    Stage2_Profit, Time, Balance2, Balance;
```

By specifying an indexing-expression after the problem name, you can define an indexed collection of problems, such as these in `multi2.run` (another web site example):

```
problem SubII {p in PROD}: Reduced_Cost[p],
    {i in ORIG, j in DEST} Trans[i,j,p],
    {i in ORIG} Supply[i,p], {j in DEST} Demand[j,p];
```

For each `p` in the set `PROD`, a problem `SubII[p]` is defined. Its components include the objective `Reduced_Cost[p]`, the variables `Trans[i,j,p]` for each combination of `i` in `ORIG` and `j` in `DEST`, and the constraints `Supply[i,p]` and `Demand[j,p]` for each `i` in `ORIG` and each `j` in `DEST`, respectively.

A problem declaration's form and interpretation naturally resemble those of other AMPL statements that specify lists of model components. The declaration begins with the keyword `problem`, a problem name not previously used for any other model component, an optional indexing expression (to define an indexed collection of problems), and a colon. Following the colon is the comma-separated list of variables, objectives and constraints to be included in the problem. This list may contain items of any of the following forms, where ''component'' refers to any variable, objective or constraint:

- A component name, such as `Cut` or `Fill`, which refers to all components having that name.

- A subscripted component name, such as `Reduced_Cost[p]`, which refers to that component alone.

- An indexing expression followed by a subscripted component name, such as `{i in ORIG} Supply[i,p]`, which refers to one component for each member of the indexing set.

To save the trouble of repeating an indexing expression when several components are indexed in the same way, the `problem` statement also allows an indexing expression followed by a parenthesized list of components. Thus for example the following would be equivalent:

```
{i in ORIG} Supply1[i,p], {i in ORIG} Supply2[i,p],
{i in ORIG, j in DEST} Trans[i,j,p],
{i in ORIG, j in DEST} Use[i,j,p]

{i in ORIG} (Supply1[i,p], Supply2[i,p],
    {j in DEST} (Trans[i,j,p], Use[i,j,p]))
```

As these examples show, the list inside the parentheses may contain any item that is valid in a component list, even an indexing expression followed by another parenthesized list. This sort of recursion is also found in AMPL's `print` command, but is more general than the list format allowed in `display` commands.

Whenever a variable, objective or constraint is declared, it is automatically added to the current problem (or all current problems, if the most recent `problem` statement specified an indexed collection of problems). Thus in our cutting-stock example, all of Figure 14-2's model components are first placed by default into the problem `Initial`; then, when the script of Figure 14-3 is run, the components are divided into the problems `Cutting_Opt` and `Pattern_Gen` by use of `problem` statements. As an alternative, we can declare empty problems and then fill in their members through AMPL declarations. Figure 14-6 (`cut2.mod`) shows how this would be done for the Figure 14-2 models. This approach is sometimes clearer or easier for simpler applications.

Any use of `drop`/`restore` or `fix`/`unfix` also modifies the current problem. The `drop` command has the effect of removing constraints or objectives from the current problem, while the `restore` command has the effect of adding constraints or objectives. Similarly, the `fix` command removes variables from the current problem and the `unfix`

```
problem Cutting_Opt;

param nPAT integer >= 0, default 0;
param roll_width;
set PATTERNS = 1..nPAT;
set WIDTHS;
param orders {WIDTHS} > 0;
param nbr {WIDTHS,PATTERNS} integer >= 0;
   check {j in PATTERNS}:
      sum {i in WIDTHS} i * nbr[i,j] <= roll_width;
var Cut {PATTERNS} >= 0;
minimize Number: sum {j in PATTERNS} Cut[j];
subject to Fill {i in WIDTHS}:
   sum {j in PATTERNS} nbr[i,j] * Cut[j] >= orders[i];

problem Pattern_Gen;

param price {WIDTHS};
var Use {WIDTHS} integer >= 0;
minimize Reduced_Cost:
   1 - sum {i in WIDTHS} price[i] * Use[i];
subject to Width_Limit:
   sum {i in WIDTHS} i * Use[i] <= roll_width;
```

Figure 14-6: Alternate definition of named cutting-stock problems (cut2.mod).

command adds variables. As an example, multi1.run uses the following problem statements:

```
problem MasterI: Artificial, Weight, Excess, Multi, Convex;
problem SubI: Artif_Reduced_Cost, Trans, Supply, Demand;

problem MasterII: Total_Cost, Weight, Multi, Convex;
problem SubII: Reduced_Cost, Trans, Supply, Demand;
```

to define named problems for phases I and II of its decomposition procedure. By contrast, multi1a.run specifies

```
problem Master: Artificial, Weight, Excess, Multi, Convex;
problem Sub: Artif_Reduced_Cost, Trans, Supply, Demand;
```

to define the problems initially, and then

```
problem Master;
   drop Artificial; restore Total_Cost; fix Excess;
problem Sub;
   drop Artif_Reduced_Cost; restore Reduced_Cost;
```

when the time comes to convert the problems to a form appropriate for the second phase. Since the names Master and Sub are used throughout the procedure, one loop in the script suffices to implement both phases.

Alternatively, a `redeclare problem` statement can give a new definition for a problem. The `drop`, `restore`, and `fix` commands above could be replaced, for instance, by

```
redeclare problem Master: Total_Cost, Weight, Multi, Convex;
redeclare problem Sub: Reduced_Cost, Trans, Supply, Demand;
```

Like other declarations, however, this cannot be used within a compound statement (`if`, `for` or `repeat`) and so cannot be used in the `multi1a.run` example.

A form of the `reset` command lets you undo any changes made to the definition of a problem. For example,

```
reset problem Cutting_Opt;
```

resets the definition of `Cutting_Opt` to the list of components in the `problem` statement that most recently defined it.

Using named problems

We next describe alternatives for changing the current problem. Any change will in general cause different objectives and constraints to be dropped, and different variables to be fixed, with the result that a different optimization problem is generated for the solver. The values associated with model components are not affected simply by a change in the current problem, however. All previously declared components are accessible regardless of the current problem, and they keep the same values unless they are explicitly changed by `let` or `data` statements, or by a `solve` in the case of variable and objective values and related quantities (such as dual values, slacks, and reduced costs).

Any `problem` statement that refers to only one problem (not an indexed collection of problems) has the effect of making that problem current. As an example, at the beginning of the cutting-stock script we want to make first one and then the other named problem current, so that we can adjust certain options in the environments of the problems. The `problem` statements in `cut1.run` (Figure 14-3):

```
problem Cutting_Opt: Cut, Number, Fill;
   option relax_integrality 1;

problem Pattern_Gen: Use, Reduced_Cost, Width_Limit;
   option relax_integrality 0;
```

serve both to define the new problems and to make those problems current. The analogous statements in `cut2.run` are simpler:

```
problem Cutting_Opt;
   option relax_integrality 1;

problem Pattern_Gen;
   option relax_integrality 0;
```

These statements serve only to make the named problems current, because the problems have already been defined by `problem` statements in `cut2.mod` (Figure 14-6).

A `problem` statement may also refer to an indexed collection of problems, as in the `multi2.run` example cited previously:

```
problem SubII {p in PROD}: Reduced_Cost[p], ...
```

This form defines potentially many problems, one for each member of the set `PROD`. Subsequent problem statements can make members of a collection current one at a time, as in a loop having the form

```
for {p in PROD} {
    problem SubII[p];
    ...
}
```

or in a statement such as `problem SubII["coils"]` that refers to a particular member.

As seen in previous examples, the `solve` statement can also include a problem name, in which case the named problem is made current and then sent to the solver. The effect of a statement such as `solve Pattern_Gen` is thus exactly the same as the effect of `problem Pattern_Gen` followed by `solve`.

Displaying named problems

The command consisting of `problem` alone tells which problem is current:

```
ampl: model cut.mod;
ampl: data cut.dat;
ampl: problem;
problem Initial;

ampl: problem Cutting_Opt: Cut, Number, Fill;
ampl: problem Pattern_Gen: Use, Reduced_Cost, Width_Limit;
ampl: problem;
problem Pattern_Gen;
```

The current problem is always `Initial` until other named problems have been defined.

The `show` command can give a list of the named problems that have been defined:

```
ampl: show problems;
problems:   Cutting_Opt    Pattern_Gen
```

We can also use `show` to see the variables, objectives and constraints that make up a particular problem or indexed collection of problems:

```
ampl: show Cutting_Opt, Pattern_Gen;
problem Cutting_Opt: Fill, Number, Cut;
problem Pattern_Gen: Width_Limit, Reduced_Cost, Use;
```

and use `expand` to see the explicit objectives and constraints of the current problem, after all data values have been substituted:

```
ampl: expand Pattern_Gen;
minimize Reduced_Cost:
        -0.166667*Use[20] - 0.416667*Use[45] - 0.5*Use[50]
        - 0.5*Use[55] - 0.833333*Use[75] + 1;

subject to Width_Limit:
        20*Use[20] + 45*Use[45] + 50*Use[50] + 55*Use[55] +
        75*Use[75] <= 110;
```

See Section 12.6 for further discussion of show and expand.

Defining and using named environments

In the same way that there is a current problem at any point in an AMPL session, there is also a current *environment*. Whereas a problem is a list of non-fixed variables and non-dropped objectives and constraints, an environment records the values of all AMPL options. By naming different environments, a script can easily switch between different collections of option settings.

In the default mode of operation, which is sufficient for many purposes, the current environment always has the same name as the current problem. At the start of an AMPL session the current environment is named Initial, and each subsequent problem statement that defines a new named problem also defines a new environment having the same name as the problem. An environment initially inherits all the option settings that existed when it was created, but it retains new settings that are made while it is current. Any problem or solve statement that changes the current problem also switches to the correspondingly named environment, with options set accordingly.

As an example, our script for the cutting stock problem (Figure 14-3) sets up the model and data and then proceeds as follows:

```
option solver cplex, solution_round 6;
option display_1col 0, display_transpose -10;

problem Cutting_Opt: Cut, Number, Fill;
option relax_integrality 1;

problem Pattern_Gen: Use, Reduced_Cost, Width_Limit;
option relax_integrality 0;
```

Options solver and three others are changed (by the first two option statements) before any of the problem statements; hence their new settings are inherited by subsequently defined environments and are the same throughout the rest of the script. Next a problem statement defines a new problem and a new environment named Cutting_Opt, and makes them current. The ensuing option statement changes relax_integrality to 1. Thereafter, when Cutting_Opt is the current problem (and environment) in the script, relax_integrality will have the value 1. Finally, another problem and option statement do much the same for problem (and environment) Pattern_Gen, except that relax_integrality is set back to 0 in that environment.

The result of these initial statements is to guarantee a proper setup for each of the subsequent `solve` statements in the repeat loop. The result of `solve Cutting_Opt` is to set the current environment to `Cutting_Opt`, thereby setting `relax_integrality` to 1 and causing the linear relaxation of the cutting optimization problem to be solved. Similarly the result of `solve Pattern_Gen` is to cause the pattern generation problem to be solved as an integer program. We could instead have used `option` statements within the loop to switch the setting of `relax_integrality`, but with this approach we have kept the loop — the key part of the script — as simple as possible.

In more complex situations, you can declare named environments independently of named problems, by use of a statement that consists of the keyword `environ` followed by a name:

```
environ Master;
```

Environments have their own name space. If the name has not been used previously as an environment name, it is defined as one and is associated with all of the current option values. Otherwise, the statement has the effect of making that environment (and its associated option values) current.

A previously declared environment may also be associated with the declaration of a new named problem, by placing `environ` and the environment name before the colon in the `problem` statement:

```
problem MasterII environ Master: ...
```

The named environment is then automatically made current whenever the associated problem becomes current. The usual creation of an environment having the same name as the problem is overridden in this case.

An indexed collection of environments may be declared in an `environ` statement by placing an AMPL indexing expression after the environment name. The name is then "subscripted" in the usual way to refer to individual environments.

Named environments handle changes in the same way as named problems. If an option's value is changed while some particular environment is current, the new value is recorded and is the value that will be reinstated whenever that environment is made current again.

Bibliography

Vašek Chvátal, *Linear Programming*. Freeman (New York, NY, 1983). A general introduction to linear programming that has chapters on the cutting-stock problem sketched in Section 14.4 and on the Dantzig-Wolfe decomposition procedure that is behind the `multi` examples cited in Section 14.5.

Marshall L. Fisher, "An Applications Oriented Guide to Lagrangian Relaxation." Interfaces **15**, 2 (1985) pp. 10–21. An introduction to the Lagrangian relaxation procedures underlying the `trn-loc2` script of Section 14.3.

Robert Fourer and David M. Gay, "Experience with a Primal Presolve Algorithm." In *Large Scale Optimization: State of the Art,* W. W. Hager, D. W. Hearn and P. M. Pardalos, eds., Kluwer Academic Publishers (Dordrecht, The Netherlands, 1994) pp. 135–154. A detailed description of the presolve procedure sketched in Section 14.1.

Robert W. Haessler, "Selection and Design of Heuristic Procedures for Solving Roll Trim Problems." Management Science **34**, 12 (1988) pp. 1460–1471. Descriptions of real cutting-stock problems, some amenable to the techniques of Section 14.4 and some not.

Leon S. Lasdon, *Optimization Theory for Large Systems.* Macmillan (New York, NY, 1970), reprinted by Dover (Mineola, NY, 2002). A classic source for several of the decomposition and generation procedures behind the scripts.

15

Network Linear Programs

Models of networks have appeared in several chapters, notably in the transportation problems in Chapter 3. We now return to the formulation of these models, and AMPL's features for handling them.

Figure 15-1 shows the sort of diagram commonly used to describe a network problem. A circle represents a *node* of the network, and an arrow denotes an *arc* running from one node to another. A *flow* of some kind travels from node to node along the arcs, in the directions of the arrows.

An endless variety of models involve optimization over such networks. Many cannot be expressed in any straightforward algebraic way or are very difficult to solve. Our discussion starts with a particular class of network optimization models in which the decision variables represent the amounts of flow on the arcs, and the constraints are limited to two kinds: simple bounds on the flows, and conservation of flow at the nodes. Models restricted in this way give rise to the problems known as network linear programs. They are especially easy to describe and solve, yet are widely applicable. Some of their benefits extend to certain generalizations of the network flow form, which we also touch upon.

We begin with minimum-cost transshipment models, which are the largest and most intuitive source of network linear programs, and then proceed to other well-known cases: maximum flow, shortest path, transportation and assignment models. Examples are initially given in terms of standard AMPL variables and constraints, defined in `var` and `subject to` declarations. In later sections, we introduce `node` and `arc` declarations that permit models to be described more directly in terms of their network structure. The last section discusses formulating network models so that the resulting linear programs can be solved most efficiently.

15.1 Minimum-cost transshipment models

As a concrete example, imagine that the nodes and arcs in Figure 15-1 represent cities and intercity transportation links. A manufacturing plant at the city marked PITT will

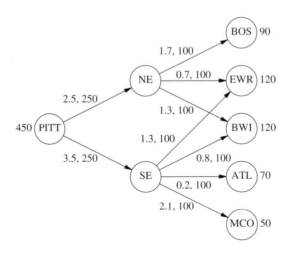

Figure 15-1: A directed network.

make 450,000 packages of a certain product in the next week, as indicated by the 450 at the left of the diagram. The cities marked NE and SE are the northeast and southeast distribution centers, which receive packages from the plant and transship them to warehouses at the cities coded as BOS, EWR, BWI, ATL and MCO. (Frequent flyers will recognize Boston, Newark, Baltimore, Atlanta, and Orlando.) These warehouses require 90, 120, 120, 70 and 50 thousand packages, respectively, as indicated by the numbers at the right. For each intercity link there is a shipping cost per thousand packages and an upper limit on the packages that can be shipped, indicated by the two numbers next to the corresponding arrow in the diagram.

The optimization problem over this network is to find the lowest-cost plan of shipments that uses only the available links, respects the specified capacities, and meets the requirements at the warehouses. We first model this as a general network flow problem, and then consider alternatives that specialize the model to the particular situation at hand. We conclude by introducing a few of the most common variations on the network flow constraints.

A general transshipment model

To write a model for any problem of shipments from city to city, we can start by defining a set of cities and a set of links. Each link is in turn defined by a start city and an end city, so we want the set of links to be a subset of the set of ordered pairs of cities:

```
set CITIES;
set LINKS within (CITIES cross CITIES);
```

Corresponding to each city there is potentially a supply of packages and a demand for packages:

```
param supply {CITIES} >= 0;
param demand {CITIES} >= 0;
```

In the case of the problem described by Figure 15-1, the only nonzero value of `supply` should be the one for PITT, where packages are manufactured and supplied to the distribution network. The only nonzero values of `demand` should be those corresponding to the five warehouses.

The costs and capacities are indexed over the links:

```
param cost {LINKS} >= 0;
param capacity {LINKS} >= 0;
```

as are the decision variables, which represent the amounts to ship over the links. These variables are nonnegative and bounded by the capacities:

```
var Ship {(i,j) in LINKS} >= 0, <= capacity[i,j];
```

The objective is

```
minimize Total_Cost:
    sum {(i,j) in LINKS} cost[i,j] * Ship[i,j];
```

which represents the sum of the shipping costs over all of the links.

It remains to describe the constraints. At each city, the packages supplied plus packages shipped in must balance the packages demanded plus packages shipped out:

```
subject to Balance {k in CITIES}:
    supply[k] + sum {(i,k) in LINKS} Ship[i,k]
        = demand[k] + sum {(k,j) in LINKS} Ship[k,j];
```

Because the expression

```
sum {(i,k) in LINKS} Ship[i,k]
```

appears within the scope of definition of the dummy index `k`, the summation is interpreted to run over all cities `i` such that `(i,k)` is in `LINKS`. That is, the summation is over all links into city `k`; similarly, the second summation is over all links out of `k`. This indexing convention, which was explained in Section 6.2, is frequently useful in describing network balance constraints algebraically. Figures 15-2a and 15-2b display the complete model and data for the particular problem depicted in Figure 15-1.

If all of the variables are moved to the left of the = sign and the constants to the right, the `Balance` constraint becomes:

```
subject to Balance {k in CITIES}:
    sum {(i,k) in LINKS} Ship[i,k]
  - sum {(k,j) in LINKS} Ship[k,j]
        = demand[k] - supply[k];
```

This variation may be interpreted as saying that, at each city `k`, shipments in minus shipments out must equal "net demand". If no city has both a plant and a warehouse (as in

```
set CITIES;
set LINKS within (CITIES cross CITIES);

param supply {CITIES} >= 0;      # amounts available at cities
param demand {CITIES} >= 0;      # amounts required at cities

check: sum {i in CITIES} supply[i] = sum {j in CITIES} demand[j];

param cost {LINKS} >= 0;         # shipment costs/1000 packages
param capacity {LINKS} >= 0;     # max packages that can be shipped

var Ship {(i,j) in LINKS} >= 0, <= capacity[i,j];
                                 # packages to be shipped

minimize Total_Cost:
   sum {(i,j) in LINKS} cost[i,j] * Ship[i,j];

subject to Balance {k in CITIES}:
   supply[k] + sum {(i,k) in LINKS} Ship[i,k]
      = demand[k] + sum {(k,j) in LINKS} Ship[k,j];
```

Figure 15-2a: General transshipment model (net1.mod).

```
set CITIES := PITT   NE SE   BOS EWR BWI ATL MCO ;

set LINKS := (PITT,NE) (PITT,SE)
             (NE,BOS) (NE,EWR) (NE,BWI)
             (SE,EWR) (SE,BWI) (SE,ATL) (SE,MCO);

param supply  default 0 := PITT 450 ;

param demand  default 0 :=
  BOS   90,   EWR 120,   BWI 120,   ATL   70,   MCO   50;

param:        cost  capacity   :=
  PITT NE     2.5    250
  PITT SE     3.5    250

  NE  BOS     1.7    100
  NE  EWR     0.7    100
  NE  BWI     1.3    100

  SE  EWR     1.3    100
  SE  BWI     0.8    100
  SE  ATL     0.2    100
  SE  MCO     2.1    100  ;
```

Figure 15-2b: Data for general transshipment model (net1.dat).

our example), then positive net demand always indicates warehouse cities, negative net demand indicates plant cities, and zero net demand indicates transshipment cities. Thus we could have gotten by with just one parameter net_demand in place of demand and supply, with the sign of net_demand[k] indicating what goes on at city k. Alternative formulations of this kind are often found in descriptions of network flow models.

Specialized transshipment models

The preceding general approach has the advantage of being able to accommodate any pattern of supplies, demands, and links between cities. For example, a simple change in the data would suffice to model a plant at one of the distribution centers, or to allow shipment links between some of the warehouses.

The disadvantage of a general formulation is that it fails to show clearly what arrangement of supplies, demands and links is expected, and in fact will allow inappropriate arrangements. If we know that the situation will be like the one shown in Figure 15-1, with supply at one plant, which ships to distribution centers, which then ship to warehouses that satisfy demand, the model can be specialized to exhibit and enforce such a structure.

To show explicitly that there are three different kinds of cities in the specialized model, we can declare them separately. We use a symbolic parameter rather than a set to hold the name of the plant, to specify that only one plant is expected:

```
param p_city symbolic;
set D_CITY;
set W_CITY;
```

There must be a link between the plant and each distribution center, so we need a subset of pairs only to specify which links connect distribution centers to warehouses:

```
set DW_LINKS within (D_CITY cross W_CITY);
```

With the declarations organized in this way, it is impossible to specify inappropriate kinds of links, such as ones between two warehouses or from a warehouse back to the plant.

One parameter represents the supply at the plant, and a collection of demand parameters is indexed over the warehouses:

```
param p_supply >= 0;
param w_demand {W_CITY} >= 0;
```

These declarations allow supply and demand to be defined only where they belong.

At this juncture, we can define the sets CITIES and LINKS and the parameters supply and demand as they would be required by our previous model:

```
set CITIES = {p_city} union D_CITY union W_CITY;
set LINKS = ({p_city} cross D_CITY) union DW_LINKS;

param supply {k in CITIES} =
   if k = p_city then p_supply else 0;

param demand {k in CITIES} =
   if k in W_CITY then w_demand[k] else 0;
```

The rest of the model can then be exactly as in the general case, as indicated in Figures 15-3a and 15-3b.

```
param p_city symbolic;

set D_CITY;
set W_CITY;
set DW_LINKS within (D_CITY cross W_CITY);

param p_supply >= 0;              # amount available at plant
param w_demand {W_CITY} >= 0;   # amounts required at warehouses

   check: p_supply = sum {k in W_CITY} w_demand[k];

set CITIES = {p_city} union D_CITY union W_CITY;
set LINKS = ({p_city} cross D_CITY) union DW_LINKS;

param supply {k in CITIES} =
   if k = p_city then p_supply else 0;

param demand {k in CITIES} =
   if k in W_CITY then w_demand[k] else 0;

### Remainder same as general transshipment model ###

param cost {LINKS} >= 0;       # shipment costs/1000 packages
param capacity {LINKS} >= 0;   # max packages that can be shipped

var Ship {(i,j) in LINKS} >= 0, <= capacity[i,j];
                             # packages to be shipped

minimize Total_Cost:
   sum {(i,j) in LINKS} cost[i,j] * Ship[i,j];

subject to Balance {k in CITIES}:
   supply[k] + sum {(i,k) in LINKS} Ship[i,k]
      = demand[k] + sum {(k,j) in LINKS} Ship[k,j];
```

Figure 15-3a: Specialized transshipment model (net2.mod).

Alternatively, we can maintain references to the different types of cities and links throughout the model. This means that we must declare two types of costs, capacities and shipments:

```
param pd_cost {D_CITY} >= 0;
param dw_cost {DW_LINKS} >= 0;

param pd_cap {D_CITY} >= 0;
param dw_cap {DW_LINKS} >= 0;

var PD_Ship {i in D_CITY} >= 0, <= pd_cap[i];
var DW_Ship {(i,j) in DW_LINKS} >= 0, <= dw_cap[i,j];
```

The ''pd'' quantities are associated with shipments from the plant to distribution centers; because they all relate to shipments from the same plant, they need only be indexed over D_CITY. The ''dw'' quantities are associated with shipments from distribution centers to warehouses, and so are naturally indexed over DW_LINKS.

The total shipment cost can now be given as the sum of two summations:

```
param p_city := PITT ;

set D_CITY := NE SE ;
set W_CITY := BOS EWR BWI ATL MCO ;

set DW_LINKS := (NE,BOS)  (NE,EWR)  (NE,BWI)
                (SE,EWR)  (SE,BWI)  (SE,ATL)  (SE,MCO);

param p_supply := 450 ;

param w_demand :=
  BOS   90,  EWR 120,  BWI 120,   ATL  70,   MCO  50;

param:        cost  capacity  :=
  PITT NE     2.5     250
  PITT SE     3.5     250

  NE  BOS     1.7     100
  NE  EWR     0.7     100
  NE  BWI     1.3     100

  SE  EWR     1.3     100
  SE  BWI     0.8     100
  SE  ATL     0.2     100
  SE  MCO     2.1     100 ;
```

Figure 15-3b: Data for specialized transshipment model (net2.dat).

```
minimize Total_Cost:
   sum {i in D_CITY} pd_cost[i] * PD_Ship[i]
 + sum {(i,j) in DW_LINKS} dw_cost[i,j] * DW_Ship[i,j];
```

Finally, there must be three kinds of balance constraints, one for each kind of city. Shipments from the plant to the distribution centers must equal the supply at the plant:

```
subject to P_Bal:  sum {i in D_CITY} PD_Ship[i] = p_supply;
```

At each distribution center, shipments in from the plant must equal shipments out to all the warehouses:

```
subject to D_Bal {i in D_CITY}:
   PD_Ship[i] = sum {(i,j) in DW_LINKS} DW_Ship[i,j];
```

And at each warehouse, shipments in from all distribution centers must equal the demand:

```
subject to W_Bal {j in W_CITY}:
   sum {(i,j) in DW_LINKS} DW_Ship[i,j] = w_demand[j];
```

The whole model, with appropriate data, is shown in Figures 15-4a and 15-4b.

The approaches shown in Figures 15-3 and 15-4 are equivalent, in the sense that they cause the same linear program to be solved. The former is more convenient for experimenting with different network structures, since any changes affect only the data for the initial declarations in the model. If the network structure is unlikely to change, however,

```
set D_CITY;
set W_CITY;
set DW_LINKS within (D_CITY cross W_CITY);

param p_supply >= 0;              # amount available at plant
param w_demand {W_CITY} >= 0;     # amounts required at warehouses

    check: p_supply = sum {j in W_CITY} w_demand[j];

param pd_cost {D_CITY} >= 0;      # shipment costs/1000 packages
param dw_cost {DW_LINKS} >= 0;

param pd_cap {D_CITY} >= 0;       # max packages that can be shipped
param dw_cap {DW_LINKS} >= 0;

var PD_Ship {i in D_CITY} >= 0, <= pd_cap[i];
var DW_Ship {(i,j) in DW_LINKS} >= 0, <= dw_cap[i,j];
                                  # packages to be shipped

minimize Total_Cost:
    sum {i in D_CITY} pd_cost[i] * PD_Ship[i] +
    sum {(i,j) in DW_LINKS} dw_cost[i,j] * DW_Ship[i,j];

subject to P_Bal:  sum {i in D_CITY} PD_Ship[i] = p_supply;

subject to D_Bal {i in D_CITY}:
    PD_Ship[i] = sum {(i,j) in DW_LINKS} DW_Ship[i,j];

subject to W_Bal {j in W_CITY}:
    sum {(i,j) in DW_LINKS} DW_Ship[i,j] = w_demand[j];
```

Figure 15-4a: Specialized transshipment model, version 2 (net3.mod).

the latter form facilitates alterations that affect only particular kinds of cities, such as the generalizations we describe next.

Variations on transshipment models

Some balance constraints in a network flow model may have to be inequalities rather than equations. In the example of Figure 15-4, if production at the plant can sometimes exceed total demand at the warehouses, we should replace = by <= in the P_Bal constraints.

A more substantial modification occurs when the quantity of flow that comes out of an arc does not necessarily equal the quantity that went in. As an example, a small fraction of the packages shipped from the plant may be damaged or stolen before the packages reach the distribution center. Suppose that a parameter pd_loss is introduced to represent the loss rate:

```
param pd_loss {D_CITY} >= 0, < 1;
```

Then the balance constraints at the distribution centers must be adjusted accordingly:

```
set D_CITY := NE SE ;

set W_CITY := BOS EWR BWI ATL MCO ;

set DW_LINKS := (NE,BOS)  (NE,EWR)  (NE,BWI)
                (SE,EWR)  (SE,BWI)  (SE,ATL)  (SE,MCO);

param p_supply := 450 ;

param w_demand :=
  BOS   90,   EWR 120,   BWI 120,   ATL   70,   MCO   50;

param:   pd_cost   pd_cap :=
  NE        2.5      250
  SE        3.5      250 ;

param:     dw_cost   dw_cap :=
  NE BOS      1.7      100
  NE EWR      0.7      100
  NE BWI      1.3      100

  SE EWR      1.3      100
  SE BWI      0.8      100
  SE ATL      0.2      100
  SE MCO      2.1      100 ;
```

Figure 15-4b: Data for specialized transshipment model, version 2 (`net3.dat`).

```
subject to D_Bal {i in D_CITY}:
   (1-pd_loss[i]) * PD_Ship[i]
      = sum {(i,j) in DW_LINKS} DW_Ship[i,j];
```

The expression to the left of the = sign has been modified to reflect the fact that only `(1-pd_loss[i]) * PD_Ship[i]` packages arrive at city `i` when `PD_Ship[i]` packages are shipped from the plant.

A similar variation occurs when the flow is not measured in the same units throughout the network. If demand is reported in cartons rather than thousands of packages, for example, the model will require a parameter to represent packages per carton:

```
param ppc integer > 0;
```

Then the demand constraints at the warehouses are adjusted as follows:

```
subject to W_Bal {j in W_CITY}:
   sum {(i,j) in DW_LINKS} (1000/ppc) * DW_Ship[i,j]
      = w_demand[j];
```

The term `(1000/ppc) * DW_Ship[i,j]` represents the number of cartons received at warehouse `j` when `DW_Ship[i,j]` thousand packages are shipped from distribution center `i`.

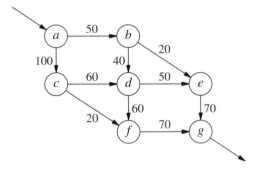

Figure 15-5: Traffic flow network.

15.2 Other network models

Not all network linear programs involve the transportation of things or the minimization of costs. We describe here three well-known model classes — maximum flow, shortest path, and transportation/assignment — that use the same kinds of variables and constraints for different purposes.

Maximum flow models

In some network design applications the concern is to send as much flow as possible through the network, rather than to send flow at lowest cost. This alternative is readily handled by dropping the balance constraints at the origins and destinations of flow, while substituting an objective that stands for total flow in some sense.

As a specific example, Figure 15-5 presents a diagram of a simple traffic network. The nodes and arcs represent intersections and roads; capacities, shown as numbers next to the roads, are in cars per hour. We want to find the maximum traffic flow that can enter the network at *a* and leave at *g*.

A model for this situation begins with a set of intersections, and symbolic parameters to indicate the intersections that serve as entrance and exit to the road network:

```
set INTER;

param entr symbolic in INTER;
param exit symbolic in INTER, <> entr;
```

The set of roads is defined as a subset of the pairs of intersections:

```
set ROADS within (INTER diff {exit}) cross (INTER diff {entr});
```

This definition ensures that no road begins at the exit or ends at the entrance.

Next, the capacity and traffic load are defined for each road:

```
set INTER;  # intersections

param entr symbolic in INTER;              # entrance to road network
param exit symbolic in INTER, <> entr;  # exit from road network

set ROADS within (INTER diff {exit}) cross (INTER diff {entr});

param cap {ROADS} >= 0;                              # capacities
var Traff {(i,j) in ROADS} >= 0, <= cap[i,j];  # traffic loads

maximize Entering_Traff: sum {(entr,j) in ROADS} Traff[entr,j];

subject to Balance {k in INTER diff {entr,exit}}:
   sum {(i,k) in ROADS} Traff[i,k] = sum {(k,j) in ROADS} Traff[k,j];

data;

set INTER := a b c d e f g ;

param entr := a ;
param exit := g ;

param:  ROADS:  cap :=
            a b    50,    a c    100
            b d    40,    b e    20
            c d    60,    c f    20
            d e    50,    d f    60
            e g    70,    f g    70 ;
```

Figure 15-6: Maximum traffic flow model and data (`netmax.mod`).

```
param cap {ROADS} >= 0;
var Traff {(i,j) in ROADS} >= 0, <= cap[i,j];
```

The constraints say that except for the entrance and exit, flow into each intersection equals flow out:

```
subject to Balance {k in INTER diff {entr,exit}}:
    sum {(i,k) in ROADS} Traff[i,k]
        = sum {(k,j) in ROADS} Traff[k,j];
```

Given these constraints, the flow out of the entrance must be the total flow through the network, which is to be maximized:

```
maximize Entering_Traff: sum {(entr,j) in ROADS} Traff[entr,j];
```

We could equally well maximize the total flow into the exit. The entire model, along with data for the example shown in Figure 15-5, is presented in Figure 15-6. Any linear programming solver will find a maximum flow of 130 cars per hour.

Shortest path models

If you were to use the optimal solution to any of our models thus far, you would have to send each of the packages, cars, or whatever along some path from a supply (or entrance) node to a demand (or exit) node. The values of the decision variables do not directly say what the optimal paths are, or how much flow must go on each one. Usually it is not too hard to deduce these paths, however, especially when the network has a regular or special structure.

If a network has just one unit of supply and one unit of demand, the optimal solution assumes a quite different nature. The variable associated with each arc is either 0 or 1, and the arcs whose variables have value 1 comprise a minimum-cost path from the supply node to the demand node. Often the ''costs'' are in fact times or distances, so that the optimum gives a shortest path.

Only a few changes need be made to the maximum flow model of Figure 15-6 to turn it into a shortest path model. There are still a parameter and a variable associated with each road from i to j, but we call them `time[i,j]` and `Use[i,j]`, and the sum of their products yields the objective:

```
param time {ROADS} >= 0;            # times to travel roads
var Use {(i,j) in ROADS} >= 0;   # 1 iff (i,j) in shortest path

minimize Total_Time: sum {(i,j) in ROADS} time[i,j] * Use[i,j];
```

Since only those variables `Use[i,j]` on the optimal path equal 1, while the rest are 0, this sum does correctly represent the total time to traverse the optimal path. The only other change is the addition of a constraint to ensure that exactly one unit of flow is available at the entrance to the network:

```
subject to Start:  sum {(entr,j) in ROADS} Use[entr,j] = 1;
```

The complete model is shown in Figure 15-7. If we imagine that the numbers on the arcs in Figure 15-5 are travel times in minutes rather than capacities, the data are the same; AMPL finds the solution as follows:

```
ampl: model netshort.mod;
ampl: solve;
MINOS 5.5: optimal solution found.
1 iterations, objective 140

ampl: option omit_zero_rows 1;
ampl: display Use;
Use :=
a b   1
b e   1
e g   1
;
```

The shortest path is a $\rightarrow$ b $\rightarrow$ e $\rightarrow$ g, which takes 140 minutes.

Transportation and assignment models

The best known and most widely used special network structure is the ''bipartite'' structure depicted in Figure 15-8. The nodes fall into two groups, one serving as origins of flow and the other as destinations. Each arc connects an origin to a destination.

```
set INTER;  # intersections

param entr symbolic in INTER;          # entrance to road network
param exit symbolic in INTER, <> entr;  # exit from road network

set ROADS within (INTER diff {exit}) cross (INTER diff {entr});

param time {ROADS} >= 0;           # times to travel roads
var Use {(i,j) in ROADS} >= 0;    # 1 iff (i,j) in shortest path

minimize Total_Time: sum {(i,j) in ROADS} time[i,j] * Use[i,j];

subject to Start:  sum {(entr,j) in ROADS} Use[entr,j] = 1;

subject to Balance {k in INTER diff {entr,exit}}:
   sum {(i,k) in ROADS} Use[i,k] = sum {(k,j) in ROADS} Use[k,j];

data;

set INTER := a b c d e f g ;

param entr := a ;
param exit := g ;

param:  ROADS:  time :=
          a b     50,     a c     100
          b d     40,     b e      20
          c d     60,     c f      20
          d e     50,     d f      60
          e g     70,     f g      70 ;
```

Figure 15-7: Shortest path model and data (`netshort.mod`).

The minimum-cost transshipment model on this network is known as the transportation model. The special case in which every origin is connected to every destination was introduced in Chapter 3; an AMPL model and sample data are shown in Figures 3-1a and 3-1b. A more general example analogous to the models developed earlier in this chapter, where a set LINKS specifies the arcs of the network, appears in Figures 6-2a and 6-2b.

Every path from an origin to a destination in a bipartite network consists of one arc. Or, to say the same thing another way, the optimal flow along an arc of the transportation model gives the actual amount shipped from some origin to some destination. This property permits the transportation model to be viewed alternatively as a so-called assignment model, in which the optimal flow along an arc is the amount of something from the origin that is assigned to the destination. The meaning of assignment in this context can be broadly construed, and in particular need not involve a shipment in any sense.

One of the more common applications of the assignment model is matching people to appropriate targets, such as jobs, offices or even other people. Each origin node is associated with one person, and each destination node with one of the targets — for example, with one project. The sets might then be defined as follows:

```
set PEOPLE;
set PROJECTS;
set ABILITIES within (PEOPLE cross PROJECTS);
```

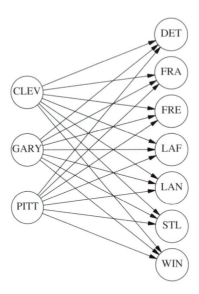

Figure 15-8: Bipartite network.

The set `ABILITIES` takes the role of `LINKS` in our earlier models; a pair `(i,j)` is placed in this set if and only if person `i` can work on project `j`.

As one possibility for continuing the model, the supply at node `i` could be the number of hours that person `i` is available to work, and the demand at node `j` could be the number of hours required for project `j`. Variables `Assign[i,j]` would represent the number of hours of person `i`'s time assigned to project `j`. Also associated with each pair `(i,j)` would be a cost per hour, and a maximum number of hours that person `i` could contribute to job `j`. The resulting model is shown in Figure 15-9.

Another possibility is to make the assignment in terms of people rather than hours. The supply at every node `i` is 1 (person), and the demand at node `j` is the number of people required for project `j`. The supply constraints ensure that `Assign[i,j]` is not greater than 1; and it will equal 1 in an optimal solution if and only if person `i` is assigned to project `j`. The coefficient `cost[i,j]` could be some kind of cost of assigning person `i` to project `j`, in which case the objective would still be to minimize total cost. Or the coefficient could be the ranking of person `i` for project `j`, perhaps on a scale from 1 (highest) to 10 (lowest). Then the model would produce an assignment for which the total of the rankings is the best possible.

Finally, we can imagine an assignment model in which the demand at each node `j` is also 1; the problem is then to match people to projects. In the objective, `cost[i,j]` could be the number of hours that person `i` would need to complete project `j`, in which case the model would find the assignment that minimizes the total hours of work needed to finish all the projects. You can create a model of this kind by replacing all references

```
set PEOPLE;
set PROJECTS;

set ABILITIES within (PEOPLE cross PROJECTS);

param supply {PEOPLE} >= 0;   # hours each person is available
param demand {PROJECTS} >= 0; # hours each project requires

check: sum {i in PEOPLE} supply[i] = sum {j in PROJECTS} demand[j];

param cost {ABILITIES} >= 0;   # cost per hour of work
param limit {ABILITIES} >= 0;  # maximum contributions to projects

var Assign {(i,j) in ABILITIES} >= 0, <= limit[i,j];

minimize Total_Cost:
   sum {(i,j) in ABILITIES} cost[i,j] * Assign[i,j];

subject to Supply {i in PEOPLE}:
   sum {(i,j) in ABILITIES} Assign[i,j] = supply[i];

subject to Demand {j in PROJECTS}:
   sum {(i,j) in ABILITIES} Assign[i,j] = demand[j];
```

Figure 15-9: Assignment model (`netasgn.mod`).

to `supply[i]` and `demand[j]` by 1 in Figure 15-9. Objective coefficients representing rankings are an option for this model as well, giving rise to the kind of assignment model that we used as an example in Section 3.3.

15.3 Declaring network models by `node` and `arc`

AMPL's algebraic notation has great power to express a variety of network linear programs, but the resulting constraint expressions are often not as natural as we would like. While the idea of constraining ''flow out minus flow in'' at each node is easy to describe and understand, the corresponding algebraic constraints tend to involve terms like

```
sum {(i,k) in LINKS} Ship[i,k]
```

that are not so quickly comprehended. The more complex and realistic the network, the worse the problem. Indeed, it can be hard to tell whether a model's algebraic constraints represent a valid collection of flow balances on a network, and consequently whether specialized network optimization software (described later in this chapter) can be used.

Algebraic formulations of network flows tend to be problematical because they are constructed explicitly in terms of variables and constraints, while the nodes and arcs are merely implicit in the way that the constraints are structured. People prefer to approach network flow problems in the opposite way. They imagine giving an explicit definition of nodes and arcs, from which flow variables and balance constraints implicitly arise. To deal with this situation, AMPL provides an alternative that allows network concepts to be declared directly in a model.

The network extensions to AMPL include two new kinds of declaration, `node` and `arc`, that take the place of the `subject to` and `var` declarations in an algebraic constraint formulation. The `node` declarations name the nodes of a network, and characterize the flow balance constraints at the nodes. The `arc` declarations name and define the arcs, by specifying the nodes that arcs connect, and by providing optional information such as bounds and costs that are associated with arcs.

This section introduces `node` and `arc` by showing how they permit various examples from earlier in this chapter to be reformulated conveniently. The following section presents the rules for these declarations more systematically.

A general transshipment model

In rewriting the model of Figure 15-2a using `node` and `arc`, we can retain all of the `set` and `param` declarations and associated data. The changes affect only the three declarations — `minimize`, `var`, and `subject to` — that define the linear program.

There is a node in the network for every member of the set `CITIES`. Using a `node` declaration, we can say this directly:

```
node Balance {k in CITIES}: net_in = demand[k] - supply[k];
```

The keyword `net_in` stands for ''net input'', that is, the flow in minus the flow out, so this declaration says that net flow in must equal net demand at each node `Balance[k]`. Thus it says the same thing as the constraint named `Balance[k]` in the algebraic version, except that it uses the concise term `net_in` in place of the lengthy expression

```
sum {(i,k) in LINKS} Ship[i,k] - sum {(k,j) in LINKS} Ship[k,j]
```

Indeed, the syntax of `subject to` and `node` are practically the same except for the way that the conservation-of-flow constraint is stated. (The keyword `net_out` may also be used to stand for flow out minus flow in, so that we could have written `net_out = supply[k] - demand[k]`.)

There is an arc in the network for every pair in the set `LINKS`. This too can be said directly, using an `arc` declaration:

```
arc Ship {(i,j) in LINKS} >= 0, <= capacity[i,j],
    from Balance[i], to Balance[j], obj Total_Cost cost[i,j];
```

An arc `Ship[i,j]` is defined for each pair in `LINKS`, with bounds of 0 and `capacity[i,j]` on its flow; to this extent, the `arc` and `var` declarations are the same. The `arc` declaration contains additional phrases, however, to say that the arc runs from the node named `Balance[i]` to the node named `Balance[j]`, with a linear coefficient of `cost[i,j]` in the objective function named `Total_Cost`. These phrases use the keywords `from`, `to`, and `obj`.

Since the information about the objective function is included in the `arc` declaration, it is not needed in the `minimize` declaration, which reduces to:

```
minimize Total_Cost;
```

```
set CITIES;
set LINKS within (CITIES cross CITIES);

param supply {CITIES} >= 0;    # amounts available at cities
param demand {CITIES} >= 0;    # amounts required at cities

   check: sum {i in CITIES} supply[i] = sum {j in CITIES} demand[j];

param cost {LINKS} >= 0;        # shipment costs/1000 packages
param capacity {LINKS} >= 0;   # max packages that can be shipped

minimize Total_Cost;

node Balance {k in CITIES}: net_in = demand[k] - supply[k];

arc Ship {(i,j) in LINKS} >= 0, <= capacity[i,j],
   from Balance[i], to Balance[j], obj Total_Cost cost[i,j];
```

Figure 15-10: General transshipment model with node and arc (net1node.mod).

The whole model is shown in Figure 15-10.

As this description suggests, arc and node take the place of var and subject to, respectively. In fact AMPL treats an arc declaration as a definition of variables, so that you would still say display Ship to look at the optimal flows in the network model of Figure 15-10; it treats a node declaration as a definition of constraints. The difference is that node and arc present the model in a way that corresponds more directly to its appearance in a network diagram. The description of the nodes always comes first, followed by a description of how the arcs connect the nodes.

A specialized transshipment model

The node and arc declarations make it easy to define a linear program for a network that has several different kinds of nodes and arcs. For an example we return to the specialized model of Figure 15-4a.

The network has a plant node, a distribution center node for each member of D_CITY, and a warehouse node for each member of W_CITY. Thus the model requires three node declarations:

```
        node Plant: net_out = p_supply;
        node Dist {i in D_CITY};
        node Whse {j in W_CITY}: net_in = w_demand[j];
```

The balance conditions say that flow out of node Plant must be p_supply, while flow into node Whse[j] is w_demand[j]. (The network has no arcs into the plant or out of the warehouses, so net_out and net_in are just the flow out and flow in, respectively.) The conditions at node Dist[i] could be written either net_in = 0 or net_out = 0, but since these are assumed by default we need not specify any condition at all.

```
set D_CITY;
set W_CITY;
set DW_LINKS within (D_CITY cross W_CITY);

param p_supply >= 0;              # amount available at plant
param w_demand {W_CITY} >= 0;   # amounts required at warehouses

    check: p_supply = sum {j in W_CITY} w_demand[j];

param pd_cost {D_CITY} >= 0;    # shipment costs/1000 packages
param dw_cost {DW_LINKS} >= 0;

param pd_cap {D_CITY} >= 0;      # max packages that can be shipped
param dw_cap {DW_LINKS} >= 0;

minimize Total_Cost;

node Plant: net_out = p_supply;

node Dist {i in D_CITY};

node Whse {j in W_CITY}: net_in = w_demand[j];

arc PD_Ship {i in D_CITY} >= 0, <= pd_cap[i],
    from Plant, to Dist[i], obj Total_Cost pd_cost[i];

arc DW_Ship {(i,j) in DW_LINKS} >= 0, <= dw_cap[i,j],
    from Dist[i], to Whse[j], obj Total_Cost dw_cost[i,j];
```

Figure 15-11: Specialized transshipment model with node and arc (net3node.mod).

This network has two kinds of arcs. There is an arc from the plant to each member of D_CITY, which can be declared by:

```
        arc PD_Ship {i in D_CITY} >= 0, <= pd_cap[i],
            from Plant, to Dist[i], obj Total_Cost pd_cost[i];
```

And there is an arc from distribution center i to warehouse j for each pair (i,j) in DW_LINKS:

```
        arc DW_Ship {(i,j) in DW_LINKS} >= 0, <= dw_cap[i,j],
            from Dist[i], to Whse[j], obj Total_Cost dw_cost[i,j];
```

The arc declarations specify the relevant bounds and objective coefficients, as in our previous example. The whole model is shown in Figure 15-11.

Variations on transshipment models

The balance conditions in node declarations may be inequalities, like ordinary algebraic balance constraints. If production at the plant can sometimes exceed total demand at the warehouses, it would be appropriate to give the condition in the declaration of node Plant as net_out <= p_supply.

An arc declaration can specify losses in transit by adding a factor at the end of the to phrase:

```
arc PD_Ship {i in D_CITY} >= 0, <= pd_cap[i],
    from Plant, to Dist[i] 1-pd_loss[i],
    obj Total_Cost pd_cost[i];
```

This is interpreted as saying that `PD_Ship[i]` is the number of packages that leave node `Plant`, but `(1-pd_loss[i]) * PD_Ship[i]` is the number that enter node `Dist[i]`.

The same option can be used to specify conversions. To use our previous example, if shipments are measured in thousands of packages but demands are measured in cartons, the arcs from distribution centers to warehouses should be declared as:

```
arc DW_Ship {(i,j) in DW_LINKS} >= 0, <= dw_cap[i,j],
    from Dist[i], to Whse[j] (1000/ppc),
    obj Total_Cost dw_cost[i,j];
```

If the shipments to warehouses are also measured in cartons, the factor should be applied at the distribution center:

```
arc DW_Ship {(i,j) in DW_LINKS} >= 0, <= dw_cap[i,j],
    from Dist[i] (ppc/1000), to Whse[j],
    obj Total_Cost dw_cost[i,j];
```

A loss factor could also be applied to the `to` phrase in these examples.

Maximum flow models

In the diagram of Figure 15-5 that we have used to illustrate the maximum flow problem, there are three kinds of intersections represented by nodes: the one where traffic enters, the one where traffic leaves, and the others where traffic flow is conserved. Thus a model of the network could have three corresponding node declarations:

```
node Entr_Int: net_out >= 0;
node Exit_Int: net_in >= 0;

node Intersection {k in INTER diff {entr,exit}};
```

The condition `net_out >= 0` implies that the flow out of node `Entr_Int` may be any amount at all; this is the proper condition, since there is no balance constraint on the entrance node. An analogous comment applies to the condition for node `Exit_Int`.

There is one arc in this network for each pair `(i,j)` in the set ROADS. Thus the declaration should look something like this:

```
arc Traff {(i,j) in ROADS} >= 0, <= cap[i,j],   # NOT RIGHT
    from Intersection[i], to Intersection[j],
    obj Entering_Traff (if i = entr then 1);
```

Since the aim is to maximize the total traffic leaving the entrance node, the arc is given a coefficient of 1 in the objective if and only if i takes the value `entr`. When i does take this value, however, the arc is specified to be from `Intersection[entr]`, a node that does not exist; the arc should rather be from node `Entr_Int`. Similarly, when j takes the value `exit`, the arc should not be to `Intersection[exit]`, but to

Exit_Int. AMPL will catch these errors and issue a message naming one of the nonexistent nodes that has been referenced.

It might seem reasonable to use an if-then-else to get around this problem, in the following way:

```
arc Traff {(i,j) in ROADS} >= 0, <= cap[i,j],   # SYNTAX ERROR
    from (if i = entr then Entr_Int else Intersection[i]),
    to   (if j = exit then Exit_Int else Intersection[j]),
    obj Entering_Traff (if i = entr then 1);
```

However, the if-then-else construct in AMPL does not apply to model components such as Entr_Int and Intersection[i]; this version will be rejected as a syntax error. Instead you need to use from and to phrases qualified by indexing expressions:

```
arc Traff {(i,j) in ROADS} >= 0, <= cap[i,j],
    from {if i = entr} Entr_Int,
    from {if i <> entr} Intersection[i],
    to   {if j = exit} Exit_Int,
    to   {if j <> exit} Intersection[j],
    obj Entering_Traff (if i = entr then 1);
```

The special indexing expression beginning with if works much the same way here as it does for constraints (Section 8.4); the from or to phrase is processed if the condition following if is true. Thus Traff[i,j] is declared to be from Entr_Int if i equals entr, and to be from Intersection[i] if i is not equal to entr, which is what we intend.

As an alternative, we can combine the declarations of the three different kinds of nodes into one. Observing that net_out is positive or zero for Entr_Int, negative or zero for Exit_Int, and zero for all other nodes Intersection[i], we can declare:

```
node Intersection {k in INTER}:
    (if k = exit then -Infinity)
        <= net_out <= (if k = entr then Infinity);
```

The nodes that were formerly declared as Entr_Int and Exit_Int are now just Intersection[entr] and Intersection[exit], and consequently the arc declaration that we previously marked ''not right'' now works just fine. The choice between this version and the previous one is entirely a matter of convenience and taste. (Infinity is a predefined AMPL parameter that may be used to specify any ''infinitely large'' bound; its technical definition is given in Section A.7.2.)

Arguably the AMPL formulation that is most convenient and appealing is neither of the above, but rather comes from interpreting the network diagram of Figure 15-5 in a slightly different way. Suppose that we view the arrows into the entrance node and out of the exit node as representing additional arcs, which happen to be adjacent to only one node rather than two. Then flow in equals flow out at every intersection, and the node declaration simplifies to:

```
node Intersection {k in INTER};
```

```
set INTER;  # intersections

param entr symbolic in INTER;            # entrance to road network
param exit symbolic in INTER, <> entr;  # exit from road network

set ROADS within (INTER diff {exit}) cross (INTER diff {entr});

param cap {ROADS} >= 0;  # capacities of roads

node Intersection {k in INTER};

arc Traff_In >= 0, to Intersection[entr];
arc Traff_Out >= 0, from Intersection[exit];

arc Traff {(i,j) in ROADS} >= 0, <= cap[i,j],
    from Intersection[i], to Intersection[j];

maximize Entering_Traff: Traff_In;

data;

set INTER := a b c d e f g ;

param entr := a ;
param exit := g ;

param:  ROADS:  cap :=
            a b     50,     a c     100
            b d     40,     b e      20
            c d     60,     c f      20
            d e     50,     d f      60
            e g     70,     f g      70 ;
```

Figure 15-12: Maximum flow model with node and arc (netmax3.mod).

The two arcs ''hanging'' at the entrance and exit are defined in the obvious way, but include only a to or a from phrase:

```
        arc Traff_In >= 0, to Intersection[entr];
        arc Traff_Out >= 0, from Intersection[exit];
```

The arcs that represent roads within the network are declared as before:

```
        arc Traff {(i,j) in ROADS} >= 0, <= cap[i,j],
            from Intersection[i], to Intersection[j];
```

When the model is represented in this way, the objective is to maximize Traff_In (or equivalently Traff_Out). We could do this by adding an obj phrase to the arc declaration for Traff_In, but in this case it is perhaps clearer to define the objective algebraically:

```
        maximize Entering_Traff: Traff_In;
```

This version is shown in full in Figure 15-12.

15.4 Rules for `node` and `arc` declarations

Having defined `node` and `arc` by example, we now describe more comprehensively the required and optional elements of these declarations, and comment on their interaction with the conventional declarations `minimize` or `maximize`, `subject to`, and `var` when both kinds appear in the same model.

node declarations

A `node` declaration begins with the keyword `node`, a name, an optional indexing expression, and a colon. The expression following the colon, which describes the balance condition at the node, may have any of the following forms:

> *net-expr* =   *arith-expr*
> *net-expr* <= *arith-expr*
> *net-expr* >= *arith-expr*
>
> *arith-expr* =   *net-expr*
> *arith-expr* <= *net-expr*
> *arith-expr* >= *net-expr*
>
> *arith-expr* <= *net-expr* <= *arith-expr*
> *arith-expr* >= *net-expr* >= *arith-expr*

where an *arith-expr* may be any arithmetic expression that uses previously declared model components and currently defined dummy indices. A *net-expr* is restricted to one of the following:

> ± `net_in`   ± `net_out`
> ± `net_in` + *arith-expr*   ± `net_out` + *arith-expr*
> *arith-expr* ± `net_in`   *arith-expr* ± `net_out`

(and a unary + may be omitted). Each node defined in this way induces a constraint in the resulting linear program. A node name is treated like a constraint name in the AMPL command environment, for example in a `display` statement.

For declarations that use `net_in`, AMPL generates the constraint by substituting, at the place where `net_in` appears in the balance conditions, a linear expression that represents flow into the node minus flow out of the node. Declarations that use `net_out` are handled the same way, except that AMPL substitutes flow out minus flow in. The expressions for flow in and flow out are deduced from the `arc` declarations.

arc declarations

An `arc` declaration consists of the keyword `arc`, a name, an optional indexing expression, and a series of optional qualifying phrases. Each arc creates a variable in the resulting linear program, whose value is the amount of flow over the arc; the arc name may be used to refer to this variable elsewhere. All of the phrases that may appear in a `var` definition have the same significance in an `arc` definition; most commonly, the >=

and <= phrases are used to specify values for lower and upper bounds on the flow along the arc.

The `from` and `to` phrases specify the nodes connected by an arc. Usually these consist of the keyword `from` or `to` followed by a node name. An arc is interpreted to contribute to the flow out of the `from` node, and to the flow into the `to` node; these interpretations are what permit the inference of the constraints associated with the nodes.

Typically one `from` and one `to` phrase are specified in an `arc` declaration. Either may be omitted, however, as in Figure 15-12. Either may also be followed by an optional indexing expression, which should be one of two kinds:

- An indexing expression that specifies — depending on the data — an empty set (in which case the `from` or `to` phrase is ignored) or a set with one member (in which case the `from` or `to` phrase is used).

- An indexing expression of the special form {if *logical-expr*}, which causes the `from` or `to` phrase to be used if and only if the *logical-expr* evaluates to true.

It is possible to specify that an arc carries flow out of or into two or more nodes, by giving more than one `from` or `to` phrase, or by using an indexing expression that specifies a set having more than one member. The result is not a network linear program, however, and AMPL displays an appropriate warning message.

At the end of a `from` or `to` phrase, you may add an arithmetic expression representing a factor to multiply the flow, as shown in our examples of shipping-loss and change-of-unit variations in Section 15.3. If the factor is in the `to` phrase, it multiplies the arc variable in determining the flow into the specified node; that is, for a given flow along the arc, an amount equal to the `to`-factor times the flow is considered to enter the `to` node. A factor in the `from` phrase is interpreted analogously. The default factor is 1.

An optional `obj` phrase specifies a coefficient that will multiply the arc variable to create a linear term in a specified objective function. Such a phrase consists of the keyword `obj`, the name of an objective that has previously been defined in a `minimize` or `maximize` declaration, and an arithmetic expression for the coefficient value. The keyword may be followed by an indexing expression, which is interpreted as for the `from` and `to` phrases.

Interaction with objective declarations

If all terms in the objective function are specified through `obj` phrases in `arc` declarations, the declaration of the objective is simply `minimize` or `maximize` followed by an optional indexing expression and a name. This declaration must come before the `arc` declarations that refer to the objective.

Alternatively, arc names may be used as variables to specify the objective function in the usual algebraic way. In this case the objective must be declared after the arcs, as in Figure 15-12.

```
set CITIES;
set LINKS within (CITIES cross CITIES);

set PRODS;

param supply {CITIES,PRODS} >= 0;  # amounts available at cities
param demand {CITIES,PRODS} >= 0;  # amounts required at cities
   check {p in PRODS}:
      sum {i in CITIES} supply[i,p] = sum {j in CITIES} demand[j,p];
param cost {LINKS,PRODS} >= 0;      # shipment costs/1000 packages
param capacity {LINKS,PRODS} >= 0; # max packages shipped
param cap_joint {LINKS} >= 0;       # max total packages shipped/link

minimize Total_Cost;

node Balance {k in CITIES, p in PRODS}:
   net_in = demand[k,p] - supply[k,p];

arc Ship {(i,j) in LINKS, p in PRODS} >= 0, <= capacity[i,j,p],
   from Balance[i,p], to Balance[j,p], obj Total_Cost cost[i,j,p];

subject to Multi {(i,j) in LINKS}:
   sum {p in PRODS} Ship[i,j,p] <= cap_joint[i,j];
```

Figure 15-13: Multicommodity flow with side constraints (`netmulti.mod`).

Interaction with constraint declarations

The components defined in `arc` declarations may be used as variables in additional `subject to` declarations. The latter represent ''side constraints'' that are imposed in addition to balance of flow at the nodes.

As an example, consider how a multicommodity flow problem can be built from the node-and-arc network formulation in Figure 15-10. Following the approach in Section 4.1, we introduce a set PRODS of different products, and add it to the indexing of all parameters, nodes and arcs. The result is a separate network linear program for each product, with the objective function being the sum of the costs for all products. To tie these networks together, we provide for a joint limit on the total shipments along any link:

```
param cap_joint {LINKS} >= 0;

subject to Multi {(i,j) in LINKS}:
   sum {p in PRODS} Ship[p,i,j] <= cap_joint[i,j];
```

The final model, shown in Figure 15-13, is not a network linear program, but the network and non-network parts of it are cleanly separated.

Interaction with variable declarations

Just as an `arc` variable may be used in a `subject to` declaration, an ordinary `var` variable may be used in a `node` declaration. That is, the balance condition in a `node` declaration may contain references to variables that were defined by preceding `var` declarations. These references define ''side variables'' to the network linear program.

As an example, we again replicate the formulation of Figure 15-10 over the set `PRODS`. This time we tie the networks together by introducing a set of feedstocks and associated data:

```
set FEEDS;
param yield {PRODS,FEEDS} >= 0;
param limit {FEEDS,CITIES} >= 0;
```

We imagine that at city `k`, in addition to the amounts `supply[p,k]` of products available to be shipped, up to `limit[f,k]` of feedstock `f` can be converted into products; one unit of feedstock `f` gives rise to `yield[p,f]` units of each product `p`. A variable `Feed[f,k]` represents the amount of feedstock `f` used at city `k`:

```
var Feed {f in FEEDS, k in CITIES} >= 0, <= limit[f,k];
```

The balance condition for product `p` at city `k` can now say that the net flow out equals net supply plus the sum of the amounts derived from the various feedstocks:

```
node Balance {p in PRODS, k in CITIES}:
   net_out = supply[p,k] - demand[p,k]
       + sum {f in FEEDS} yield[p,f] * Feed[f,k];
```

The arcs are unchanged, leading to the model shown in Figure 15-14. At a given city `k`, the variables `Feed[f,k]` appear in the node balance conditions for all the different products, bringing together the product networks into a single linear program.

15.5 Solving network linear programs

All of the models that we have described in this chapter give rise to linear programs that have a ''network'' property of some kind. AMPL can send these linear programs to an LP solver and retrieve the optimal values, much as for any other class of LPs. If you use AMPL in this way, the network structure is helpful mainly as a guide to formulating the model and interpreting the results.

Many of the models that we have described belong as well to a more restricted class of problems that (confusingly) are also known as ''network linear programs.'' In modeling terms, the variables of a network LP must represent flows on the arcs of a network, and the constraints must be only of two types: bounds on the flows along the arcs, and limits on flow out minus flow in at the nodes. A more technical way to say the same thing is that each variable of a network linear program must appear in at most two constraints (aside from lower or upper bounds on the variables), such that the variable has a coefficient of +1 in at most one constraint, and a coefficient of −1 in at most one constraint.

''Pure'' network linear programs of this restricted kind have some very strong properties that make their use particularly desirable. So long as the supplies, demands, and

```
set CITIES;

set LINKS within (CITIES cross CITIES);

set PRODS;

param supply {PRODS,CITIES} >= 0;   # amounts available at cities

param demand {PRODS,CITIES} >= 0;   # amounts required at cities
   check {p in PRODS}:
       sum {i in CITIES} supply[p,i] = sum {j in CITIES} demand[p,j];

param cost {PRODS,LINKS} >= 0;      # shipment costs/1000 packages
param capacity {PRODS,LINKS} >= 0;  # max packages shipped of product

set FEEDS;

param yield {PRODS,FEEDS} >= 0;     # amounts derived from feedstocks
param limit {FEEDS,CITIES} >= 0;    # feedstocks available at cities

minimize Total_Cost;

var Feed {f in FEEDS, k in CITIES} >= 0, <= limit[f,k];

node Balance {p in PRODS, k in CITIES}:
   net_out = supply[p,k] - demand[p,k]
       + sum {f in FEEDS} yield[p,f] * Feed[f,k];

arc Ship {p in PRODS, (i,j) in LINKS} >= 0, <= capacity[p,i,j],
   from Balance[p,i], to Balance[p,j],
   obj Total_Cost cost[p,i,j];
```

Figure 15-14: Multicommodity flow with side variables (`netfeeds.mod`).

bounds are integers, a network linear program must have an optimal solution in which all flows are integers. Moreover, if the solver is of a kind that finds "extreme" solutions (such as those based on the simplex method) it will always find one of the all-integer optimal solutions. We have taken advantage of this property, without explicitly mentioning it, in assuming that the variables in the shortest path problem and in certain assignment problems come out to be either zero or one, and never some fraction in between.

Network linear programs can also be solved much faster than other linear programs of comparable size, through the use of solvers that are specialized to take advantage of the network structure. If you write your model in terms of `node` and `arc` declarations, AMPL automatically communicates the network structure to the solver, and any special network algorithms available in the solver can be applied automatically. On the other hand, a network expressed algebraically using `var` and `subject to` may or may not be recognized by the solver, and certain options may have to be set to ensure that it is recognized. For example, when using the algebraic model of Figure 15-4a, you may see the usual response from the general LP algorithm:

```
ampl: model net3.mod; data net3.dat; solve;
CPLEX 8.0.0: optimal solution; objective 1819
1 dual simplex iterations (0 in phase I)
```

But when using the equivalent node and arc formulation of Figure 15-11, you may get a somewhat different response to reflect the application of a special network LP algorithm:

```
ampl: model net3node.mod
ampl: data net3.dat
ampl: solve;
CPLEX 8.0.0: optimal solution; objective 1819
Network extractor found 7 nodes and 7 arcs.
7 network simplex iterations.
0 simplex iterations (0 in phase I)
```

To determine how your favorite solver behaves in this situation, consult the solver-specific documentation that is supplied with your AMPL installation.

Because network linear programs are much easier to solve, especially with integer data, the success of a large-scale application may depend on whether a pure network formulation is possible. In the case of the multicommodity flow model of Figure 15-13, for example, the joint capacity constraints disrupt the network structure — they represent a third constraint in which each variable figures — but their presence cannot be avoided in a correct representation of the problem. Multicommodity flow problems thus do not necessarily have integer solutions, and are generally much harder to solve than single-commodity flow problems of comparable size.

In some cases, a judicious reformulation can turn what appears to be a more general model into a pure network model. Consider, for instance, a generalization of Figure 15-10 in which capacities are defined at the nodes as well as along the arcs:

```
param city_cap {CITIES} >= 0;
param link_cap {LINKS} >= 0;
```

The arc capacities represent, as before, upper limits on the shipments between cities. The node capacities limit the throughput, or total flow handled at a city, which may be written as the supply at the city plus the sum of the flows in, or equivalently as the demand at the city plus the sum of the flows out. Using the former, we arrive at the following constraint:

```
subject to through_limit {k in CITIES}:
    supply[k] + sum {(i,k) in LINKS} Ship[i,k] <= node_cap[k];
```

Viewed in this way, the throughput limit is another example of a "side constraint" that disrupts the network structure by adding a third coefficient for each variable. But we can achieve the same effect without a side constraint, by using two nodes to represent each city; one receives flow into a city plus any supply, and the other sends flow out of a city plus any demand:

```
node Supply {k in CITIES}: net_out = supply[k];
node Demand {k in CITIES}: net_in = demand[k];
```

A shipment link between cities i and j is represented by an arc that connects the node Demand[i] to node Supply[j]:

```
set CITIES;
set LINKS within (CITIES cross CITIES);

param supply {CITIES} >= 0;   # amounts available at cities
param demand {CITIES} >= 0;   # amounts required at cities

   check: sum {i in CITIES} supply[i] = sum {j in CITIES} demand[j];

param cost {LINKS} >= 0;        # shipment costs per ton

param city_cap {CITIES} >= 0; # max throughput at cities
param link_cap {LINKS} >= 0;  # max shipment over links

minimize Total_Cost;

node Supply {k in CITIES}: net_out = supply[k];
node Demand {k in CITIES}: net_in = demand[k];

arc Ship {(i,j) in LINKS} >= 0, <= link_cap[i,j],
   from Demand[i], to Supply[j], obj Total_Cost cost[i,j];

arc Through {k in CITIES} >= 0, <= city_cap[k],
   from Supply[k], to Demand[k];
```

Figure 15-15: Transshipment model with node capacities (netthru.mod).

```
        arc Ship {(i,j) in LINKS} >= 0, <= link_cap[i,j],
            from Demand[i], to Supply[j], obj Total_Cost cost[i,j];
```

The throughput at city k is represented by a new kind of arc, from Supply[k] to Demand[k]:

```
        arc Through {k in cities} >= 0, <= city_cap[k],
            from Supply[k], to Demand[k];
```

The throughput limit is now represented by an upper bound on this arc's flow, rather than by a side constraint, and the network structure of the model is preserved. A complete listing appears in Figure 15-15.

The preceding example exhibits an additional advantage of using the node and arc declarations when developing a network model. If you use only node and arc in their simple forms — no variables in the node conditions, and no optional factors in the from and to phrases — your model is guaranteed to give rise only to pure network linear programs. By contrast, if you use var and subject to, it is your responsibility to ensure that the resulting linear program has the necessary network structure.

Some of the preceding comments can be extended to "generalized network" linear programs in which each variable still figures in at most two constraints, but not necessarily with coefficients of +1 and –1. We have seen examples of generalized networks in the cases where there is a loss of flow or change of units on the arcs. Generalized network LPs do not necessarily have integer optimal solutions, but fast algorithms for them do exist. A solver that promises a "network" algorithm may or may not have an extension to generalized networks; check the solver-specific documentation before you make any assumptions.

Bibliography

Ravindra K. Ahuja, Thomas L. Magnanti and James B. Orlin, *Network Flows: Theory, Algorithms, and Applications.* Prentice-Hall (Englewood Cliffs, NJ, 1993).

Dimitri P. Bertsekas *Network Optimization: Continuous and Discrete Models.* Athena Scientific (Princeton, NJ, 1998).

L. R. Ford, Jr. and D. R. Fulkerson, *Flows in Networks.* Princeton University Press (Princeton, NJ, 1962). A highly influential survey of network linear programming and related topics, which stimulated much subsequent study.

Fred Glover, Darwin Klingman and Nancy V. Phillips, *Network Models in Optimization and their Applications in Practice.* John Wiley & Sons (New York, 1992).

Walter Jacobs, "The Caterer Problem." Naval Research Logistics Quarterly **1** (1954) pp. 154–165. The origin of the network problem described in Exercise 15-8.

Katta G. Murty, *Network Programming.* Prentice-Hall (Englewood Cliffs, NJ, 1992).

Exercises

15-1. The following diagram can be interpreted as representing a network transshipment problem:

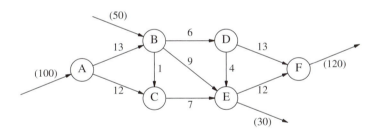

The arrows into nodes A and B represent supply in the indicated amounts, 100 and 50; the arrows out of nodes E and F similarly represent demand in the amounts 30 and 120. The remaining arrows indicate shipment possibilities, and the numbers on them are the unit shipping costs. There is a capacity of 80 on every arc.

(a) Solve this problem by developing appropriate data statements to go along with the model of Figure 15-2a.

(b) Check that you get the same solution using the node and arc formulation of Figure 15-10. Does the solver appear to be using the same algorithm as in (a)? Try this comparison with each LP solver available to you.

15-2. Reinterpret the numbers on the arcs *between* nodes, in the diagram of the preceding exercise, to solve the following problems. (Ignore the numbers on the arrows into A and B and on the arrows out of E and F.)

(a) Regarding the numbers on the arcs between nodes as lengths, use a model such as the one in Figure 15-7 to find the shortest path from A to F.

(b) Regarding the numbers on the arcs between nodes as capacities, use a model such as the one in Figure 15-6 to find the maximum flow from A to F.

(c) Generalize the model from (b) so that it can find the maximum flow from any subset of nodes to any other subset of nodes, in some meaningful sense. Use your generalization to find the maximum flow from A and B to E and F.

15-3. Section 4.2 showed how a multiperiod model could be constructed by replicating a static model over time periods, and using inventories to tie the periods together. Consider applying the same approach to the specialized transshipment model of Figure 15-4a.

(a) Construct a multi-week version of Figure 15-4a, with the inventories kept at the distribution centers (the members of D_CITY).

(b) Now consider expanding the data of Figure 15-4b for this model. Suppose that the initial inventory is 200 at NE and 75 at SE, and that the inventory carrying cost per 1000 packages is 0.15 per week at NE and 0.12 per week at SE. Let the supplies and demands over 5 weeks be as follows:

		Demand				
Week	Supply	BOS	EWR	BWI	ATL	MCO
1	450	50	90	95	50	20
2	450	65	100	105	50	20
3	400	70	100	110	50	25
4	250	70	110	120	50	40
5	325	80	115	120	55	45

Leave the cost and capacity data unchanged, the same in all weeks. Develop an appropriate data file for this situation, and solve the multi-week problem.

(c) The multi-week model in this case can be viewed as a pure network model, with arcs representing inventories as well as shipments. To show that this is the case, reformulate the model from (a) using only node and arc declarations for the constraints and variables.

15-4. For each of the following network problems, construct a data file that permits it to be solved using the general transshipment model of Figure 15-2a.

(a) The transportation problem of Figure 3-1.

(b) The assignment problem of Figure 3-2.

(c) The maximum flow problem of Figure 15-6.

(d) The shortest path problem of Figure 15-7.

15-5. Reformulate each of the following network models using node and arc declarations as much as possible:

(a) The transportation model of Figure 6-2a.

(b) The shortest path model of Figure 15-7.

(c) The production/transportation model of Figure 4-6.

(d) The multicommodity transportation model of Figure 6-5.

15-6. The professor in charge of an industrial engineering design course is faced with the problem of assigning 28 students to eight projects. Each student must be assigned to one project, and each project group must have 3 or 4 students. The students have been asked to rank the projects, with 1 being the best ranking and higher numbers representing lower rankings.

(a) Formulate an algebraic assignment model, using `var` and `subject to` declarations, for this problem.

(b) Solve the assignment problem for the following table of rankings:

	A	ED	EZ	G	H1	H2	RB	SC		A	ED	EZ	G	H1	H2	RB	SC
Allen	1	3	4	7	7	5	2	6	Knorr	7	4	1	2	2	5	6	3
Black	6	4	2	5	5	7	1	3	Manheim	4	7	2	1	1	3	6	5
Chung	6	2	3	1	1	7	5	4	Morris	7	5	4	6	6	3	1	2
Clark	7	6	1	2	2	3	5	4	Nathan	4	7	5	6	6	3	1	2
Conners	7	6	1	3	3	4	5	2	Neuman	7	5	4	6	6	3	1	2
Cumming	6	7	4	2	2	3	5	1	Patrick	1	7	5	4	4	2	3	6
Demming	2	5	4	6	6	1	3	7	Rollins	6	2	3	1	1	7	5	4
Eng	4	7	2	1	1	6	3	5	Schuman	4	7	3	5	5	1	2	6
Farmer	7	6	5	2	2	1	3	4	Silver	4	7	3	1	1	2	5	6
Forest	6	7	2	5	5	1	3	4	Stein	6	4	2	5	5	7	1	3
Goodman	7	6	2	4	4	5	1	3	Stock	5	2	1	6	6	7	4	3
Harris	4	7	5	3	3	1	2	6	Truman	6	3	2	7	7	5	1	4
Holmes	6	7	4	2	2	3	5	1	Wolman	6	7	4	2	2	3	5	1
Johnson	7	2	4	6	6	5	3	1	Young	1	3	4	7	7	6	2	5

How many students are assigned second or third choice?

(c) Some of the projects are harder than others to reach without a car. Thus it is desirable that at least a certain number of students assigned to each project must have a car; the numbers vary by project as follows:

        A 1   ED 0   EZ 0   G 2   H1 2   H2 2   RB 1   SC 1

The students who have cars are:

        Chung   Eng   Manheim   Nathan   Rollins
        Demming Holmes   Morris   Patrick   Young

Modify the model to add this car constraint, and solve the problem again. How many more students than before must be assigned second or third choice?

(d) Your formulation in (c) can be viewed as a transportation model with side constraints. By defining appropriate network nodes and arcs, reformulate it as a "pure" network flow model, as discussed in Section 15.5. Write the formulation in AMPL using only `node` and `arc` declarations for the constraints and variables. Solve with the same data as in (c), to show that the optimal value is the same.

15-7. To manage its excess cash over the next 12 months, a company may purchase 1-month, 2-month or 3-month certificates of deposit from any of several different banks. The current cash on hand and amounts invested are known, while the company must estimate the cash receipts and expenditures for each month, and the returns on the different certificates.

The company's problem is to determine the best investment strategy, subject to cash requirements. (As a practical matter, the company would use the first month of the optimal solution as a guide to its current purchases, and then re-solve with updated estimates at the beginning of the next month.)

(a) Draw a network diagram for this situation. Show each month as a node, and the investments, receipts and expenditures as arcs.

(b) Formulate the relevant optimization problem as an AMPL model, using `node` and `arc` declarations. Assume that any cash from previously-purchased certificates coming due in the early months is included in data for the receipts.

There is more than one way to describe the objective function for this model. Explain your choice.

(c) Suppose that the company's estimated receipts and expenses (in thousands of dollars) over the next 12 months are as follows:

	receipt	expense
1	3200	200
2	3600	200
3	3100	400
4	1000	800
5	1000	2100
6	1000	4500
7	1200	3300
8	1200	1800
9	1200	600
10	1500	200
11	1800	200
12	1900	200

The two banks competing for the business are estimating the following rates of return for the next 12 months:

CIT:	1	2	3	NBD:	1	2	3
1	0.00433	0.01067	0.01988	1	0.00425	0.01067	0.02013
2	0.00437	0.01075	0.02000	2	0.00429	0.01075	0.02025
3	0.00442	0.01083	0.02013	3	0.00433	0.01083	0.02063
4	0.00446	0.01092	0.02038	4	0.00437	0.01092	0.02088
5	0.00450	0.01100	0.02050	5	0.00442	0.01100	0.02100
6	0.00458	0.01125	0.02088	6	0.00450	0.01125	0.02138
7	0.00467	0.01142	0.02113	7	0.00458	0.01142	0.02162
8	0.00487	0.01183	0.02187	8	0.00479	0.01183	0.02212
9	0.00500	0.01217	0.02237	9	0.00492	0.01217	0.02262
10	0.00500	0.01217	0.02250	10	0.00492	0.01217	0.02275
11	0.00492	0.01217	0.02250	11	0.00483	0.01233	0.02275
12	0.00483	0.01217	0.02275	12	0.00475	0.01250	0.02312

Construct an appropriate data file, and solve the resulting linear program. Use `display` to produce a summary of the indicated purchases.

(d) Company policy prohibits investing more than 70% of its cash in new certificates of any one bank in any month. Devise a side constraint on the model from (b) to impose this restriction.

Again solve the resulting linear program, and summarize the indicated purchases. How much income is lost due to the restrictive policy?

15-8. A caterer has booked dinners for the next T days, and has as a result a requirement for a certain number of napkins each day. He has a certain initial stock of napkins, and can buy new ones each day at a certain price. In addition, used napkins can be laundered either at a slow service that takes 4 days, or at a faster but more expensive service that takes 2 days. The caterer's problem is to find the most economical combination of purchase and laundering that will meet the forthcoming demand.

(a) It is not hard to see that the decision variables for this problem should be something like the following:

Buy[t]	clean napkins bought for day t
Carry[t]	clean napkins still on hand at the end of day t
Wash2[t]	used napkins sent to the fast laundry after day t
Wash4[t]	used napkins sent to the slow laundry after day t
Trash[t]	used napkins discarded after day t

There are two collections of constraints on these variables, which can be described as follows:

– The number of clean napkins acquired through purchase, carryover and laundering on day t must equal the number sent to laundering, discarded or carried over after day t.

– The number of used napkins laundered or discarded after day t must equal the number that were required for that day's catering.

Formulate an AMPL linear programming model for this problem.

(b) Formulate an alternative network linear programming model for this problem. Write it in AMPL using node and arc declarations.

(c) The ''caterer problem'' was introduced in a 1954 paper by Walter Jacobs of the U.S. Air Force. Although it has been presented in countless books on linear and network programming, it does not seem to have ever been used by any caterer. In what application do you suppose it really originated?

(d) Since this is an artificial problem, you might as well make up your own data for it. Use your data to check that the formulations in (a) and (b) give the same optimal value.

16

Columnwise Formulations

Because the fundamental idea of an optimization problem is to minimize or maximize a function of the decision variables, subject to constraints on them, AMPL is oriented toward explicit descriptions of variables and constraints. This is why `var` declarations tend to come first, followed by the `minimize` or `maximize` and `subject to` declarations that use the variables. A wide variety of optimization problems can be formulated with this approach, as the examples in this book demonstrate.

For certain kinds of linear programs, however, it can be preferable to declare the objective and constraints before the variables. Usually in these cases, there is a much simpler pattern or interpretation to the coefficients of a single variable down all the constraints than to the coefficients of a single constraint across all the variables. In the jargon of linear programming, it is easier to describe the matrix of constraint coefficients "columnwise" than "row-wise". As a result, the formulation is simplified by first declaring the constraints and objective, then listing the nonzero coefficients in the declarations of the variables.

One example of this phenomenon is provided by the network linear programs described in Chapter 15. Each variable has at most two nonzero coefficients in the constraints, a +1 and a −1. Rather than trying to describe the constraints algebraically, you may find it easier to specify, in each variable's declaration, the one or two constraints that the variable figures in. In fact, this is exactly what you do by using the special `node` and `arc` declarations introduced by Section 15.3. The `node` declarations come first to describe the nature of the constraints at the network nodes. Then the `arc` declarations define the network flow variables, using `from` and `to` phrases to locate their nonzero coefficients among the node constraints. This approach is particularly appealing because it corresponds directly to the way most people think about a network flow problem.

It would be impractical for AMPL to offer special declarations and phrases for every kind of linear program that you might want to declare by columns rather than by rows. Instead, additional options to the `var` and `subject to` declarations permit any linear program to be given a columnwise declaration. This chapter introduces AMPL's columnwise features through two contrasting examples — an input-output production model, and a work-shift scheduling model — and concludes with a summary of the language extensions that may be used in columnwise formulations.

16.1 An input-output model

In simple maximum-profit production models such as the examples in Chapter 1, the goods produced are distinct from the resources consumed, so that overall production is limited in an obvious way by resources available. In a more realistic model of a complex operation such as a steel mill or refinery, however, production is carried out at a series of units; as a result, some of a production unit's inputs may be the outputs from other units. For this situation we need a model that deals more generally with materials that may be inputs or outputs, and with production activities that may involve several inputs and outputs each.

We begin by developing an AMPL formulation in the usual row-wise (or constraint-oriented) way. Then we explain the columnwise (or variable-oriented) alternative, and discuss refinements of the model.

Formulation by constraints

The definition of our model starts with a set of materials and a set of activities:

```
set MAT;
set ACT;
```

The key data values are the input-output coefficients for all material-activity combinations:

```
param io {MAT,ACT};
```

If $io[i,j] > 0$, it is interpreted as the amount of material i produced (as an output) by a unit of activity j. On the other hand, if $io[i,j] < 0$, it represents minus the amount of material i consumed (as an input) by a unit of activity j. For example, a value of 10 represents 10 units of i produced per unit of j, while a value of -10 represents 10 units consumed per unit of j. Of course, we can expect that for many combinations of i and j we will have $io[i,j] = 0$, signifying that material i does not figure in activity j at all.

To see why we want to interpret $io[i,j]$ in this manner, suppose we define $Run[j]$ to be the level at which we operate (run) activity j:

```
param act_min {ACT} >= 0;
param act_max {j in ACT} >= act_min[j];

var Run {j in ACT} >= act_min[j], <= act_max[j];
```

Then $io[i,j] * Run[j]$ is the total amount of material i produced (if $io[i,j] > 0$) or minus the amount of material i consumed (if $io[i,j] < 0$) by activity j. Summing over all activities, we see that

```
set MAT;                # materials
set ACT;                # activities
param io {MAT,ACT};     # input-output coefficients

param revenue {ACT};
param act_min {ACT} >= 0;
param act_max {j in ACT} >= act_min[j];

var Run {j in ACT} >= act_min[j], <= act_max[j];

maximize Net_Profit:  sum {j in ACT} revenue[j] * Run[j];

subject to Balance {i in MAT}:
   sum {j in ACT} io[i,j] * Run[j] = 0;
```

Figure 16-1: Input-output model by rows (`iorow.mod`).

```
   sum {j in ACT} io[i,j] * Run[j]
```

represents the amount of material `i` produced in the operation minus the amount consumed. These amounts must balance, as expressed by the following constraint:

```
   subject to Balance {i in MAT}:
      sum {j in ACT} io[i,j] * Run[j] = 0;
```

What about the availability of resources, or the requirements for finished goods? These are readily modeled through additional activities that represent the purchase or sale of materials. A purchase activity for material `i` has no inputs and just `i` as an output; the upper limit on `Run[i]` represents the amount of this resource available. Similarly, a sale activity for material `i` has no outputs and just `i` as an input, and the lower limit on `Run[i]` represents the amount of this good that must be produced for sale.

We complete the model by associating unit revenues with the activities. Sale activities necessarily have positive revenues, while purchase and production activities have negative revenues — that is, costs. The sum of unit revenues times activity levels gives the total net profit of the operation:

```
   param revenue {ACT};
   maximize Net_Profit: sum {j in ACT} revenue[j] * Run[j];
```

The completed model is shown in Figure 16-1.

A columnwise formulation

As our discussion of purchase and sale activities suggests, everything in this model can be organized by activity. Specifically, for each activity `j` we have a decision variable `Run[j]`, a cost or income represented by `revenue[j]`, limits `act_min[j]` and `act_max[j]`, and a collection of input-output coefficients `io[i,j]`. Changes such as improving the yield of a unit, or acquiring a new source of supply, are accommodated by adding an activity or by modifying the data for an activity.

In the formulation by rows, the activities' importance to this model is somewhat hidden. While act_min[j] and act_max[j] appear in the declaration of the variables, revenue[j] is in the objective, and the io[i,j] values are in the constraint declaration. The columnwise alternative brings all of this information together, by adding obj and coeff phrases to the var declaration:

```
var Run {j in ACT} >= act_min[j], <= act_max[j],
   obj Net_Profit revenue[j],
   coeff {i in MAT} Balance[i] io[i,j];
```

The obj phrase says that in the objective function named Net_Profit, the variable Run[j] has the coefficient revenue[j]; that is, the term revenue[j] * Run[j] should be added in. The coeff phrase is a bit more complicated, because it is indexed over a set. It says that for each material i, in the constraint Balance[i] the variable Run[j] should have the coefficient io[i,j], so that the term io[i,j] * Run[j] is added in. Together, these phrases describe all the coefficients of all the variables in the linear program.

Since we have placed all the coefficients in the var declaration, we must remove them from the other declarations:

```
maximize Net_Profit;
subject to Balance {i in MAT}: to_come = 0;
```

The keyword to_come indicates where the terms io[i,j] * Run[j] generated by the var declaration are to be "added in." You can think of to_come = 0 as a template for the constraint, which will be filled out as the coefficients are declared. No template is needed for the objective in this example, however, since it is exclusively the sum of the terms revenue[j] * Run[j]. Templates may be written in a limited variety of ways, as shown in Section 16.3 below.

Because the obj and coeff phrases refer to Net_Profit and Balance, the var declaration must come after the maximize and subject to declarations in the columnwise formulation. The complete model is shown in Figure 16-2.

```
set MAT;              # materials
set ACT;              # activities
param io {MAT,ACT};   # input-output coefficients

param revenue {ACT};
param act_min {ACT} >= 0;
param act_max {j in ACT} >= act_min[j];

maximize Net_Profit;

subject to Balance {i in MAT}: to_come = 0;

var Run {j in ACT} >= act_min[j], <= act_max[j],
   obj Net_Profit revenue[j],
   coeff {i in MAT} Balance[i] io[i,j];
```

Figure 16-2: Columnwise formulation (iocol1.mod).

Refinements of the columnwise formulation

The advantages of a columnwise approach become more evident as the model becomes more complicated. As one example, consider what happens if we want to have separate variables to represent sales of finished materials. We declare a subset of materials that can be sold, and use it to index new collections of bounds, revenues and variables:

```
set MATF within MAT;    # finished materials

param revenue {MATF} >= 0;

param sell_min {MATF} >= 0;
param sell_max {i in MATF} >= sell_min[i];

var Sell {i in MATF} >= sell_min[i], <= sell_max[i];
```

We may now dispense with the special sale activities previously described. Since the remaining members of ACT represent purchase or production activities, we can introduce a nonnegative parameter cost associated with them:

```
param cost {ACT} >= 0;
```

In the row-wise approach, the new objective is written as

```
maximize Net_Profit:
    sum {i in MATF} revenue[i] * Sell[i]
    - sum {j in ACT} cost[j] * Run[j];
```

to represent total sales revenue minus total raw material and production costs.

So far we seem to have improved upon the model in Figure 16-1. The composition of net profit is more clearly modeled, and sales are restricted to explicitly designated finished materials; also the optimal amounts sold are more easily examined apart from the other variables, by a command such as display Sell. It remains to fix up the constraints. We would like to say that the net output of material i from all activities, represented as

```
sum {j in ACT} io[i,j] * Run[j]
```

in Figure 16-1, must balance the amount sold — either Sell[i] if i is a finished material, or zero. Thus the constraint declaration must be written:

```
subject to Balance {i in MAT}:
    sum {j in ACT} io[i,j] * Run[j]
        = if i in MATF then Sell[i] else 0;
```

Unfortunately this constraint seems less clear than our original one, due to the complication introduced by the if-then-else expression.

In the columnwise alternative, the objective and constraints are the same as in Figure 16-2, while all the changes are reflected in the declarations of the variables:

```
set MAT;              # materials
set ACT;              # activities

param io {MAT,ACT};   # input-output coefficients

set MATF within MAT; # finished materials

param revenue {MATF} >= 0;

param sell_min {MATF} >= 0;
param sell_max {i in MATF} >= sell_min[i];

param cost {ACT} >= 0;
param act_min {ACT} >= 0;
param act_max {j in ACT} >= act_min[j];

maximize Net_Profit;

subject to Balance {i in MAT}: to_come = 0;

var Run {j in ACT} >= act_min[j], <= act_max[j],
   obj Net_Profit -cost[j],
   coeff {i in MAT} Balance[i] io[i,j];

var Sell {i in MATF} >= sell_min[i], <= sell_max[i],
   obj Net_Profit revenue[i],
   coeff Balance[i] -1;
```

Figure 16-3: Columnwise formulation, with sales activities (iocol2.mod).

```
var Run {j in ACT} >= act_min[j], <= act_max[j],
   obj Net_Profit -cost[j],
   coeff {i in MAT} Balance[i] io[i,j];


var Sell {i in MATF} >= sell_min[i], <= sell_max[i],
   obj Net_Profit revenue[i],
   coeff Balance[i] -1;
```

In this view, the variable Sell[i] represents the kind of sale activity that we previously described, with only material i as input and no materials as output — hence the single coefficient of –1 in constraint Balance[i]. We need not specify all the zero coefficients for Sell[i]; a zero is assumed for any constraint not explicitly cited in a coeff phrase in the declaration. The whole model is shown in Figure 16-3.

This example suggests that a columnwise approach is particularly suited to refinements of the input-output model that distinguish different kinds of activities. It would be easy to add another group of variables that represent purchases of raw materials, for instance.

On the other hand, versions of the input-output model that involve numerous specialized constraints would lend themselves more to a formulation by rows.

16.2 A scheduling model

In Section 2.4 we observed that the general form of a blending model was applicable to certain scheduling problems. Here we describe a related scheduling model for which the columnwise approach is particularly attractive.

Suppose that a factory's production for the next week is divided into fixed time periods, or shifts. You want to assign employees to shifts, so that the required number of people are working on each shift. You cannot fill each shift independently of the others, however, because only certain weekly schedules are allowed; for example, a person cannot work five shifts in a row. Your problem is thus more properly viewed as one of assigning employees to schedules, so that each shift is covered and the overall assignment is the most economical.

We can conveniently represent the schedules for this problem by an indexed collection of subsets of shifts:

```
set SHIFTS;                  # shifts

param Nsched;                # number of schedules;
set SCHEDS = 1..Nsched;      # set of schedules

set SHIFT_LIST {SCHEDS} within SHIFTS;
```

For each schedule j in the set SCHEDS, the shifts that a person works on schedule j are contained in the set SHIFT_LIST[j]. We also specify a pay rate per person on each schedule, and the number of people required on each shift:

```
param rate {SCHEDS} >= 0;
param required {SHIFTS} >= 0;
```

We let the variable Work[j] represent the number of people assigned to work each schedule j, and minimize the sum of rate[j] * Work[j] over all schedules:

```
var Work {SCHEDS} >= 0;
minimize Total_Cost: sum {j in SCHEDS} rate[j] * Work[j];
```

Finally, our constraints say that the total of employees assigned to each shift i must be at least the number required:

```
subject to Shift_Needs {i in SHIFTS}:
   sum {j in SCHEDS: i in SHIFT_LIST[j]} Work[j]
      >= required[i];
```

On the left we take the sum of Work[j] over all schedules j such that i is in SHIFT_LIST[j]. This sum represents the total employees who are assigned to schedules that contain shift i, and hence equals the total employees covering shift i.

The awkward description of the constraint in this formulation motivates us to try a columnwise formulation. As in our previous examples, we declare the objective and constraints first, but with the variables left out:

```
minimize Total_Cost;
subject to Shift_Needs {i in SHIFTS}: to_come >= required[i];
```

The coefficients of `Work[j]` appear instead in its `var` declaration. In the objective, it has a coefficient of `rate[j]`. In the constraints, the membership of `SHIFT_LIST[j]` tells us exactly what we need to know: `Work[j]` has a coefficient of 1 in constraint `Shift_Needs[i]` for each `i` in `SHIFT_LIST[j]`, and a coefficient of 0 in the other constraints. This leads us to the following concise declaration:

```
var Work {j in SCHEDS} >= 0,
   obj Total_Cost rate[j],
   coeff {i in SHIFT_LIST[j]} Shift_Needs[i] 1;
```

The full model is shown in Figure 16-4.

As a specific instance of this model, imagine that you have three shifts a day on Monday through Friday, and two shifts on Saturday. Each day you need 100, 78 and 52 employees on the first, second and third shifts, respectively. To keep things simple, suppose that the cost per person is the same regardless of schedule, so that you may just minimize the total number of employees by setting `rate[j]` to 1 for all `j`.

As for the schedules, a reasonable scheduling rule might be that each employee works five shifts in a week, but never more than one shift in any 24-hour period. Part of the data file is shown in Figure 16-5; we don't show the whole file, because there are 126 schedules that satisfy the rule! The resulting 126-variable linear program is not hard to solve, however:

```
ampl: model sched.mod; data sched.dat; solve;
MINOS 5.5: optimal solution found.
19 iterations, objective 265.6

ampl: option display_eps .000001;
ampl: option omit_zero_rows 1;
ampl: option display_1col 0, display_width 60;

ampl: display Work;
Work [*] :=
 10 28.8     30 14.4     71 35.6    106 23.2    123 35.6
 18  7.6     35  6.8     73 28      109 14.4
 24  6.8     66 35.6     87 14.4    113 14.4
 ;
```

As you can see, this optimal solution makes use of 13 of the schedules, some in fractional amounts. (There exist many other optimal solutions to this problem, so the results you get may differ.) If you round each fraction in this solution up to the next highest value, you get a pretty good feasible solution using 271 employees. To determine whether this is the best whole-number solution, however, it is necessary to use integer programming techniques, which are the subject of Chapter 20.

The convenience of the columnwise formulation in this case follows directly from how we have chosen to represent the data. We imagine that the modeler will be thinking in terms of schedules, and will want to try adding, dropping or modifying different schedules to see what solutions can be obtained. The subsets `SHIFT_LIST[j]` provide a convenient and concise way of maintaining the schedules in the data. Since the data are then organized by schedules, and there is also a variable for each schedule, it proves to be

```
set SHIFTS;                  # shifts

param Nsched;                # number of schedules;
set SCHEDS = 1..Nsched;      # set of schedules

set SHIFT_LIST {SCHEDS} within SHIFTS;

param rate {SCHEDS} >= 0;
param required {SHIFTS} >= 0;

minimize Total_Cost;
subject to Shift_Needs {i in SHIFTS}: to_come >= required[i];

var Work {j in SCHEDS} >= 0,
   obj Total_Cost rate[j],
   coeff {i in SHIFT_LIST[j]} Shift_Needs[i] 1;
```

Figure 16-4: Columnwise scheduling model (`sched.mod`).

```
set SHIFTS := Mon1 Tue1 Wed1 Thu1 Fri1 Sat1
              Mon2 Tue2 Wed2 Thu2 Fri2 Sat2
              Mon3 Tue3 Wed3 Thu3 Fri3 ;

param Nsched := 126 ;

set SHIFT_LIST[  1] := Mon1 Tue1 Wed1 Thu1 Fri1 ;
set SHIFT_LIST[  2] := Mon1 Tue1 Wed1 Thu1 Fri2 ;
set SHIFT_LIST[  3] := Mon1 Tue1 Wed1 Thu1 Fri3 ;
set SHIFT_LIST[  4] := Mon1 Tue1 Wed1 Thu1 Sat1 ;
set SHIFT_LIST[  5] := Mon1 Tue1 Wed1 Thu1 Sat2 ;
```

(117 lines omitted)

```
set SHIFT_LIST[123] := Tue1 Wed1 Thu1 Fri2 Sat2 ;
set SHIFT_LIST[124] := Tue1 Wed1 Thu2 Fri2 Sat2 ;
set SHIFT_LIST[125] := Tue1 Wed2 Thu2 Fri2 Sat2 ;
set SHIFT_LIST[126] := Tue2 Wed2 Thu2 Fri2 Sat2 ;

param rate  default 1 ;

param required := Mon1 100   Mon2 78   Mon3 52
                  Tue1 100   Tue2 78   Tue3 52
                  Wed1 100   Wed2 78   Wed3 52
                  Thu1 100   Thu2 78   Thu3 52
                  Fri1 100   Fri2 78   Fri3 52
                  Sat1 100   Sat2 78 ;
```

Figure 16-5: Partial data for scheduling model (`sched.dat`).

simpler — and for larger examples, more efficient — to specify the coefficients by variable.

Models of this kind are used for a variety of scheduling problems. As a convenience, the keyword `cover` may be used (in the manner of `from` and `to` for networks) to specify a coefficient of 1:

```
        var Work {j in SCHEDS} >= 0,
           obj Total_Cost rate[j],
           cover {i in SHIFT_LIST[j]} Shift_Needs[i];
```

Some of the best known and largest examples are in airline crew scheduling, where the variables may represent the assignment of crews rather than individuals, the shifts become flights, and the requirement is one crew for each flight. We then have what is known as a set covering problem, in which the objective is to most economically cover the set of all flights with subsets representing crew schedules.

16.3 Rules for columnwise formulations

The algebraic description of an AMPL constraint can be written in any of the following ways:

> *arith-expr* `<=` *arith-expr*
> *arith-expr* `=` *arith-expr*
> *arith-expr* `>=` *arith-expr*
>
> *const-expr* `<=` *arith-expr* `<=` *const-expr*
> *const-expr* `>=` *arith-expr* `>=` *const-expr*

Each *const-expr* must be an arithmetic expression not containing variables, while an *arith-expr* may be any valid arithmetic expression — though it must be linear in the variables (Section 8.2) if the result is to be a linear program. To permit a columnwise formulation, one of the *arith-expr*s may be given as:

> `to_come`
> `to_come` + *arith-expr*
> *arith-expr* + `to_come`

Most often a "template" constraint of this kind consists, as in our examples, of `to_come`, a relational operator and a *const-expr*; the constraint's linear terms are all provided in subsequent `var` declarations, and `to_come` shows where they should go. If the template constraint does contain variables, they must be from previous `var` declarations, and the model becomes a sort of hybrid between row-wise and columnwise forms.

The expression for an objective function may also incorporate `to_come` in one of the ways shown above. If the objective is a sum of linear terms specified entirely by subsequent `var` declarations, as in our examples, the expression for the objective is just `to_come` and may be omitted.

In a `var` declaration, constraint coefficients may be specified by one or more phrases consisting of the keyword `coeff`, an optional indexing expression, a constraint name, and an *arith-expr*. If an indexing expression is present, a coefficient is generated for each member of the indexing set; otherwise, one coefficient is generated. The indexing expression may also take the special form `{if` *logical-expr*`}` as seen in Section 8.4 or 15.3, in which case a coefficient is generated only if the *logical-expr* evaluates to true. Our simple examples have required just one `coeff` phrase in each `var` declaration, but

```
set CITIES;
set LINKS within (CITIES cross CITIES);

set PRODS;

param supply {CITIES,PRODS} >= 0;  # amounts available at cities
param demand {CITIES,PRODS} >= 0;  # amounts required at cities

    check {p in PRODS}:
        sum {i in CITIES} supply[i,p] = sum {j in CITIES} demand[j,p];

param cost {LINKS,PRODS} >= 0;      # shipment costs/1000 packages
param capacity {LINKS,PRODS} >= 0; # max packages shipped
param cap_joint {LINKS} >= 0;      # max total packages shipped/link

minimize Total_Cost;

node Balance {k in CITIES, p in PRODS}:
    net_in = demand[k,p] - supply[k,p];

subject to Multi {(i,j) in LINKS}:
    to_come <= cap_joint[i,j];

arc Ship {(i,j) in LINKS, p in PRODS} >= 0, <= capacity[i,j,p],
    from Balance[i,p], to Balance[j,p],
    coeff Multi[i,j] 1.0,
    obj Total_Cost cost[i,j,p];
```

Figure 16-6: Columnwise formulation of Figure 15-13 (`netmcol.mod`).

in general a separate `coeff` phrase is needed for each different indexed collection of constraints in which a variable appears.

Objective function coefficients may be specified in the same way, except that the keyword `obj` is used instead of `coeff`.

The `obj` phrase in a `var` declaration is the same as the `obj` phrase used in `arc` declarations for networks (Section 15.4). The constraint coefficients for the network variables defined by an `arc` declaration are normally given in `from` and `to` phrases, but `coeff` phrases may be present as well; they can be useful if you want to give a columnwise description of ''side'' constraints that apply in addition to the balance-of-flow constraints. As an example, Figure 16-6 shows how a `coeff` phrase can be used to rewrite the multicommodity flow model of Figure 15-13 in an entirely columnwise manner.

Bibliography

Gerald Kahan, ''Walking Through a Columnar Approach to Linear Programming of a Business.'' Interfaces **12**, 3 (1982) pp. 32–39. A brief for the columnwise approach to linear programming, with a small example.

Exercises

16-1. (a) Construct a columnwise formulation of the diet model of Figure 2-1.

(b) Construct a columnwise formulation of the diet model of Figure 5-1. Since there are two separate collections of constraints in this diet model, you will need two `coeff` phrases in the `var` declaration.

16-2. Expand the columnwise production model of Figure 16-3 to incorporate variables `Buy[i]` that represent purchases of raw materials.

16-3. Formulate a multiperiod version of the model of Figure 16-3, using the approach introduced in Section 4.2: first replicate the model over a set of weeks, then introduce inventory variables to tie the weeks together. Use only columnwise declarations for all of the variables, including those for inventory.

16-4. The "roll trim" or "cutting stock" problem has much in common with the scheduling problem described in this chapter. Review the description of the roll trim problem in Exercise 2-6, and use it to answer the following questions.

(a) What is the relationship between the available cutting patterns in the roll trim problem, and the coefficients of the variables in the linear programming formulation?

(b) Formulate an AMPL model for the roll trim problem that uses only columnwise declarations of the variables.

(c) Solve the roll trim problem for the data given in Exercise 2-6. As a test of your formulation, show that it gives the same optimal value as a row-wise formulation of the model.

16-5. The set covering problem, mentioned at the end of Section 16.2, can be stated in a general way as follows. You are given a set S, and certain subsets $T_1, T_2, \ldots, T_n$ of S; a cost is associated with each of the subsets. A selection of some of the subsets T_j is said to *cover* S if every member of S is also a member of at least one of the selected subsets. For example, if

        S = {1,2,3,4}

and

        T1 = {1,2,4}   T2 = {2,3}   T3 = {1}   T4 = {3,4}   T5 = {1,3}

the selections (T1,T2) and (T2,T4,T5) cover S, but the selection (T3,T4,T5) does not.

The goal of the set covering problem is to find the least costly selection of subsets that covers S. Formulate a columnwise linear program that solves this problem for any given set and subsets.

17

Piecewise-Linear Programs

Several kinds of linear programming problems use functions that are not really linear, but are pieced together from connected linear segments:

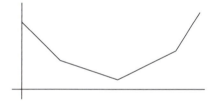

These "piecewise-linear" terms are easy to imagine, but can be hard to describe in conventional algebraic notation. Hence AMPL provides a special, concise way of writing them.

This chapter introduces AMPL's piecewise-linear notation through examples of piecewise-linear objective functions. In Section 17.1, terms of potentially many pieces are used to describe costs more accurately than a single linear relationship. Section 17.2 shows how terms of two or three pieces can be valuable for such purposes as penalizing deviations from constraints, dealing with infeasibilities, and modeling "reversible" activities. Finally, Section 17.3 describes piecewise-linear functions that can be written with other AMPL operators and functions; some are most effectively handled by converting them to the piecewise-linear notation, while others can be accommodated only through more extensive transformations.

Although the piecewise-linear examples in this chapter are all easy to solve, seemingly similar examples can be much more difficult. The last section of this chapter thus offers guidelines for forming and using piecewise-linear terms. We explain how the easy cases can be characterized by the convexity or concavity of the piecewise-linear terms.

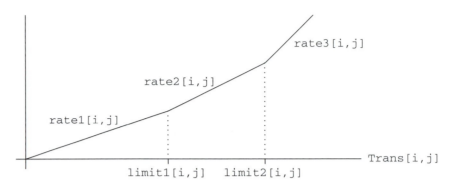

Figure 17-1: Piecewise-linear function, with three slopes.

17.1 Cost terms

Piecewise-linearities are often employed to give a more realistic description of costs than can be achieved by linear terms alone. In this kind of application, piecewise-linear terms serve much the same purpose as nonlinear ones, but without some of the difficulties to be described in Chapter 18.

To make the comparison explicit, we will use the same transportation example as in Chapter 18. We introduce AMPL's notation for piecewise-linear terms with a simple example that has a fixed number of cost levels (and linear pieces) for each shipping link. Then we show how an extension of the notation can use indexing expressions to specify a varying number of pieces controlled through the data.

Fixed numbers of pieces

In a linear transportation model like Figure 3-1a, any number of units can be shipped from a given origin to a given destination at the same cost per unit. More realistically, however, the most favorable rate may be available for only a limited number of units; shipments beyond this limit pay higher rates. As an example, imagine that three cost rate levels are specified for each origin-destination pair. Then the total cost of shipments along a link increases with the amount shipped in a piecewise-linear fashion, with three pieces as shown in Figure 17-1.

To model the three-piece costs, we replace the parameter cost of Figure 3-1a by three rates and two limits:

```
param rate1 {i in ORIG, j in DEST} >= 0;
param rate2 {i in ORIG, j in DEST} >= rate1[i,j];
param rate3 {i in ORIG, j in DEST} >= rate2[i,j];

param limit1 {i in ORIG, j in DEST} > 0;
param limit2 {i in ORIG, j in DEST} > limit1[i,j];
```

Shipments from i to j are charged at rate1[i,j] per unit up to limit1[i,j] units, then at rate2[i,j] per unit up to limit2[i,j], and then at rate3[i,j]. Normally rate2[i,j] would be greater than rate1[i,j] and rate3[i,j] would be greater than rate2[i,j], but they may be equal if the link from i to j does not have three distinct rates.

In the linear transportation model, the objective is expressed in terms of the variables and the parameter cost as follows:

```
var Trans {ORIG,DEST} >= 0;

minimize Total_Cost:
    sum {i in ORIG, j in DEST} cost[i,j] * Trans[i,j];
```

We could express a piecewise-linear objective analogously, by introducing three collections of variables, one to represent the amount shipped at each rate:

```
var Trans1 {i in ORIG, j in DEST} >= 0, <= limit1[i,j];
var Trans2 {i in ORIG, j in DEST} >= 0, <= limit2[i,j]
                                             - limit1[i,j];
var Trans3 {i in ORIG, j in DEST} >= 0;

minimize Total_Cost:
    sum {i in ORIG, j in DEST} (rate1[i,j] * Trans1[i,j]
        + rate2[i,j] * Trans2[i,j] + rate3[i,j] * Trans3[i,j]);
```

But then the new variables would have to be introduced into all the constraints, and we would also have to deal with these variables whenever we wanted to display the optimal results. Rather than go to all this trouble, we would much prefer to describe the piecewise-linear cost function explicitly in terms of the original variables. Since there is no standard way to describe piecewise-linear functions in algebraic notation, AMPL supplies its own syntax for this purpose.

The piecewise-linear function depicted in Figure 17-1 is written in AMPL as follows:

```
<<limit1[i,j], limit2[i,j];
  rate1[i,j], rate2[i,j], rate3[i,j]>> Trans[i,j]
```

The expression between << and >> describes the piecewise-linear function, and is followed by the name of the variable to which it applies. (You can think of it as "multiplying" Trans[i,j], but by a series of coefficients rather than just one.) There are two parts to the expression, a list of *breakpoints* where the slope of the function changes, and a list of the *slopes* — which in this case are the cost rates. The lists are separated by a semicolon, and members of each list are separated by commas. Since the first slope applies to values before the first breakpoint, and the last slope to values after the last breakpoint, the number of slopes must be one more than the number of breakpoints.

Although the lists of breakpoints and slopes are sufficient to describe the piecewise-linear cost function for optimization, they do not quite specify the function uniquely. If we added, say, 10 to the cost at every point, we would have a different cost function even though all the breakpoints and slopes would be the same. To resolve this ambiguity,

```
set ORIG;   # origins
set DEST;   # destinations

param supply {ORIG} >= 0;    # amounts available at origins
param demand {DEST} >= 0;    # amounts required at destinations

    check: sum {i in ORIG} supply[i] = sum {j in DEST} demand[j];

param rate1 {i in ORIG, j in DEST} >= 0;
param rate2 {i in ORIG, j in DEST} >= rate1[i,j];
param rate3 {i in ORIG, j in DEST} >= rate2[i,j];

param limit1 {i in ORIG, j in DEST} > 0;
param limit2 {i in ORIG, j in DEST} > limit1[i,j];

var Trans {ORIG,DEST} >= 0;     # units to be shipped

minimize Total_Cost:
   sum {i in ORIG, j in DEST}
      <<limit1[i,j], limit2[i,j];
         rate1[i,j], rate2[i,j], rate3[i,j]>> Trans[i,j];

subject to Supply {i in ORIG}:
   sum {j in DEST} Trans[i,j] = supply[i];

subject to Demand {j in DEST}:
   sum {i in ORIG} Trans[i,j] = demand[j];
```

Figure 17-2: Piecewise-linear model with three slopes (`transpl1.mod`).

AMPL assumes that a piecewise-linear function evaluates to zero at zero, as in Figure 17-1. Options for other possibilities are discussed later in this chapter.

Summing the cost over all links, the piecewise-linear objective function is now written

```
minimize Total_Cost:
   sum {i in ORIG, j in DEST}
      <<limit1[i,j], limit2[i,j];
         rate1[i,j], rate2[i,j], rate3[i,j]>> Trans[i,j];
```

The declarations of the variables and constraints stay the same as before; the complete model is shown in Figure 17-2.

Varying numbers of pieces

The approach taken in the preceding example is most useful when there are only a few linear pieces for each term. If there were, for example, 12 pieces instead of three, a model defining `rate1[i,j]` through `rate12[i,j]` and `limit1[i,j]` through `limit11[i,j]` would be unwieldy. Realistically, moreover, there would more likely be up to 12 pieces, rather than exactly 12, for each term; a term with fewer than 12 pieces could be handled by making some rates equal, but for large numbers of pieces this would be a cumbersome device that would require many unnecessary data values and would obscure the actual number of pieces in each term.

A much better approach is to let the number of pieces (that is, the number of shipping rates) itself be a parameter of the model, indexed over the links:

```
param npiece {ORIG,DEST} integer >= 1;
```

We can then index the rates and limits over all combinations of links and pieces:

```
param rate {i in ORIG, j in DEST, p in 1..npiece[i,j]}
    >= if p = 1 then 0 else rate[i,j,p-1];

param limit {i in ORIG, j in DEST, p in 1..npiece[i,j]-1}
    > if p = 1 then 0 else limit[i,j,p-1];
```

For any particular origin i and destination j, the number of linear pieces in the cost term is given by npiece[i,j]. The slopes are rate[i,j,p] for p ranging from 1 to npiece[i,j], and the intervening breakpoints are limit[i,j,p] for p from 1 to npiece[i,j]-1. As before, there is one more slope than there are breakpoints.

To use AMPL's piecewise-linear function notation with these data values, we have to give indexed lists of breakpoints and slopes, rather than the explicit lists of the previous example. This is done by placing indexing expressions in front of the slope and breakpoint values:

```
minimize Total_Cost:
    sum {i in ORIG, j in DEST}
        <<{p in 1..npiece[i,j]-1} limit[i,j,p];
          {p in 1..npiece[i,j]} rate[i,j,p]>> Trans[i,j];
```

Once again, the rest of the model is the same. Figure 17-3a shows the whole model and Figure 17-3b illustrates how the data would be specified. Notice that since npiece["PITT","STL"] is 1, Trans["PITT","STL"] has only one slope and no breakpoints; this implies a one-piece linear term for Trans["PITT","STL"] in the objective function.

17.2 Common two-piece and three-piece terms

Simple piecewise-linear terms have a variety of uses in otherwise linear models. In this section we present three cases: allowing limited violations of the constraints, analyzing infeasibility, and representing costs for variables that are meaningful at negative as well as positive levels.

Penalty terms for "soft" constraints

Linear programs most easily express "hard" constraints: that production must be at least at a certain level, for example, or that resources used must not exceed those available. Real situations are often not nearly so definite. Production and resource use may

```
set ORIG;    # origins
set DEST;    # destinations

param supply {ORIG} >= 0;    # amounts available at origins
param demand {DEST} >= 0;    # amounts required at destinations

    check: sum {i in ORIG} supply[i] = sum {j in DEST} demand[j];

param npiece {ORIG,DEST} integer >= 1;

param rate {i in ORIG, j in DEST, p in 1..npiece[i,j]}
  >= if p = 1 then 0 else rate[i,j,p-1];

param limit {i in ORIG, j in DEST, p in 1..npiece[i,j]-1}
  > if p = 1 then 0 else limit[i,j,p-1];

var Trans {ORIG,DEST} >= 0;    # units to be shipped

minimize Total_Cost:
   sum {i in ORIG, j in DEST}
      <<{p in 1..npiece[i,j]-1} limit[i,j,p];
        {p in 1..npiece[i,j]} rate[i,j,p]>> Trans[i,j];

subject to Supply {i in ORIG}:
   sum {j in DEST} Trans[i,j] = supply[i];

subject to Demand {j in DEST}:
   sum {i in ORIG} Trans[i,j] = demand[j];
```

Figure 17-3a: Piecewise-linear model with indexed slopes (`transp12.mod`).

have certain preferred levels, yet we may be allowed to violate these levels by accepting some extra costs or reduced profits. The resulting ''soft'' constraints can be modeled by adding piecewise-linear ''penalty'' terms to the objective function.

For an example, we return to the multi-week production model developed in Chapter 4. As seen in Figure 4-4, the constraints say that, in each of weeks 1 through T, total hours used to make all products may not exceed hours available:

```
subject to Time {t in 1..T}:
   sum {p in PROD} (1/rate[p]) * Make[p,t] <= avail[t];
```

Suppose that, in reality, a larger number of hours may be used in each week, but at some penalty per hour to the total profit. Specifically, we replace the parameter `avail[t]` by two availability levels and an hourly penalty rate:

```
param avail_min {1..T} >= 0;
param avail_max {t in 1..T} >= avail_min[t];

param time_penalty {1..T} > 0;
```

Up to `avail_min[t]` hours are available without penalty in week t, and up to `avail_max[t]` hours are available at a loss of `time_penalty[t]` in profit for each hour above `avail_min[t]`.

To model this situation, we introduce a new variable `Use[t]` to represent the hours used by production. Clearly `Use[t]` may not be less than zero, or greater than

```
param: ORIG:  supply :=
       GARY 1400  CLEV 2600  PITT 2900 ;

param: DEST:  demand :=
       FRA  900    DET 1200   LAN  600    WIN  400
       STL 1700    FRE 1100   LAF 1000 ;

param npiece:  FRA DET LAN WIN STL FRE LAF :=
       GARY      3   3   3   2   3   2   3
       CLEV      3   3   3   3   3   3   3
       PITT      2   2   2   2   1   2   1 ;

param rate :=
  [GARY,FRA,*] 1 39   2 50   3 70     [GARY,DET,*] 1 14   2  17   3  33
  [GARY,LAN,*] 1 11   2 12   3 23     [GARY,WIN,*] 1 14   2  17
  [GARY,STL,*] 1 16   2 23   3 40     [GARY,FRE,*] 1 82   2  98
  [GARY,LAF,*] 1  8   2 16   3 24

  [CLEV,FRA,*] 1 27   2 37   3 47     [CLEV,DET,*] 1  9   2  19   3  24
  [CLEV,LAN,*] 1 12   2 32   3 39     [CLEV,WIN,*] 1  9   2  14   3  21
  [CLEV,STL,*] 1 26   2 36   3 47     [CLEV,FRE,*] 1 95   2 105   3 129
  [CLEV,LAF,*] 1  8   2 16   3 24

  [PITT,FRA,*] 1 24   2 34            [PITT,DET,*] 1 14   2  24
  [PITT,LAN,*] 1 17   2 27            [PITT,WIN,*] 1 13   2  23
  [PITT,STL,*] 1 28                   [PITT,FRE,*] 1 99   2 140
  [PITT,LAF,*] 1 20 ;

param limit :=
  [GARY,*,*] FRA 1  500   FRA 2 1000   DET 1  500   DET 2 1000
             LAN 1  500   LAN 2 1000   WIN 1 1000
             STL 1  500   STL 2 1000   FRE 1 1000
             LAF 1  500   LAF 2 1000

  [CLEV,*,*] FRA 1  500   FRA 2 1000   DET 1  500   DET 2 1000
             LAN 1  500   LAN 2 1000   WIN 1  500   WIN 2 1000
             STL 1  500   STL 2 1000   FRE 1  500   FRE 2 1000
             LAF 1  500   LAF 2 1000

  [PITT,*,*] FRA 1 1000   DET 1 1000   LAN 1 1000   WIN 1 1000
             FRE 1 1000 ;
```

Figure 17-3b: Data for piecewise-linear model (`transp12.dat`).

avail_max[t]. In place of our previous constraint, we say that the total hours used to make all products must equal Use[t]:

```
    var Use {t in 1..T} >= 0, <= avail_max[t];

    subject to Time {t in 1..T}:
       sum {p in PROD} (1/rate[p]) * Make[p,t] = Use[t];
```

We can now describe the hourly penalty in terms of this new variable. If Use[t] is between 0 and avail_min[t], there is no penalty; if Use[t] is between avail_min[t] and avail_max[t], the penalty is time_penalty[t] per hour

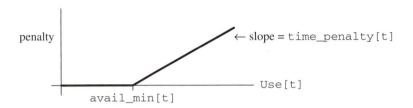

Figure 17-4: Piecewise-linear penalty function for hours used.

that it exceeds `avail_min[t]`. That is, the penalty is a piecewise-linear function of `Use[t]` as shown in Figure 17-4, with slopes of 0 and `time_penalty[t]` surrounding a breakpoint at `avail_min[t]`. Using the syntax previously introduced, we can rewrite the expression for the objective function as:

```
maximize Net_Profit:
    sum {p in PROD, t in 1..T} (revenue[p,t]*Sell[p,t] -
        prodcost[p]*Make[p,t] - invcost[p]*Inv[p,t])
    - sum {t in 1..T} <<avail_min[t]; 0,time_penalty[t]>> Use[t];
```

The first summation is the same expression for total profit as before, while the second is the sum of the piecewise-linear penalty functions over all weeks. Between << and >> are the breakpoint `avail_min[t]` and a list of the surrounding slopes, 0 and `time_penalty[t]`; this is followed by the argument `Use[t]`.

The complete revised model is shown in Figure 17-5a, and our small data set from Chapter 4 is expanded with the new availabilities and penalties in Figure 17-5b. In the optimal solution, we find that the hours used are as follows:

```
ampl: model steelp11.mod; data steelp11.dat; solve;
MINOS 5.5: optimal solution found.
21 iterations, objective 457572.8571

ampl: display avail_min,Use,avail_max;
: avail_min  Use avail_max    :=
1      35       35     42
2      35       42     42
3      30       30     40
4      35       42     42
;
```

In weeks 1 and 3 we use only the unpenalized hours available, while in weeks 2 and 4 we also use the penalized hours. Solutions to piecewise-linear programs usually display this sort of solution, in which many (though not necessarily all) of the variables ''stick'' at one of the breakpoints.

```
set PROD;        # products
param T > 0;     # number of weeks

param rate {PROD} > 0;             # tons per hour produced
param inv0 {PROD} >= 0;            # initial inventory
param commit {PROD,1..T} >= 0;     # minimum tons sold in week
param market {PROD,1..T} >= 0;     # limit on tons sold in week

param avail_min {1..T} >= 0;       # unpenalized hours available
param avail_max {t in 1..T} >= avail_min[t]; # total hours avail
param time_penalty {1..T} > 0;

param prodcost {PROD} >= 0;        # cost/ton produced
param invcost {PROD} >= 0;         # carrying cost/ton of inventory
param revenue {PROD,1..T} >= 0;    # revenue/ton sold

var Make {PROD,1..T} >= 0;              # tons produced
var Inv {PROD,0..T} >= 0;               # tons inventoried
var Sell {p in PROD, t in 1..T}
   >= commit[p,t], <= market[p,t];      # tons sold

var Use {t in 1..T} >= 0, <= avail_max[t];   # hours used

maximize Total_Profit:
   sum {p in PROD, t in 1..T} (revenue[p,t]*Sell[p,t] -
      prodcost[p]*Make[p,t] - invcost[p]*Inv[p,t])
 - sum {t in 1..T} <<avail_min[t]; 0,time_penalty[t]>> Use[t];

                # Objective: total revenue less costs in all weeks

subject to Time {t in 1..T}:
   sum {p in PROD} (1/rate[p]) * Make[p,t] = Use[t];

                # Total of hours used by all products
                # may not exceed hours available, in each week

subject to Init_Inv {p in PROD}:  Inv[p,0] = inv0[p];

                # Initial inventory must equal given value

subject to Balance {p in PROD, t in 1..T}:
   Make[p,t] + Inv[p,t-1] = Sell[p,t] + Inv[p,t];

                # Tons produced and taken from inventory
                # must equal tons sold and put into inventory
```

Figure 17-5a: Piecewise-linear objective with penalty function (`steelpl1.mod`).

Dealing with infeasibility

The parameters commit[p,t] in Figure 17-5b represent the minimum production amounts for each product in each week. If we change the data to raise these commitments:

```
param commit:     1       2       3       4 :=
        bands   3500    5900    3900    6400
        coils   2500    2400    3400    4100 ;
```

```
param T := 4;
set PROD := bands coils;

param:     rate  inv0  prodcost  invcost :=
  bands     200    10      10       2.5
  coils     140     0      11        3 ;

param: avail_min  avail_max  time_penalty :=
    1        35        42         3100
    2        35        42         3000
    3        30        40         3700
    4        35        42         3100 ;

param revenue:     1     2     3     4 :=
          bands   25    26    27    27
          coils   30    35    37    39 ;

param commit:     1     2     3     4 :=
         bands  3000  3000  3000  3000
         coils  2000  2000  2000  2000 ;

param market:     1     2     3     4 :=
         bands  6000  6000  4000  6500
         coils  4000  2500  3500  4200 ;
```

Figure 17-5b: Data for Figure 17-5a (`steelp11.dat`).

then there are not enough hours to produce even these minimum amounts, and the solver reports that the problem is infeasible:

```
ampl: model steelp11.mod;
ampl: data steelp12.dat;

ampl: solve;
MINOS 5.5: infeasible problem.
13 iterations
```

In the solution that is returned, the inventory of coils in the last period is negative:

```
ampl: option display_1col 0;
ampl: display Inv;
Inv [*,*] (tr)
: bands   coils    :=
0    10       0
1     0     937
2     0     287
3     0       0
4     0   -2700
;
```

and production of coils in several periods is below the minimum required:

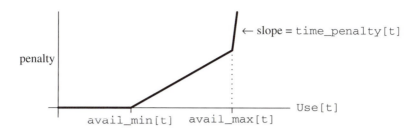

Figure 17-6: Penalty function for hours used, with two breakpoints.

```
ampl: display commit,Make,market;
:          commit    Make market     :=
bands 1     3500     3490    6000
bands 2     5900     5900    6000
bands 3     3900     3900    4000
bands 4     6400     6400    6500
coils 1     2500     3437    4000
coils 2     2400     1750    2500
coils 3     3400     2870    3500
coils 4     4100     1400    4200
;
```

These are typical of the infeasible results that solvers return. The infeasibilities are scattered around the solution, so that it is hard to tell what changes might be necessary to achieve feasibility. By extending the idea of penalties, we can better concentrate the infeasibility where it can be understood.

Suppose that we want to view the infeasibility in terms of a shortage of hours. Imagine that we extend the piecewise-linear penalty function of Figure 17-4 to the one shown in Figure 17-6. Now Use[t] is allowed to increase past avail_max[t], but only with an extremely steep penalty per hour — so that the solution will use hours above avail_max[t] only to the extent absolutely necessary.

In AMPL, the new penalty function is introduced through the following changes:

```
var Use {t in 1..T} >= 0;

maximize Total_Profit:
    sum {p in PROD, t in 1..T} (revenue[p,t]*Sell[p,t] -
        prodcost[p]*Make[p,t] - invcost[p]*Inv[p,t])
    - sum {t in 1..T} <<avail_min[t],avail_max[t];
                        0,time_penalty[t],100000>> Use[t];
```

The former bound avail_max[t] has become a breakpoint, and to its right a very large slope of 100,000 has been introduced. Now we get a feasible solution, which uses hours as follows:

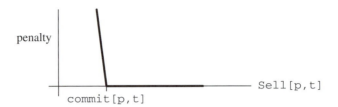

Figure 17-7: Penalty function for sales.

```
ampl: model steelpl2.mod; data steelpl2.dat; solve;
MINOS 5.5: optimal solution found.
19 iterations, objective -1576814.857

ampl: display avail_max,Use;
: avail_max  Use      :=
1     42        42
2     42        42
3     40        41.7357
4     42        61.2857
;
```

This table implies that the commitments can be met only by adding about 21 hours, mostly in the last week.

Alternatively, we may view the infeasibility in terms of an excess of commitments. For this purpose we subtract a very large penalty from the objective for each unit that Sell[p,t] falls below commit[p,t]; the penalty as a function of Sell[p,t] is depicted in Figure 17-7.

Since this function has a breakpoint at commit[p,t], with a slope of 0 to the right and a very negative value to the left, it would seem that the AMPL representation could be

```
<<commit[p,t]; -100000,0>> Sell[p,t]
```

Recall, however, AMPL's convention that such a function takes the value zero at zero. Figure 17-7 clearly shows that we want our penalty function to take a positive value at zero, so that it will fall to zero at commit[p,t] and beyond. In fact we want the function to take a value of 100000 * commit[p,t] at zero, and we could express the function properly by adding this constant to the penalty expression:

```
<<commit[p,t]; -100000,0>> Sell[p,t] + 100000*commit[p,t]
```

The same thing may be said more concisely by using a second argument that states explicitly where the piecewise-linear function should evaluate to zero:

```
<<commit[p,t]; -100000,0>> (Sell[p,t],commit[p,t])
```

This says that the function should be zero at commit[p,t], as Figure 17-7 shows. In the completed model, we have:

```
    var Sell {p in PROD, t in 1..T} >= 0, <= market[p,t];

    maximize Total_Profit:
        sum {p in PROD, t in 1..T} (revenue[p,t]*Sell[p,t] -
            prodcost[p]*Make[p,t] - invcost[p]*Inv[p,t])
      - sum {t in 1..T} <<avail_min[t]; 0,time_penalty[t]>> Use[t]
      - sum {p in PROD, t in 1..T}
            <<commit[p,t]; -100000,0>> (Sell[p,t],commit[p,t]);
```

The rest of the model is the same as in Figure 17-5a. Notice that `Sell[p,t]` appears in both a linear and a piecewise-linear term within the objective function; AMPL automatically recognizes that the sum of these terms is also piecewise-linear.

This version, using the same data, produces a solution in which the amounts sold are as follows:

```
ampl: model steelp13.mod; data steelp12.dat; solve;
MINOS 5.5: optimal solution found.
24 iterations, objective -293856347

ampl: display Sell,commit;
:         Sell commit    :=
bands 1   3500   3500
bands 2   5900   5900
bands 3   3900   3900
bands 4   6400   6400
coils 1      0   2500
coils 2   2400   2400
coils 3   3400   3400
coils 4   3657   4100
;
```

To get by with the given number of hours, commitments to deliver coils are cut by 2500 tons in the first week and 443 tons in the fourth week.

Reversible activities

Almost all of the linear programs in this book are formulated in terms of nonnegative variables. Sometimes a variable makes sense at negative as well as positive values, however, and in many such cases the associated cost is piecewise-linear with a breakpoint at zero.

One example is provided by the inventory variables in Figure 17-5a. We have defined `Inv[p,t]` to represent the tons of product p inventoried at the end of week t. That is, after week t there are `Inv[p,t]` tons of product p that have been made but not sold. A negative value of `Inv[p,t]` could thus reasonably be interpreted as representing tons of product p that have been sold but not made — tons backordered, in effect. The material balance constraints,

```
    subject to Balance {p in PROD, t in 1..T}:
        Make[p,t] + Inv[p,t-1] = Sell[p,t] + Inv[p,t];
```

remain valid under this interpretation.

This analysis suggests that we remove the `>= 0` from the declaration of `Inv` in our model. Then backordering might be especially attractive if the sales price were expected to drop in later weeks, like this:

```
param revenue:    1      2      3      4 :=
         bands   25     26     23     20
         coils   30     35     31     25 ;
```

When we re-solve with appropriately revised model and data files, however, the results are not what we expect:

```
ampl: model steelpl4.mod; data steelpl4.dat; solve;
MINOS 5.5: optimal solution found.
15 iterations, objective 1194250

ampl: display Make,Inv,Sell;
:        Make     Inv     Sell     :=
bands 0     .       10        .
bands 1     0    -5990     6000
bands 2     0   -11990     6000
bands 3     0   -15990     4000
bands 4     0   -22490     6500
coils 0     .        0        .
coils 1     0    -4000     4000
coils 2     0    -6500     2500
coils 3     0   -10000     3500
coils 4     0   -14200     4200
;
```

The source of difficulty is in the objective function, where `invcost[p] * Inv[p,t]` is subtracted from the sales revenue. When `Inv[p,t]` is negative, a negative amount is subtracted, increasing the apparent total profit. The greater the amount backordered, the more the total profit is increased — hence the odd solution in which the maximum possible sales are backordered, while nothing is produced!

A proper inventory cost function for this model looks like the one graphed in Figure 17-8. It increases both as `Inv[p,t]` becomes more positive (greater inventories) and as `Inv[p,t]` becomes more negative (greater backorders). We represent this piecewise-linear function in AMPL by declaring a backorder cost to go with the inventory cost:

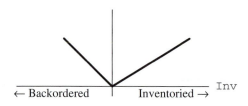

Figure 17-8: Inventory cost function.

```
param invcost {PROD} >= 0;
param backcost {PROD} >= 0;
```

Then the slopes for the `Inv[p,t]` term in the objective are `-backcost[p]` and `invcost[p]`, with the breakpoint at zero, and the correct objective function is:

```
maximize Total_Profit:
   sum {p in PROD, t in 1..T}
      (revenue[p,t]*Sell[p,t] - prodcost[p]*Make[p,t]
       - <<0; -backcost[p],invcost[p]>> Inv[p,t])
   - sum {t in 1..T} <<avail_min[t]; 0,time_penalty[t]>> Use[t];
```

In contrast to our first example, the piecewise-linear function is subtracted rather than added. The result is still piecewise-linear, though; it's the same as if we had added the expression `<<0; backcost[p], -invcost[p]>> Inv[p,t]`.

When we make this change, and add some backorder costs to the data, we get a more reasonable-looking solution. Nevertheless, there remains a tendency to make nothing and backorder everything in the later periods; this is an "end effect" that occurs because the model does not account for the eventual production cost of items backordered past the last period. As a quick fix, we can rule out any remaining backorders at the end, by adding a constraint that final-week inventory must be nonnegative:

```
subject to Final {p in PROD}:  Inv[p,T] >= 0;
```

Solving with this constraint, and with `backcost` values of 1.5 for band and 2 for coils:

```
ampl: model steelp15.mod; data steelp15.dat; solve;
MINOS 5.5: optimal solution found.
20 iterations, objective 370752.8571

ampl: display Make,Inv,Sell;
:           Make     Inv    Sell       :=
bands 0        .       10      .
bands 1    4142.86      0    4152.86
bands 2    6000         0    6000
bands 3    3000         0    3000
bands 4    3000         0    3000
coils 0        .        0      .
coils 1    2000         0    2000
coils 2    1680      -820    2500
coils 3    2100      -800    2080
coils 4    2800         0    2000
;
```

About 800 tons of coils for weeks 2 and 3 will be delivered a week late under this plan.

17.3 Other piecewise-linear functions

Many simple piecewise-linear functions can be modeled in several equivalent ways in AMPL. The function of Figure 17-4, for example, could be written as

```
if Use[t] > avail_min[t]
    then time_penalty[t] * (Use[t] - avail_min[t]) else 0
```

or more concisely as

```
max(0, time_penalty[t] * (Use[t] - avail_min[t]))
```

The current version of AMPL does not detect that these expressions are piecewise-linear, so you are unlikely to get satisfactory results if you try to solve a model that has expressions like these in its objective. To take advantage of linear programming techniques that can be applied for piecewise-linear terms, you need to use the piecewise-linear terminology

```
<<avail_min[t]; 0,time_penalty[t]>> Use[t]
```

so the structure can be noted and passed to a solver.

The same advice applies to the function `abs`. Imagine that we would like to encourage the number of hours used to be close to `avail_min[t]`. Then we would want the penalty term to equal `time_penalty[t]` times the amount that `Use[t]` deviates from `avail_min[t]`, either above or below. Such a term can be written as

```
time_penalty[t] * abs(Use[t] - avail_min[t])
```

To express it in an explicitly piecewise-linear fashion, however, you should write it as

```
time_penalty[t] * <<avail_min[t]; -1,1>> Use[t]
```

or equivalently,

```
<<avail_min[t]; -time_penalty[t],time_penalty[t]>> Use[t]
```

As this example shows, multiplying a piecewise-linear function by a constant is the same as multiplying each of its slopes individually.

As a final example of a common piecewise-linearity in the objective, we return to the kind of assignment model that was discussed in Chapter 15. Recall that, for `i` in the set `PEOPLE` and `j` in the set `PROJECTS`, `cost[i,j]` is the cost for person `i` to work an hour on project `j`, and the decision variable `Assign[i,j]` is the number of hours that person `i` is assigned to work on project `j`:

```
set PEOPLE;
set PROJECTS;

param cost {PEOPLE,PROJECTS} >= 0;
var Assign {PEOPLE,PROJECTS} >= 0;
```

We originally formulated the objective as the total cost of all assignments,

```
sum {i in PEOPLE, j in PROJECTS} cost[i,j] * Assign[i,j]
```

What if we want the fairest assignment instead of the cheapest? Then we might minimize the maximum cost of any one person's assignments:

```
set PEOPLE;
set PROJECTS;

param supply {PEOPLE} >= 0;    # hours each person is available
param demand {PROJECTS} >= 0; # hours each project requires

   check: sum {i in PEOPLE} supply[i]
          = sum {j in PROJECTS} demand[j];

param cost {PEOPLE,PROJECTS} >= 0;    # cost per hour of work
param limit {PEOPLE,PROJECTS} >= 0;   # maximum contributions
                                      # to projects

var M;
var Assign {i in PEOPLE, j in PROJECTS} >= 0, <= limit[i,j];

minimize Max_Cost: M;

subject to M_def {i in PEOPLE}:
   M >= sum {j in PROJECTS} cost[i,j] * Assign[i,j];

subject to Supply {i in PEOPLE}:
   sum {j in PROJECTS} Assign[i,j] = supply[i];

subject to Demand {j in PROJECTS}:
   sum {i in PEOPLE} Assign[i,j] = demand[j];
```

Figure 17-9: Min-max assignment model (`minmax.mod`).

```
minimize Max_Cost:
   max {i in PEOPLE}
      sum {j in PROJECTS} cost[i,j] * Assign[i,j];
```

This function is also piecewise-linear, in a sense; it is pieced together from the linear functions `sum {j in PROJECTS} cost[i,j] * Assign[i,j]` for different people `i`. However, it is not piecewise-linear in the individual variables — in mathematical jargon, it is not separable — and hence it cannot be written using the `<< ... >>` notation.

This is a case in which piecewise-linearity can only be handled by rewriting the model as a linear program. We introduce a new variable `M` to represent the maximum. Then we write constraints to guarantee that `M` is greater than or equal to each cost of which it is the maximum:

```
var M;
minimize Max_Cost: M;

subject to M_def {i in PEOPLE}:
   M >= sum {j in PROJECTS} cost[i,j] * Assign[i,j];
```

Because `M` is being minimized, at the optimal solution it will in fact equal the maximum of `sum {j in PROJECTS} cost[i,j] * Assign[i,j]` over all `i` in `PEOPLE`. The other constraints are the same as in any assignment problem, as shown in Figure 17-9.

This kind of reformulation can be applied to any problem that has a ''min-max'' objective. The same idea works for the analogous ''max-min'' objective, with `maximize` instead of `minimize` and with `M <= ...` in the constraints.

17.4 Guidelines for piecewise-linear optimization

AMPL's piecewise-linear notation has the power to specify a variety of useful functions. We summarize below its various forms, most of which have been illustrated earlier in this chapter.

Because this notation is so general, it can be used to specify many functions that are not readily optimized by any efficient and reliable algorithms. We conclude by describing the kinds of piecewise-linear functions that are most likely to appear in tractable models, with particular emphasis on the property of convexity or concavity.

Forms for piecewise-linear expressions

An AMPL piecewise-linear term has the general form

> <<*breakpoint-list*; *slope-list*>> *pl-argument*

where *breakpoint-list* and *slope-list* each consist of a comma-separated list of one or more items. An item may be an individual arithmetic expression, or an indexing expression followed by an arithmetic expression. In the latter case, the indexing expression must be an ordered set; the item is expanded to a list by evaluating the arithmetic expression once for each set member (as in the example of Figure 17-3a).

After any indexed items are expanded, the number of slopes must be one more than the number of breakpoints, and the breakpoints must be nondecreasing. The resulting piecewise-linear function is constructed by interleaving the slopes and breakpoints in the order given, with the first slope to the left of the first breakpoint, and the last slope to the right of the last breakpoint. By indexing breakpoints over an empty set, it is possible to specify no breakpoints and one slope, in which case the function is linear.

The *pl-argument* may have one of the forms

> *var-ref*
> (*arg-expr*)
> (*arg-expr*, *zero-expr*)

The *var-ref* (a reference to a previously declared variable) or the *arg-expr* (an arithmetic expression) specifies the point where the piecewise-linear function is to be evaluated. The *zero-expr* is an arithmetic expression that specifies a place where the function is zero; when the *zero-expr* is omitted, the function is assumed to be zero at zero.

Suggestions for piecewise-linear models

As seen in all of our examples, AMPL's terminology for piecewise-linear functions of variables is limited to describing functions of individual variables. In model declarations, no variables may appear in the *breakpoint-list*, *slope-list* and *zero-expr* (if any), while an *arg-expr* can only be a reference to an individual variable. (Piecewise-linear expressions in commands like `display` may use variables without limitation, however.)

A piecewise-linear function of an individual variable remains such a function when multiplied or divided by an arithmetic expression without variables. AMPL also treats a sum or difference of piecewise-linear and linear functions of the same variable as representing one piecewise-linear function of that variable. A *separable* piecewise-linear function of a model's variables is a sum or difference (using +, – or `sum`) of piecewise-linear or linear functions of the individual variables. Optimizers can effectively handle these separable functions, which are the ones that appear in our examples.

A piecewise-linear function is convex if successive slopes are nondecreasing (along with the breakpoints), and is concave if the slopes are nonincreasing. The two kinds of piecewise-linear optimization most easily handled by solvers are minimizing a separable convex piecewise-linear function, and maximizing a separable concave piecewise-linear function, subject to linear constraints. You can easily check that all of this chapter's examples are of these kinds. AMPL can obtain solutions in these cases by translating to an equivalent linear program, applying any LP solver, and then translating the solution back; the whole sequence occurs automatically when you type `solve`.

Outside of these two cases, optimizing a separable piecewise-linear function must be viewed as an application of integer programming — the topic of Chapter 20 — and AMPL must translate piecewise-linear terms to equivalent integer programming forms. This, too, is done automatically, for solution by an appropriate solver. Because integer programs are usually much harder to solve than similar linear programs of comparable size, however, you should not assume that just any separable piecewise-linear function can be readily optimized; a degree of experimentation may be necessary to determine how large an instance your solver can handle. The best results are likely to be obtained by solvers that accept an option known (mysteriously) as ''special ordered sets of type 2''; check the solver-specific documentation for details.

The situation for the constraints can be described in a similar way. However, a separable piecewise-linear function in a constraint can be handled through linear programming only under a restrictive set of circumstances:

- If it is convex and on the left-hand side of a $\leq$ constraint (or equivalently, the right-hand side of a $\geq$ constraint);
- If it is concave and on the left-hand side of a $\geq$ constraint (or equivalently, the right-hand side of a $\leq$ constraint).

Other piecewise-linearities in the constraints must be dealt with through integer programming techniques, and the preceding comments for the case of the objective apply.

If you have access to a solver that can handle piecewise-linearities directly, you can turn off AMPL's translation to the linear or integer programming form by setting the option `pl_linearize` to 0. The case of minimizing a convex or maximizing a concave separable piecewise-linear function can in particular be handled very efficiently by piecewise-linear generalizations of LP techniques. A solver intended for nonlinear programming may also accept piecewise-linear functions, but it is unlikely to handle them reliably unless it has been specially designed for ''nondifferentiable'' optimization.

The differences between hard and easy piecewise-linear cases can be slight. This chapter's transportation example is easy, in particular because the shipping rates increase along with shipping volume. The same example would be hard if economies of scale caused shipping rates to decrease with volume, since then we would be minimizing a concave rather than a convex function. We cannot say definitively that shipping rates ought to be one way or the other; their behavior depends upon the specifics of the situation being modeled.

In all cases, the difficulty of piecewise-linear optimization gradually increases with the total number of pieces. Thus piecewise-linear cost functions are most effective when the costs can be described or approximated by relatively few pieces. If you need more than about a dozen pieces to describe a cost accurately, you may be better off using a nonlinear function as described in Chapter 18.

Bibliography

Robert Fourer, ''A Simplex Algorithm for Piecewise-Linear Programming III: Computational Analysis and Applications.'' Mathematical Programming **53** (1992) pp. 213–235. A survey of conversions from piecewise-linear to linear programs, and of applications.

Robert Fourer and Roy E. Marsten, ''Solving Piecewise-Linear Programs: Experiments with a Simplex Approach.'' ORSA Journal on Computing **4** (1992) pp. 16–31. Descriptions of varied applications and of experience in solving them.

Spyros Kontogiorgis, ''Practical Piecewise-Linear Approximation for Monotropic Optimization.'' INFORMS Journal on Computing **12** (2000) pp. 324–340. Guidelines for choosing the breakpoints when approximating a nonlinear function by a piecewise-linear one.

Exercises

17-1. Piecewise-linear models are sometimes an alternative to the nonlinear models described in Chapter 18, replacing a smooth curve by a series of straight-line segments. This exercise deals with the model shown in Figure 18-4.

(a) Reformulate the model of Figure 18-4 so that it approximates each nonlinear term

```
Trans[i,j] / (1 - Trans[i,j]/limit[i,j])
```

by a piecewise-linear term having three pieces. Set the breakpoints at `(1/3) * limit[i,j]` and `(2/3) * limit[i,j]`. Pick the slopes so that the approximation equals the original nonlinear term when `Trans[i,j]` is 0, `1/3 * limit[i,j]`, `2/3 * limit[i,j]`, or `11/12 * limit[i,j]`; you should find that the three slopes are 3/2, 9/2 and 36 in every term, regardless of the size of `limit[i,j]`. Finally, place an explicit upper limit of `0.99 * limit[i,j]` on `Trans[i,j]`.

(b) Solve the approximation with the data given in Figure 18-5, and compare the optimal shipment amounts to the amounts recommended by the nonlinear model.

(c) Formulate a more sophisticated version in which the number of linear pieces for each term is given by a parameter `nsl`. Pick the breakpoints to be at `(k/nsl) * limit[i,j]` for k from 1 to `nsl-1`. Pick the slopes so that the piecewise-linear function equals the original nonlinear function when `Trans[i,j]` is `(k/nsl) * limit[i,j]` for any k from 0 to `nsl-1`, or when `Trans[i,j]` is `(nsl-1/4)/nsl * limit[i,j]`.

Check your model by showing that you get the same results as in (b) when `nsl` is 3. Then, by trying higher values of `nsl`, determine how many linear pieces the approximation requires in order to determine all shipment amounts to within about 10% of the amounts recommended by the original nonlinear model.

17-2. This exercise asks how you might convert the demand constraints in the transportation model of Figure 3-1a into the kind of ''soft'' constraints described in Section 17.2.

Suppose that instead of a single parameter called `demand[j]` at each destination j, you are given the following four parameters that describe a more complicated situation:

`dem_min_abs[j]`	absolute minimum that must be shipped to j
`dem_min_ask[j]`	preferred minimum amount shipped to j
`dem_max_ask[j]`	preferred maximum amount shipped to j
`dem_max_abs[j]`	absolute maximum that may be shipped to j

There are also two penalty costs for shipment amounts outside of the preferred limits:

`dem_min_pen`	penalty per unit that shipments fall below `dem_min_ask[j]`
`dem_max_pen`	penalty per unit that shipments exceed `dem_max_ask[j]`

Because the total shipped to j is no longer fixed, a new variable `Receive[j]` is introduced to represent the amount received at j.

(a) Modify the model of Figure 3-1a to use this new information. The modifications will involve declaring `Receive[j]` with the appropriate lower and upper bounds, adding a three-piece piecewise-linear penalty term to the objective function, and substituting `Receive[j]` for `demand[j]` in the constraints.

(b) Add the following demand information to the data of Figure 3-1b:

	dem_min_abs	dem_min_ask	dem_max_ask	dem_max_abs
FRA	800	850	950	1100
DET	975	1100	1225	1250
LAN	600	600	625	625
WIN	350	375	450	500
STL	1200	1500	1800	2000
FRE	1100	1100	1100	1125
LAF	800	900	1050	1175

Let `dem_min_pen` and `dem_max_pen` be 2 and 4, respectively. Find the new optimal solution. In the solution, which destinations receive shipments that are outside the preferred levels?

17-3. When the diet model of Figure 2-1 is run with the data of Figure 2-3, there is no feasible solution. This exercise asks you to use the ideas of Section 17.2 to find some good near-feasible solutions.

(a) Modify the model so that it is possible, at a very high penalty, to purchase more than the specified maximum of a food. In the resulting solution, which maximums are exceeded?

(b) Modify the model so that it is possible, at a very high penalty, to supply more than the specified maximum of a nutrient. In the resulting solution, which maximums are exceeded?

(c) Using extremely large penalties, such as 10^{20} may give the solver numerical difficulties. Experiment to see how available solvers behave when you use penalty terms like 10^{20} and 10^{30}.

17-4. In the model of Exercise 4-4(b), the change in crews from one period to the next is limited to some number M. As an alternative to imposing this limit, suppose that we introduce a new variable D_t that represents the change in number of crews (in all shifts) at period t. This variable may be positive, indicating an increase in crews over the previous period, or negative, indicating a decrease in crews.

To make use of this variable, we introduce a defining constraint,

$$D_t = \sum_{s \in S} (Y_{st} - Y_{s,t-1}),$$

for each $t = 1, \ldots, T$. We then estimate costs of c^+ per crew added from period to period, and c^- per crew dropped from period to period; as a result, the following cost must be included in the objective for each month t:

$$c^- D_t, \qquad \text{if } D_t < 0;$$
$$c^+ D_t, \qquad \text{if } D_t > 0.$$

Reformulate the model in AMPL accordingly, using a piecewise-linear function to represent this extra cost.

Solve using $c^- = -20000$ and $c^+ = 100000$, together with the data previously given. How does this solution compare to the one from Exercise 4-4(b)?

17-5. The following "credit scoring" problem appears in many contexts, including the processing of credit card applications. A set APPL of people apply for a card, each answering a set QUES of questions on the application. The response of person i to question j is converted to a number, ans[i,j]; typical numbers are years at current address, monthly income, and a home ownership indicator (say, 1 if a home is owned and 0 otherwise).

To summarize the responses, the card issuer chooses weights Wt[j], from which a score for each person i in APPL is computed by the linear formula

```
sum {j in QUES} ans[i,j] * Wt[j]
```

The issuer also chooses a cutoff, Cut; credit is granted when an applicant's score is greater than or equal to the cutoff, and is denied when the score is less than the cutoff. In this way the decision can be made objectively (if not always most wisely).

To choose the weights and the cutoff, the card issuer collects a sample of previously accepted applications, some from people who turned out to be good customers, and some from people who never paid their bills. If we denote these two collections of people by sets GOOD and BAD, then the ideal weights and cutoff (for this data) would satisfy

```
sum {j in QUES} ans[i,j] * Wt[j] >= Cut    for each i in GOOD
sum {j in QUES} ans[i,j] * Wt[j] <  Cut    for each i in BAD
```

Since the relationship between answers to an application and creditworthiness is imprecise at best, however, no values of Wt[j] and Cut can be found to satisfy all of these inequalities. Instead, the issuer has to choose values that are merely the best possible, in some sense. There are any number of ways to make such a choice; here, naturally, we consider an optimization approach.

(a) Suppose that we define a new variable `Diff[i]` that equals the difference between person i's score and the cutoff:

```
Diff[i] = sum {j in QUES} ans[i,j] * Wt[j] - Cut
```

Clearly the undesirable cases are where `Diff[i]` is negative for i in GOOD, and where it is non-negative for i in BAD. To discourage these cases, we can tell the issuer to minimize the function

```
sum {i in GOOD} max(0,-Diff[i]) + sum {i in BAD} max(0,Diff[i])
```

Explain why minimizing this function tends to produce a desirable choice of weights and cutoff.

(b) The expression above is a piecewise-linear function of the variables `Diff[i]`. Rewrite it using AMPL's notation for piecewise-linear functions.

(c) Incorporate the expression from (b) into an AMPL model for finding the weights and cutoff.

(d) Given this approach, any positive value for `Cut` is as good as any other. We can fix it at a convenient round number — say, 100. Explain why this is the case.

(e) Using a `Cut` of 100, apply the model to the following imaginary credit data:

```
set GOOD := _17 _18 _19 _22 _24 _26 _28 _29 ;
set BAD  := _15 _16 _20 _21 _23 _25 _27 _30 ;

set QUES := Q1 Q2 R1 R2 R3 S2 T4 ;

param ans:   Q1    Q2    R1    R2    R3    S2    T4 :=
      _15   1.0    10    15    20    10     8    10
      _16   0.0     5    15    40     8    10     8
      _17   0.5    10    25    35     8    10    10
      _18   1.5    10    25    30     8     6    10
      _19   1.5     5    20    25     8     8     8
      _20   1.0     5     5    30     8     8     6
      _21   1.0    10    20    30     8    10    10
      _22   0.5    10    25    40     8     8    10
      _23   0.5    10    25    25     8     8    14
      _24   1.0    10    15    40     8    10    10
      _25   0.0     5    15    15    10    12    10
      _26   0.5    10    15    20     8    10    10
      _27   1.0     5    10    25    10     8     6
      _28   0.0     5    15    40     8    10     8
      _29   1.0     5    15    40     8     8    10
      _30   1.5     5    20    25    10    10    14 ;
```

What are the chosen weights? Using these weights, how many of the good customers would be denied a card, and how many of the bad risks would be granted one?

You should find that a lot of the bad risks have scores right at the cutoff. Why does this happen in the solution? How might you adjust the cutoff to deal with it?

(f) To force scores further away from the cutoff (in the desired direction), it might be preferable to use the following objective,

```
sum {i in GOOD} max(0,-Diff[i]+offset) +
sum {i in BAD} max(0,Diff[i]+offset)
```

where `offset` is a positive parameter whose value is supplied. Explain why this change has the desired effect. Try offset values of 2 and 10 and compare the results with those in (e).

(g) Suppose that giving a card to a bad credit risk is considered much more undesirable than refusing a card to a good credit risk. How would you change the model to take this into account?

(h) Suppose that when someone's application is accepted, his or her score is also used to suggest an initial credit limit. Thus it is particularly important that bad credit risks not receive very large scores. How would you add pieces to the piecewise-linear objective function terms to account for this concern?

17-6. In Exercise 18-3, we suggest a way to estimate position, velocity and acceleration values from imprecise data, by minimizing a nonlinear "sum of squares" function:

$$\sum_{j=1}^{n} [h_j - (a_0 - a_1 t_j - \tfrac{1}{2} a_2 t_j^2)]^2.$$

An alternative approach instead minimizes a sum of absolute values:

$$\sum_{j=1}^{n} |h_j - (a_0 - a_1 t_j - \tfrac{1}{2} a_2 t_j^2)|.$$

(a) Substitute the sum of absolute values directly for the sum of squares in the model from Exercise 18-3, first with the `abs` function, and then with AMPL's explicit piecewise-linear notation.

Explain why neither of these formulations is likely to be handled effectively by any solver.

(b) To model this situation effectively, we introduce variables e_j to represent the individual formulas $h_j - (a_0 - a_1 t_j - \tfrac{1}{2} a_2 t_j^2)$ whose absolute values are being taken. Then we can express the minimization of the sum of absolute values as the following constrained optimization problem:

$$\text{Minimize } \sum_{j=1}^{n} |e_j|$$

$$\text{Subject to } \quad e_j = h_j - (a_0 - a_1 t_j - \tfrac{1}{2} a_2 t_j^2), \; j = 1, \ldots, n$$

Write an AMPL model for this formulation, using the piecewise-linear notation for the terms $|e_j|$.

(c) Solve for a_0, a_1, and a_2 using the data from Exercise 18-3. How much difference is there between this estimate and the least-squares one?

Use `display` to print the e_j values for both the least-squares and the least-absolute-values solutions. What is the most obvious qualitative difference?

(d) Yet another possibility is to focus on the greatest absolute deviation, rather than the sum:

$$\max_{j=1,\ldots,n} |h_j - (a_0 - a_1 t_j - \tfrac{1}{2} a_2 t_j^2)|.$$

Formulate an AMPL linear program that will minimize this quantity, and test it on the same data as before. Compare the resulting estimates and e_j values. Which of the three estimates would you choose in this case?

17-7. A planar *structure* consists of a set of *joints* connected by *bars*. For example, in the following diagram, the joints are represented by circles, and the bars by lines between two circles:

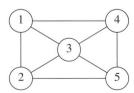

Consider the problem of finding a minimum-weight structure to meet certain external forces. We let J be the set of joints, and $B \subseteq J \times J$ be the set of *admissible* bars; for the diagram above, we could take $J = \{1,2,3,4,5\}$, and

$$B = \{(1,2), (1,3), (1,4), (2,3), (2,5), (3,4), (3,5), (4,5)\}.$$

The "origin" and "destination" of a bar are arbitrary. The bar between joints 1 and 2, for example, could be represented in B by either $(1,2)$ or $(2,1)$, but it need not be represented by both.

We can use two-dimensional Euclidean coordinates to specify the position of each joint in the plane, taking some arbitrary point as the origin:

a_i^x horizontal position of joint i relative to the origin

a_i^y vertical position of joint i relative to the origin

For the example, if the origin lies exactly at joint 2, we might have

$$(a_1^x, a_1^y) = (0, 2), \ (a_2^x, a_2^y) = (0, 0), \ (a_3^x, a_3^y) = (2, 1),$$
$$(a_4^x, a_4^y) = (4, 2), \ (a_5^x, a_5^y) = (4, 0).$$

The remaining data consist of the external forces on the joints:

f_i^x horizontal component of the external force on joint i

f_i^y vertical component of the external force on joint i

To resist this force, a subset $S \subseteq J$ of joints is fixed in position. (It can be proved that fixing two joints is sufficient to guarantee a solution.)

The external forces induce stresses on the bars, which we can represent as

F_{ij} if > 0, tension on bar (i,j)

        if < 0, compression of bar (i,j)

A set of stresses is in *equilibrium* if the external forces, tensions and compressions balance at all joints, in both the horizontal and vertical components — except at the fixed joints. That is, for each joint $k \notin S$,

$$\sum_{i \in J:(i,k) \in B} c_{ik}^x F_{ik} - \sum_{j \in J:(k,j) \in B} c_{kj}^x F_{kj} = f_k^x$$

$$\sum_{i \in J:(i,k) \in B} c_{ik}^y F_{ik} - \sum_{j \in J:(k,j) \in B} c_{kj}^y F_{kj} = f_k^y,$$

where c_{st}^x and c_{st}^y are the cosines of the direction from joint s to joint t with the horizontal and vertical axes,

$$c_{st}^x = (a_t^x - a_s^x)/l_{st},$$
$$c_{st}^y = (a_t^y - a_s^y)/l_{st},$$

and l_{st} is the length of the bar (s,t):

$$l_{st} = \sqrt{(a_t^x - a_s^x)^2 + (a_t^y - a_s^y)^2}.$$

In general, there are infinitely many different sets of equilibrium stresses. However, it can be shown that a given system of stresses will be realized in a structure of minimum weight if and only if the cross-sectional areas of the bars are proportional to the absolute values of the stresses. Since the weight of a bar is proportional to the cross section times length, we can take the (suitably scaled) weight of bar (i,j) to be

$$w_{ij} = l_{ij} \cdot |F_{ij}|.$$

The problem is then to find a system of stresses F_{ij} that meet the equilibrium conditions, and that minimize the sum of the weights w_{ij} over all bars $(i,j) \in B$.

(a) The indexing sets for this linear program can be declared in AMPL as:

```
set joints;
set fixed within joints;
set bars within {i in joints, j in joints: i <> j};
```

Using these set declarations, formulate an AMPL model for the minimum-weight structural design problem. Use the piecewise-linear notation of this chapter to represent the absolute-value terms in the objective function.

(b) Now consider in particular a structure that has the following joints:

Assume that there is one unit horizontally and vertically between joints, and that the origin is at the lower left; thus $(a_1^x, a_1^y) = (0, 2)$ and $(a_{15}^x, a_{15}^y) = (5, 0)$.

Let there be external forces of 3.25 and 1.75 units straight downward on joints 1 and 7, so that $f_1^y = -3.25$, $f_7^y = -1.75$, and otherwise all $f_i^x = 0$ and $f_i^y = 0$. Let $S = \{6, 15\}$. Finally, let the admissible bars consist of all possible bars that do not go directly through a joint; for example, $(1, 2)$ or $(1, 9)$ or $(1, 13)$ would be admissible, but not $(1, 3)$ or $(1, 12)$ or $(1, 14)$.

Determine all the data for the problem that is needed by the linear program, and represent it as AMPL data statements.

(c) Use AMPL to solve the linear program and to examine the minimum-weight structure that is determined.

Draw a diagram of the optimal structure, indicating the cross sections of the bars and the nature of the stresses. If there is zero force on a bar, it has a cross section of zero, and may be left out of your diagram.

(d) Repeat parts (b) and (c) for the case in which all possible bars are admissible. Is the resulting structure different? Is it any lighter?

18

Nonlinear Programs

Although any model that violates the linearity rules of Chapter 8 is ''not linear'', the term ''nonlinear program'' is traditionally used in a more narrow sense. For our purposes in this chapter, a nonlinear program, like a linear program, has continuous (rather than integer or discrete) variables; the expressions in its objective and constraints need not be linear, but they must represent ''smooth'' functions. Intuitively, a smooth function of one variable has a graph like that in Figure 18-1a, for which there is a well-defined slope at every point; there are no jumps as in Figure 18-1b, or kinks as in Figure 18-1c. Mathematically, a smooth function of any number of variables must be continuous and must have a well-defined gradient (vector of first derivatives) at every point; Figures 18-1b and 18-1c exhibit points at which a function is discontinuous and nondifferentiable, respectively.

Optimization problems in functions of this kind have been singled out for several reasons: because they are readily distinguished from other ''not linear'' problems, because they have many and varied applications, and because they are amenable to solution by well-established types of algorithms. Indeed, most solvers for nonlinear programming use methods that rely on the assumptions of continuity and differentiability. Even with these assumptions, nonlinear programs are typically a lot harder to formulate and solve than comparable linear ones.

This chapter begins with an introduction to sources of nonlinearity in mathematical programs. We do not try to cover the variety of nonlinear models systematically, but instead give a few examples to indicate why and how nonlinearities occur. Subsequent sections discuss the implications of nonlinearity for AMPL variables and expressions. Finally, we point out some of the difficulties that you are likely to confront in trying to solve nonlinear programs.

While the focus of this chapter is on nonlinear optimization, keep in mind that AMPL can also express systems of nonlinear equations or inequalities, even if there is no objective to optimize. There exist solvers specialized to this case, and many solvers for nonlinear optimization can also do a decent job of finding a feasible solution to an equation or inequality system.

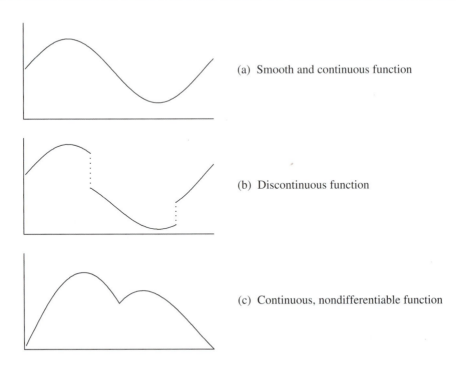

(a) Smooth and continuous function

(b) Discontinuous function

(c) Continuous, nondifferentiable function

Figure 18-1: Classes of nonlinear functions.

18.1 Sources of nonlinearity

We discuss here three ways that nonlinearities come to be included in optimization models: by dropping a linearity assumption, by constructing a nonlinear function to achieve a desired effect, and by modeling an inherently nonlinear physical process.

As an example, we describe some nonlinear variants of the linear network flow model `net1.mod` introduced in Chapter 15 (Figure 15-2a). This linear program's objective is to minimize total shipping cost,

```
minimize Total_Cost:
    sum {(i,j) in LINKS} cost[i,j] * Ship[i,j];
```

where `cost[i,j]` and `Ship[i,j]` represent the cost per unit and total units shipped between cities `i` and `j`, with `LINKS` being the set of all city pairs between which shipment routes exist. The constraints are balance of flow at each city:

```
subject to Balance {k in CITIES}:
    supply[k] + sum {(i,k) in LINKS} Ship[i,k]
        = demand[k] + sum {(k,j) in LINKS} Ship[k,j];
```

with the nonnegative parameters `supply[i]` and `demand[i]` representing the units either available or required at city `i`.

Dropping a linearity assumption

The linear network flow model assumes that each unit shipped from city `i` to city `j` incurs the same shipping cost, `cost[i,j]`. Figure 18-2a shows a typical plot of shipping cost versus amount shipped in this case; the plot is a line with slope `cost[i,j]` (hence the term linear). The other plots in Figure 18-2 show a variety of other ways, none of them linear, in which shipping cost could depend on the shipment amount.

In Figure 18-2b the cost also tends to increase linearly with the amount shipped, but at certain critical amounts the cost per unit (that is, the slope of the line) makes an abrupt change. This kind of function is called piecewise-linear. It is not linear, strictly speaking, but it is also not smoothly nonlinear. The use of piecewise-linear objectives is the topic of Chapter 17.

In Figure 18-2c the function itself jumps abruptly. When nothing is shipped, the shipping cost is zero; but when there is any shipment at all, the cost is linear starting from a value greater than zero. In this case there is a fixed cost for using the link from `i` to `j`, plus a variable cost per unit shipped. Again, this is not a function that can be handled by linear programming techniques, but it is also not a smooth nonlinear function. Fixed costs are most commonly handled by use of integer variables, which are the topic of Chapter 20.

The remaining plots illustrate the sorts of smooth nonlinear functions that we want to consider in this chapter. Figure 18-2d shows a kind of concave cost function. The incremental cost for each additional unit shipped (that is, the slope of the plot) is great at first, but becomes less as more units are shipped; after a certain point, the cost is nearly linear. This is a continuous alternative to the fixed cost function of Figure 18-2c. It could also be used to approximate the cost for a situation (resembling Figure 18-2b) in which volume discounts become available as the amount shipped increases.

Figure 18-2e shows a kind of convex cost function. The cost is more or less linear for smaller shipments, but rises steeply as shipment amounts approach some critical amount. This sort of function would be used to model a situation in which the lowest cost shippers are used first, while shipping becomes progressively more expensive as the number of units increases. The critical amount represents, in effect, an upper bound on the shipments.

These are some of the simplest functional forms. The functions that you consider will depend on the kind of situation that you are trying to model. Figure 18-2f shows a possibility that is neither concave nor convex, combining features of the previous two examples.

Whereas linear functions are essentially all the same except for the choice of coefficients (or slopes), nonlinear functions can be defined by an infinite variety of different formulas. Thus in building a nonlinear programming model, it is up to you to derive or specify nonlinear functions that properly represent the situation at hand. In the objective

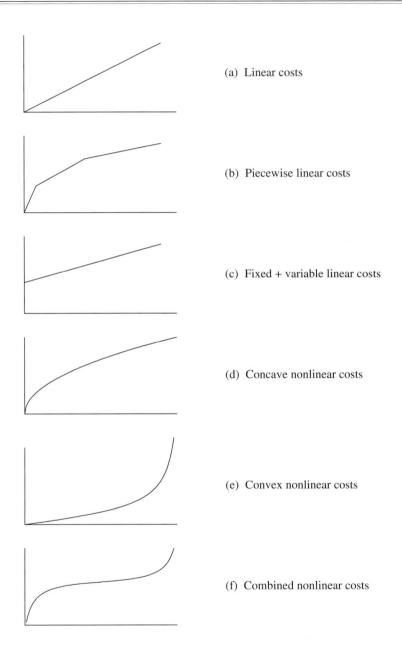

(a) Linear costs

(b) Piecewise linear costs

(c) Fixed + variable linear costs

(d) Concave nonlinear costs

(e) Convex nonlinear costs

(f) Combined nonlinear costs

Figure 18-2: Nonlinear cost functions.

of the transportation example, for instance, one possibility would be to replace the product `cost[i,j] * Ship[i,j]` by

```
(cost1[i,j] + cost2[i,j]*Ship[i,j]) / (1 + Ship[i,j])
   * Ship[i,j]
```

This function grows quickly at small shipment levels but levels off to essentially linear at
larger levels. Thus it represents one way to implement the curve shown in Figure 18-2d.

Another way to approach the specification of a nonlinear objective function is to
focus on the *slopes* of the plots in Figure 18-2. In the linear case of Figure 18-2a, the
slope of the plot is constant; that is why we can use a single parameter `cost[i,j]` to
represent the cost per unit shipped. In the piecewise-linear case of Figure 18-2b, the
slope is constant within each interval; we can express such piecewise-linear functions as
explained in Chapter 17.

In the nonlinear case, however, the slope varies continuously with the amount
shipped. This suggests that we go back to our original linear formulation of the network
flow problem, and turn the parameter `cost[i,j]` into a variable `Cost[i,j]`:

```
var Cost {ORIG,DEST};        # shipment costs per unit
var Ship {ORIG,DEST} >= 0;   # units to ship

minimize Total_Cost:
    sum {i in ORIG, j in DEST} Cost[i,j] * Ship[i,j];
```

This is no longer a linear objective, because it multiplies a variable by another variable.
We add some equations to specify how the cost relates to the amount shipped:

```
subject to Cost_Relation {i in ORIG, j in DEST}:
    Cost[i,j] =
       (cost1[i,j] + cost2[i,j]*Ship[i,j]) / (1 + Ship[i,j]);
```

These equations are also nonlinear, because they involve division by an expression that
contains a variable. It is easy to see that `Cost[i,j]` is near `cost1[i,j]` where shipments are near zero, but levels off to `cost2[i,j]` at sufficiently high shipment levels.
Thus the concave cost of Figure 18-2d is realized provided that the first cost is greater
than the second.

Assumptions of nonlinearity can be found in constraints as well. The constraints of
the network flow model embody only a weak linearity assumption, to the effect that the
total shipped out of a city is the sum of the shipments to the other cities. But in the production model of Figure 1-6a, the constraint

```
subject to Time {s in STAGE}:
    sum {p in PROD} (1/rate[p,s]) * Make[p] <= avail[s];
```

embodies a strong assumption that the number of hours used in each stage `s` of making
each product `p` grows linearly with the level of production.

Achieving a nonlinear effect

Sometimes nonlinearity arises from a need to model a situation in which a linear function could not possibly exhibit the desired behavior.

In a network model of traffic flows, as one example, it may be necessary to take congestion into account. The total time to traverse a shipment link should be essentially a constant for small shipment amounts, but should increase rapidly towards infinity as the capacity of the link is approached. No linear function has this property, so we are forced to make travel time a nonlinear function of shipment load in order to get the desired effect.

One possibility for expressing the travel time is given by the function

```
time[i,j] + (sens[i,j]*Ship[i,j]) / (1 - Ship[i,j]/cap[i,j])
```

This function is approximately `time[i,j]` for small values of `Ship[i,j]`, but goes to infinity as `Ship[i,j]` approaches `cap[i,j]`; a third parameter `sens[i,j]` governs the shape of the function between the two extremes. This function is always convex, and so has a graph resembling Figure 18-2e. (Exercise 18-4 suggests how this travel time function can be incorporated into a network model of traffic flows.)

As another example, we may want to allow demand to be satisfied only approximately. We can model this possibility by introducing a variable `Discrepancy[k]`, to represent the deviation of the amount delivered from the amount demanded. This variable, which can be either positive or negative, is added to the right-hand side of the balance constraint:

```
subject to Balance {k in CITIES}:
    supply[k] + sum {(i,k) in LINKS} Ship[i,k]
        = demand[k] + Discrepancy[k] +
            sum {(k,j) in LINKS} Ship[k,j];
```

One established approach for keeping the discrepancy from becoming too large is to add a penalty cost to the objective. If this penalty is proportional to the amount of the discrepancy, then we have a convex piecewise-linear penalty term,

```
minimize Total_Cost:
    sum {(i,j) in LINKS} cost[i,j] * Ship[i,j] +
    sum {k in CITIES} pen * <<-1,1; 0>> Discrepancy[k];
```

where `pen` is a positive parameter. AMPL readily transforms this objective to a linear one.

This form of penalty may not achieve the effect that we want, however, because it penalizes each unit of the discrepancy equally. To discourage large discrepancies, we would want the penalty to become steadily larger per unit as the discrepancy becomes worse, but this is not a property that can be achieved by linear penalty functions (or piecewise-linear ones that have a finite number of pieces). Instead a more appropriate penalty function would be quadratic:

```
minimize Total_Cost:
    sum {(i,j) in LINKS} cost[i,j] * Ship[i,j] +
    sum {k in CITIES} pen * Discrepancy[k] ^ 2;
```

Nonlinear objectives that are a sum of squares of some quantities are common in optimization problems that involve approximation or data fitting.

Modeling an inherently nonlinear process

There are many sources of nonlinearity in models of physical activities like oil refining, power transmission, and structural design. More often than not, the nonlinearities in these models cannot be traced to the relaxation of any linearity assumptions, but are a consequence of inherently nonlinear relationships that govern forces, volumes, currents and so forth. The forms of the nonlinear functions in physical models may be easier to determine, because they arise from empirical measurements and the underlying laws of physics (rather than economics). On the other hand, the nonlinearities in physical models tend to involve more complicated functional forms and interactions among the variables.

As a simple example, a model of a natural gas pipeline network must incorporate not only the shipments between cities but also the pressures at individual cities, which are subject to certain bounds. Thus in addition to the flow variables `Ship[i,j]` the model must define a variable `Press[k]` to represent the pressure at each city `k`. If the pressure is greater at city `i` than at city `j`, then the flow is from `i` to `j` and is related to the pressure by

```
Flow[i,j]^2 = c[i,j]^2 * (Press[i]^2 - Press[j]^2)
```

where `c[i,j]` is a constant determined by the length, diameter, and efficiency of the pipe and the properties of the gas. Compressors and valves along the pipeline give rise to different nonlinear flow relationships. Other kinds of networks, notably for transmission of electricity, have their own nonlinear flow relationships that are dictated by the physics of the situation.

If you know the algebraic form of a nonlinear expression that you want to include in your model, you can probably see a way to write it in AMPL. The next two sections of this chapter consider some of the specific issues and features relevant to declaring variables for nonlinear programs and to writing nonlinear expressions. Lest you get carried away by the ease of writing nonlinear expressions, however, the last section offers some cautionary advice on solving nonlinear programs.

18.2 Nonlinear variables

Although AMPL variables are declared for nonlinear programs in the same way as for linear programs, two features of variables — initial values and automatic substitution — are particularly useful in working with nonlinear models.

Initial values of variables

You may specify values for AMPL variables. Prior to optimization, these "initial" values can be displayed and manipulated by AMPL commands. When you type `solve`, they are passed to the solver, which may use them as a starting guess at the solution. After the solver has finished its work, the initial values are replaced by the computed optimal ones.

All of the AMPL features for assigning values to parameters are also available for variables. A `var` declaration may also specify initial values in an optional `:=` phrase; for the transportation example, you can write

```
var Ship {LINKS} >= 0, := 1;
```

to set every `Ship[i,j]` initially to 1, or

```
var Ship {(i,j) in LINKS} >= 0, := cap[i,j] - 1;
```

to initialize each `Ship[i,j]` to 1 less than `cap[i,j]`. Alternatively, initial values may be given in a data statement along with the other data for a model:

```
var Ship:    FRA   DET   LAN   WIN   STL   FRE   LAF :=
        GARY   800   400   400   200   400   200   200
        CLEV   800   800   800   600   600   500   600
        PITT   800   800   800   200   300   800   500 ;
```

Any of the data statements for parameters can also be used for variables, as explained in Section 9.4.

All of these features for assigning values to the regular ("primal") variables also apply to the dual variables associated with constraints (Section 12.5). AMPL interprets an assignment to a constraint name as an assignment to the associated dual variable or (in the terminology more common in nonlinear programming) to the associated Lagrange multiplier. A few solvers, such as MINOS, can make use of initial values for these multipliers.

You can often speed up the work of the solver by suggesting good initial values. This can be so even for linear programs, but the effect is much stronger in the nonlinear case. The choice of an initial guess may determine what value of the objective is found to be "optimal" by the solver, or even whether the solver finds any optimal solution at all. These possibilities are discussed further in the last section of this chapter.

If you don't give any initial value for a variable, then AMPL will tentatively set it to zero. If the solver incorporates a routine for determining initial values, then it may re-set the values of any uninitialized variables, while making use of the values of variables that have been initialized. Otherwise, uninitialized variables will be left at zero. Although zero is an obvious starting point, it has no special significance; for some of the examples that we will give in Section 18.4, the solver cannot optimize successfully unless the initial values are reset away from zero.

Automatic substitution of variables

The issue of substituting variables has already arisen in an example of the previous section, where we declared variables to represent the shipping costs, and then defined them in terms of other variables by use of a constraint:

```
subject to Cost_Relation {(i,j) in LINKS}:
   Cost[i,j] =
      (cost1[i,j] + cost2[i,j]*Ship[i,j]) / (1 + Ship[i,j]);
```

If the expression to the right of the = sign is substituted for every appearance of `Cost[i,j]`, the `Cost` variables can be eliminated from the model, and these constraints need not be passed to the solver. There are two ways in which you can tell AMPL to make such substitutions automatically.

First, by changing option `substout` from its default value of zero to one, you can tell AMPL to look for all "defining" constraints that have the form shown above: a single variable to the left of an = sign. When this alternative is employed, AMPL tries to use as many of these constraints as possible to substitute variables out of the model. After you have typed `solve` and a nonlinear program has been generated from a model and data, the constraints are scanned in the order that they appeared in the model. A constraint is identified as "defining" provided that

- it has just one variable to the left of an = sign;
- the left-hand variable's declaration did not specify any restrictions, such as integrality or bounds; and
- the left-hand variable has not already appeared in a constraint identified as defining.

The expression to the right of the = sign is then substituted for every appearance of the left-hand variable in the other constraints, and the defining constraint is dropped. These rules give AMPL an easy way to avoid circular substitutions, but they do imply that the nature and number of substitutions may depend on the ordering of the constraints.

As an alternative, if you want to specify explicitly that a certain collection of variables is to be substituted out, use an = phrase in the declarations of the variables. For the preceding example, you could write:

```
var Cost {(i,j) in LINKS}
   = (cost1[i,j] + cost2[i,j]*Ship[i,j]) / (1 + Ship[i,j]);
```

Then the variables `Cost[i,j]` would be replaced everywhere by the expression following the = sign. Declarations of this kind can appear in any order, subject to the usual requirement that every variable appearing in an = phrase must be previously defined.

Variables that can be substituted out are not mathematically necessary to the optimization problem. Nevertheless, they often serve an essential descriptive purpose; by associating names with nonlinear expressions, they permit more complicated expressions to be written understandably. Moreover, even though these variables have been removed from the problem sent to the solver, their names remain available for use in browsing through the results of optimization.

When the same nonlinear expression appears more than once in the objective and constraints, introducing a defined variable to represent it may make the model more concise as well as more readable. AMPL also processes such a substitution efficiently. In generating a representation of the nonlinear program for the solver, AMPL does not substitute a copy of the whole defining expression for each occurrence of a defined variable. Instead it breaks the expression into a linear and a nonlinear part, and saves one copy of the nonlinear part together with a list of the locations where its value is to be substituted; only the terms of the linear part are substituted explicitly in multiple locations. This separate treatment of linear terms is advantageous for solvers (such as MINOS) that handle the linear terms specially, but it may be turned off by setting option `linelim` to zero.

From the solver's standpoint, substitutions reduce the number of constraints and variables, but tend to make the constraint and objective expressions more complex. As a result, there are circumstances in which a solver will perform better if defined variables are not substituted out. When developing a new model, you may have to experiment to determine which substitutions give the best results.

18.3 Nonlinear expressions

Any of AMPL's arithmetic operators (Table 7-1) and arithmetic functions (Table 7-2) may be applied to variables as well as parameters. If any resulting objective or constraint does not satisfy the rules for linearity (Chapter 8) or piecewise-linearity (Chapter 17), AMPL treats it as ''not linear''. When you type `solve`, AMPL passes along instructions that are sufficient for your solver to evaluate every expression in every objective and constraint, together with derivatives if appropriate.

If you are using a typical nonlinear solver, it is up to you to define your objective and constraints in terms of the ''smooth'' functions that the solver requires. The generality of AMPL's expression syntax can be misleading in this regard. For example, if you are trying to use variables `Flow[i,j]` representing flow between points `i` and `j`, it is tempting to write expressions like

```
cost[i,j] * abs(Flow[i,j])
```

or

```
if Flow[i,j] = 0 then 0 else base[i,j] + cost[i,j]*Flow[i,j]
```

These are certainly not linear, but the first is not smooth (its slope changes abruptly at zero) and the second is not even continuous (its value jumps suddenly at zero). If you try to use such expressions, AMPL will not complain, and your solver may even return what it claims to be an optimal solution — but the results could be wrong.

Expressions that apply nonsmooth functions (such as `abs`, `min`, and `max`) to variables generally produce nonsmooth results; the same is true of `if-then-else` expressions in which a condition involving variables follows `if`. Nevertheless, there are useful exceptions where a carefully written expression can preserve smoothness. As an exam-

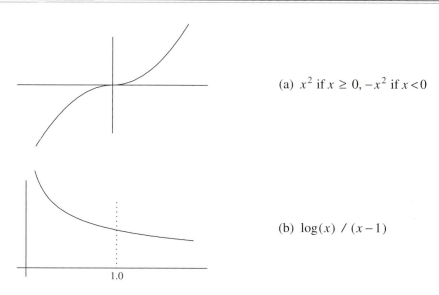

(a) x^2 if $x \geq 0$, $-x^2$ if $x < 0$

(b) $\log(x) / (x-1)$

Figure 18-3: Smooth nonlinear functions.

ple, consider again the flow-pressure relationship from Section 18.1. If the pressure is greater at city `i` than at city `j`, then the flow is from `i` to `j` and is related to the pressure by

```
Flow[i,j]^2 = c[i,j]^2 * (Press[i]^2 - Press[j]^2)
```

If instead the pressure is greater at city `j` than at city `i`, a similar equation can be written:

```
Flow[j,i]^2 = c[j,i]^2 * (Press[j]^2 - Press[i]^2)
```

But since the constants `c[i,j]` and `c[j,i]` refer to the same pipe, they are equal. Thus instead of defining a separate variable for flow in each direction, we can let `Flow[i,j]` be unrestricted in sign, with a positive value indicating flow from `i` to `j` and a negative value indicating the opposite. Using this variable, the previous pair of flow-pressure constraints can be replaced by one:

```
(if Flow[i,j] >= 0 then Flow[i,j]^2 else -Flow[i,j]^2)
   = c[i,j]^2 * (Press[i]^2 - Press[j]^2)
```

Normally the use of an `if` expression would give rise to a nonsmooth constraint, but in this case it gives a function whose two quadratic halves "meet" smoothly where `Flow[i,j]` is zero, as seen in Figure 18-3a.

As another example, the convex function in Figure 18-3b is most easily written `log(Flow[i,j]) / (Flow[i,j]-1)`, but unfortunately if `Flow[i,j]` is 1 this simplifies to 0/0, which would be reported as an error. In fact, this expression does not evaluate accurately if `Flow[i,j]` is merely very close to zero. If instead we write

```
if abs(Flow[i,j]-1) > 0.00001 then
    log(Flow[i,j])/(Flow[i,j]-1)
else
    1.5 - Flow[i,j]/2
```

a highly accurate linear approximation is substituted at small magnitudes of
Flow[i,j]. This alternative is not smooth in a literal mathematical sense, but it is
numerically close enough to being smooth to suffice for use with some solvers.

In the problem instance that it sends to a solver, AMPL distinguishes linear from nonlinear constraints and objectives, and breaks each constraint and objective expression into
a sum of linear terms plus a not-linear part. Additional terms that become linear due to
the fixing of some variables are recognized as linear. For example, in our example from
Section 18.1,

```
minimize Total_Cost:
    sum {i in ORIG, j in DEST} Cost[i,j] * Ship[i,j];
```

each fixing of a Cost[i,j] variable gives rise to a linear term; if all the Cost[i,j]
variables are fixed, then the objective is represented to the solver as linear. Variables
may be fixed by a fix command (Section 11.4) or through the actions of the presolve
phase (Section 14.1); although the presolving algorithm ignores nonlinear constraints, it
works with any linear constraints that are available, as well as any constraints that
become linear as variables are fixed.

AMPL's built-in functions are only some of the ones most commonly used in model
formulations. Libraries of other useful functions can be introduced when needed. To use
cumulative normal and inverse cumulative normal functions from a library called
statlib, for example, you would first load the library with a statement such as

```
load statlib.dll;
```

and declare the functions by use of AMPL function statements:

```
function cumnormal;
function invcumnormal;
```

Your model could then make use of these functions to form expressions such as
cumnormal(mean[i],sdev[i],Inv[i,t]) and invcumnormal(6). If these
functions are applied to variables, AMPL also arranges for function evaluations to be carried out during the solution process.

A function declaration specifies a library function's name and (optionally) its
required arguments. There may be any number of arguments, and even iterated collections of arguments. Each function's declaration must appear before its use. For your
convenience, a script containing the function declarations may be supplied along with
the library, so that a statement such as include statlib is sufficient to provide
access to all of the library's functions. Documentation for the library will indicate the
functions available and the numbers and meanings of their arguments.

Determining the correct `load` command may involve a number of details that depend on the type of system you're using and even its specific configuration. See Section A.22 for further discussion of the possibilities and the related `AMPLFUNC` option.

If you are ambitious, you can write and compile your own function libraries. Instructions and examples are available from the AMPL web site.

18.4 Pitfalls of nonlinear programming

While AMPL gives you the power to formulate diverse nonlinear optimization models, no solver can guarantee an acceptable solution every time you type `solve`. The algorithms used by solvers are susceptible to a variety of difficulties inherent in the complexities of nonlinear functions. This is in unhappy contrast to the linear case, where a well-designed solver can be relied upon to solve almost any linear program.

This section offers a brief introduction to the pitfalls of nonlinear programming. We focus on two common kinds of difficulties, function range violations and multiple local optima, and then mention several other traps more briefly.

For illustration we begin with the nonlinear transportation model shown in Figure 18-4. It is identical to our earlier transportation example (Figure 3-1a) except that the terms `cost[i,j] * Trans[i,j]` are replaced by nonlinear terms in the objective:

```
minimize Total_Cost:
    sum {i in ORIG, j in DEST}
        rate[i,j] * Trans[i,j] / (1 - Trans[i,j]/limit[i,j]);
```

Each term is a convex increasing function of `Trans[i,j]` like that depicted in Figure 18-2e; it is approximately linear with slope `rate[i,j]` at relatively small values of `Trans[i,j]`, but goes to infinity as `Trans[i,j]` approaches `limit[i,j]`. Associated data values, also similar to those used for the linear transportation example in Chapter 3, are given in Figure 18-5.

Function range violations

An attempt to solve using the model and data as given proves unsuccessful:

```
ampl: model nltrans.mod;
ampl: data nltrans.dat;

ampl: solve;
MINOS 5.5 Error evaluating objective Total_Cost
can't compute 8000/0
MINOS 5.5: solution aborted.
8 iterations, objective 0
```

The solver's message is cryptic, but strongly suggests a division by zero while evaluating the objective. That could only happen if the expression

```
1 - Trans[i,j]/limit[i,j]
```

```
set ORIG;    # origins
set DEST;    # destinations

param supply {ORIG} >= 0;    # amounts available at origins
param demand {DEST} >= 0;    # amounts required at destinations

   check: sum {i in ORIG} supply[i] = sum {j in DEST} demand[j];

param rate {ORIG,DEST} >= 0;    # base shipment costs per unit
param limit {ORIG,DEST} > 0;    # limit on units shipped

var Trans {i in ORIG, j in DEST} >= 0; # units to ship

minimize Total_Cost:
   sum {i in ORIG, j in DEST}
      rate[i,j] * Trans[i,j] / (1 - Trans[i,j]/limit[i,j]);

subject to Supply {i in ORIG}:
   sum {j in DEST} Trans[i,j] = supply[i];

subject to Demand {j in DEST}:
   sum {i in ORIG} Trans[i,j] = demand[j];
```

Figure 18-4: Nonlinear transportation model (`nltrans.mod`).

```
param: ORIG:    supply :=
          GARY    1400    CLEV    2600    PITT    2900 ;

param: DEST:    demand :=
          FRA      900    DET    1200    LAN     600
          WIN      400    STL    1700    FRE    1100
          LAF     1000 ;

param rate :    FRA    DET    LAN    WIN    STL    FRE    LAF :=
          GARY    39     14     11     14     16     82      8
          CLEV    27      9     12      9     26     95     17
          PITT    24     14     17     13     28     99     20 ;

param limit :   FRA    DET    LAN    WIN    STL    FRE    LAF :=
          GARY   500   1000   1000   1000    800    500   1000
          CLEV   500    800    800    800    500    500   1000
          PITT   800    600    600    600    500    500    900 ;
```

Figure 18-5: Data for nonlinear transportation model (`nltrans.dat`).

is zero at some point. If we use `display` to print the pairs where `Trans[i,j]` equals `limit[i,j]`:

```
ampl: display {i in ORIG, j in DEST: Trans[i,j] = limit[i,j]};
set {i in ORIG, j in DEST: Trans[i,j] == limit[i,j]}
        := (GARY,LAF) (PITT,LAN) ;

ampl: display Trans['GARY','LAF'], limit['GARY','LAF'];
Trans['GARY','LAF'] = 1000
limit['GARY','LAF'] = 1000
```

we can see the problem. The solver has allowed `Trans[GARY,LAF]` to have the value 1000, which equals `limit[GARY,LAF]`. As a result, the objective function term

```
rate[GARY,LAF] * Trans[GARY,LAF]
   / (1 - Trans[GARY,LAF]/limit[GARY,LAF])
```

evaluates to 8000/0. Since the solver is unable to evaluate the objective function, it gives up without finding an optimal solution.

Because the behavior of a nonlinear optimization algorithm can be sensitive to the choice of starting guess, we might hope that the solver will have greater success from a different start. To ensure that the comparison is meaningful, we first set

```
ampl: option send_statuses 0;
```

so that status information about variables that was returned by the previous solve will not be sent back to help determine a starting point for the next solve. Then AMPL's `let` command may be used to suggest, for example, a new initial value for each `Trans[i,j]` that is half of `limit[i,j]`:

```
ampl: let {i in ORIG, j in DEST} Trans[i,j] := limit[i,j]/2;
ampl: solve;
MINOS 5.5: the current point cannot be improved.
32 iterations, objective -7.385903389e+18
```

This time the solver runs to completion, but there is still something wrong. The objective is less than -10^{18}, or $-\infty$ for all practical purposes, and the solution is described as "cannot be improved" rather than optimal.

Examining the values of `Trans[i,j]/limit[i,j]` in the solution that the solver has returned gives a clue to the difficulty:

```
ampl: display {i in ORIG, j in DEST} Trans[i,j]/limit[i,j];
Trans[i,j]/limit[i,j] [*,*] (tr)
:         CLEV           GARY          PITT        :=
DET   -6.125e-14     0             2
FRA    0             1.5           0.1875
FRE    0.7           1             0.5
LAF    0.4           0.15          0.5
LAN    0.375         7.03288e-15   0.5
STL    2.9           0             0.5
WIN    0.125         0             0.5
;
```

These ratios show that the shipments for several pairs, such as `Trans[CLEV,STL]`, significantly exceed their limits. More seriously, `Trans[GARY,FRE]` seems to be right at `limit[GARY,FRE]`, since their ratio is given as 1. If we display them to full precision, however, we see:

```
ampl: option display_precision 0;
ampl: display Trans['GARY','FRE'], limit['GARY','FRE'];
Trans['GARY','FRE'] = 500.0000000000028
limit['GARY','FRE'] = 500
```

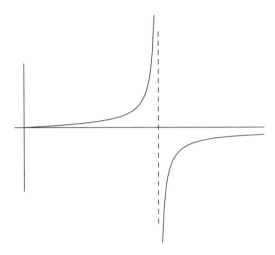

Figure 18-6: Singularity in cost function $y = x/(1 - x/c)$.

The variable is just slightly larger than the limit, so the cost term has a huge negative value. If we graph the entire cost function, as in Figure 18-6, we see that indeed the cost function goes off to $-\infty$ to the right of the singularity at limit[GARY,FRE].

The source of error in both runs above is our assumption that, since the objective goes to $+\infty$ as Trans[i,j] approaches limit[i,j] from below, the solver will keep Trans[i,j] between 0 and limit[i,j]. At least for this solver, we must enforce such an assumption by giving each Trans[i,j] an explicit upper bound that is slightly less than limit[i,j], but close enough not to affect the eventual optimal solution:

```
var Trans {i in ORIG, j in DEST} >= 0, <= .9999 * limit[i,j];
```

With this modification, the solver readily finds an optimum:

```
ampl: option display_precision 6;
ampl: model nltransb.mod; data nltrans.dat; solve;
MINOS 5.5: optimal solution found.
81 iterations, objective 1212117
ampl: display Trans;
Trans [*,*] (tr)
:        CLEV       GARY       PITT        :=
DET    586.372    187.385    426.242
FRA    294.993     81.2205   523.787
FRE    365.5      369.722    364.778
LAF    490.537      0        509.463
LAN    294.148      0        305.852
STL    469.691    761.672    468.637
WIN     98.7595     0        301.241
;
```

These values of the variables are well away from any singularity, with `Trans[i,j]/limit[i,j]` being less than 0.96 in every case. (If you change the starting guess to be `limit[i,j]/2` as before, you should find that the solution is the same but the solver needs only about half as many iterations to find it.)

The immediate lesson here is that nonlinear functions can behave quite badly outside the intended range of the variables. The incomplete graph in Figure 18-2e made this cost function look misleadingly well-behaved, whereas Figure 18-6 shows the need for a bound to keep the variable away from the singularity.

A more general lesson is that difficulties posed by a nonlinear function may lead the solver to fail in a variety of ways. When developing a nonlinear model, you need to be alert to bizarre results from the solver, and you may have to do some detective work to trace the problem back to a flaw in the model.

Multiple local optima

To illustrate a different kind of difficulty in nonlinear optimization, we consider a slightly modified objective function that has the following formula:

```
minimize Total_Cost:
  sum {i in ORIG, j in DEST}
    rate[i,j] * Trans[i,j]^0.8 / (1 - Trans[i,j]/limit[i,j]);
```

By raising the amount shipped to the power 0.8, we cause the cost function to be concave at lower shipment amounts and convex at higher amounts, in the manner of Figure 18-2f. Attempting to solve this new model, we again initially run into technical difficulties:

```
ampl: model nltransc.mod; data nltrans.dat; solve;
MINOS 5.5: Error evaluating objective Total_Cost:
                        can't evaluate pow'(0,0.8)
MINOS 5.5: solution aborted.
8 iterations, objective 0
```

This time our suspicion naturally centers upon `Trans[i,j]^0.8`, the only expression that we have changed in the model. A further clue is provided by the error message's reference to `pow'(0,0.8)`, which denotes the derivative of the exponential (power) function at zero. When `Trans[i,j]` is zero, this function has a well-defined value, but its derivative with respect to the variable — the slope of the graph in Figure 18-2f — is infinite. As a result, the partial derivative of the total cost with respect to any variable at zero cannot be returned to the solver; since the solver requires all the partial derivatives for its optimization algorithm, it gives up.

This is another variation on the range violation problem, and again it can be remedied by imposing some bounds to keep the solution away from troublesome points. In this case, we move the lower bound from zero to a very small positive number:

```
var Trans {i in ORIG, j in DEST}
    >= 1e-10, <= .9999 * limit[i,j], := 0;
```

We might also move the starting guess away from zero, but in this example the solver takes care of that automatically, since the initial values only suggest a starting point.

With the bounds adjusted, the solver runs normally and reports a solution:

```
ampl: model nltransd.mod; data nltrans.dat; solve;
MINOS 5.5: optimal solution found.
65 iterations, objective 427568.1225
ampl: display Trans;
Trans [*,*] (tr)
:        CLEV        GARY        PITT         :=
DET      689.091    1e-10        510.909
FRA      1e-10       199.005     700.995
FRE      385.326     326.135     388.54
LAF      885.965     114.035    1e-10
LAN      169.662    1e-10        430.338
STL      469.956     760.826     469.218
WIN      1e-10      1e-10        400
;
```

We can regard each 1e-10 as a zero, since such a small value is negligible in comparison with the rest of the solution.

Next we again try a starting guess at limit[i,j]/2, in the hope of speeding things up. This is the result:

```
ampl: let {i in ORIG, j in DEST} Trans[i,j] := limit[i,j]/2;
ampl: solve;
MINOS 5.5: optimal solution found.
40 iterations, objective 355438.2006

ampl: display Trans;
Trans [*,*] (tr)
:        CLEV        GARY        PITT         :=
DET      540.601     265.509     393.89
FRA      328.599    1e-10        571.401
FRE      364.639     371.628     363.732
LAF      491.262    1e-10        508.738
LAN      301.741    1e-10        298.259
STL      469.108     762.863     468.029
WIN      104.049    1e-10        295.951
;
```

Not only is the solution completely different, but the optimal value is 17% lower! The first solution could not truly have minimized the objective over all solutions that are feasible in the constraints.

Actually both solutions can be considered correct, in the sense that each is *locally* optimal. That is, each solution is less costly than any other nearby solutions. All of the classical methods of nonlinear optimization, which are the methods considered in this chapter, are designed to seek a local optimum. Started from one specified initial guess, these methods are not guaranteed to find a solution that is *globally* optimal, in the sense of giving the best objective value among all solutions that satisfy the constraints. In general, finding a global optimum is much harder than finding a local one.

Fortunately, there are many cases in which a local optimum is entirely satisfactory. When the objective and constraints satisfy certain properties, any local optimum is also global; the model considered at the beginning of this section is one example, where the convexity of the objective, together with the linearity of the constraints, guarantees that the solver will find a global optimum. (Linear programming is an even more special case with this property; that's why in previous chapters we never encountered local optima that were not global.)

Even when there is more than one local optimum, a knowledge of the situation being modeled may help you to identify the global one. Perhaps you can choose an initial solution near to the global optimum, or you can add some constraints that rule out regions known to contain local optima.

Finally, you may be content to find a very good local optimum, even if you don't have a guarantee that it is global. One straightforward approach is to try a series of starting points systematically, and take the best among the solutions. As a simple illustration, suppose that we declare the variables in our example as follows:

```
param alpha >=0, <= 1;

var Trans {i in ORIG, j in DEST}
   >= 1e-10, <= .9999 * limit[i,j], := alpha * limit[i,j];
```

For each choice of `alpha` we get a different starting guess, and potentially a different solution. Here are some resulting objective values for `alpha` ranging from 0 to 1:

alpha	Total_Cost
0.0	427568.1
0.1	366791.2
0.2	366791.2
0.3	366791.2
0.4	366791.2
0.5	355438.2
0.6	356531.5
0.7	376043.3
0.8	367014.4
0.9	402795.3
1.0	365827.2

The solution that we previously found for an `alpha` of 0.5 is still the best, but in light of these results we are now more inclined to believe that it is a very good solution. We might also observe that, although the reported objective value varies somewhat erratically with the choice of starting point — a feature of nonlinear programs generally — the second-best value of `Total_Cost` was found by setting `alpha` to 0.6. This suggests that a closer search of `alpha` values around 0.5 might be worthwhile.

Some of the more sophisticated methods for global optimization attempt to search through starting points in this way, but with a more elaborate and systematic procedure for deciding which starting points to try next. Others treat global optimization as more of a combinatorial problem, and apply solution methods motivated by those for integer programming (Chapter 20). Global optimization methods are still at a relatively early stage of development, and are likely to improve as experience accumulates, new ideas are tried, and available computing power further increases.

Other pitfalls

Many other factors can influence the efficiency and success of a nonlinear solver, including the way that the model is formulated and the choice of units (or scaling) for the variables. As a rule, nonlinearities are more easily handled when they appear in the objective function rather than in the constraints. AMPL's option to substitute variables automatically, described earlier in this chapter, may help in this regard. Another rule of thumb is that the values of the variables should differ by at most a few orders of magnitude; solvers can be misled when some variables are, say, in millions and others are in thousandths. Some solvers automatically scale a problem to try to avoid such a situation, but you can help them considerably by judiciously picking the units in which the variables are expressed.

Nonlinear solvers also have many modes of failure besides the ones we have discussed. Some methods of nonlinear optimization can get stuck at ''stationary'' points that are not optimal in any sense, can identify a maximum when a minimum is desired (or vice-versa), and can falsely give an indication that there is no feasible solution to the constraints. In these cases your only recourse may be to try a different starting guess; it can sometimes help to specify a start that is feasible for many of the nonlinear constraints. You may also improve the solver's chances of success by placing realistic bounds on the variables. If you know, for instance, that an optimal value of 80 is plausible for some variables, but a value of 800 is not, you may want to give them a bound of 400. (Once an indicated optimum is at hand, you should be sure to check whether these ''safety'' bounds have been reached by any of the variables; if so, the bounds should be relaxed and the problem re-solved.)

The intent of this section has been to illustrate that extra caution is advisable in working with nonlinear models. If you encounter a difficulty that cannot be resolved by any of the simple devices described here, you may need to consult a textbook in nonlinear programming, the documentation for the particular solver that you are using, or a numerical analyst versed in nonlinear optimization techniques.

Bibliography

Roger Fletcher, *Practical Methods of Optimization.* John Wiley & Sons (New York, NY, 1987). A concise survey of theory and methods.

Philip E. Gill, Walter Murray and Margaret H. Wright, *Practical Optimization.* Academic Press (New York, NY, 1981). Theory, algorithms and practical advice.

Jorge Nocedal and Stephen J. Wright, *Numerical Optimization.* Springer Verlag (Heidelberg, 1999). A text on methods for optimization of smooth functions.

Richard P. O'Neill, Mark Williard, Bert Wilkins and Ralph Pike, "A Mathematical Programming Model for Allocation of Natural Gas." *Operations Research* **27**, 5 (1979) pp. 857–873. A source for the nonlinear relationships in natural gas pipeline networks described in Section 18.1.

Exercises

18-1. In the last example of Section 18.4, try some more starting points to see if you can find an even better locally optimal solution. What is the best solution you can find?

18-2. The following little model `fence.mod` determines the dimensions of a rectangular field of maximum area that can be surrounded by a fence of given length:

```
param fence > 0;

var Xfield >= 0;
var Yfield >= 0;

maximize Area:  Xfield * Yfield;
subject to Enclose:  2*Xfield + 2*Yfield <= fence;
```

It's well known that the optimum field is a square.

(a) When we tried to solve this problem for a fence of 40 meters, with the default initial guess of zero for the variables, we got the following result:

```
ampl: solve;
MINOS 5.5: optimal solution found.
0 iterations, objective 0

ampl: display Xfield, Yfield;
Xfield = 0
Yfield = 0
```

What could explain this unexpected outcome? Try the same problem on any nonlinear solver available to you, and report the behavior that you observe.

(b) Using a different starting point if necessary, run your solver to confirm that the optimal dimensions for 40 meters of fence are indeed 10×10.

(c) Experiment with an analogous model for determining the dimensions of a box of maximum volume that can be wrapped by paper of a given area.

(d) Solve the same problem as in (c), but for wrapping a cylinder rather than a box.

18-3. A falling object on a nameless planet has been observed to have approximately the following heights h_j at (mostly) one-second intervals t_j:

t_j	0.0	0.5	1.5	2.5	3.5	4.5	5.5	6.5	7.5	8.5	9.5	10.0
h_j	100	95	87	76	66	56	47	38	26	15	6	0

According to the laws of physics on this planet, the height of the object at any time should be given by the formula

$$h_j = a_0 - a_1 t_j - \tfrac{1}{2} a_2 t_j^2,$$

where a_0 is the initial height, a_1 is the initial velocity, and a_2 is the acceleration due to gravity. But since the observations were not made exactly, there exists no choice of a_0, a_1, and a_2 that will cause all of the data to fit this formula exactly. Instead, we wish to estimate these three values by choosing them so as to minimize the "sum of squares"

$$\sum_{j=1}^{n} [h_j - (a_0 - a_1 t_j - \tfrac{1}{2} a_2 t_j^2)]^2.$$

where t_j and h_j are the observations from the jth entry of the table, and n is the number of observations. This expression measures the error between the ideal formula and the observed behavior.

(a) Create an AMPL model that minimizes the sum of squares for any number n of observations t_j and h_j. This model should have three variables and an objective function, but no constraints.

(b) Use AMPL data statements to represent the sample observations given above, and solve the resulting nonlinear program to determine the estimates of a_0, a_1, and a_2.

18-4. This problem involves a very simple "traffic flow" network:

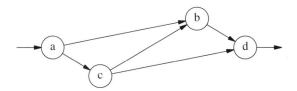

Traffic proceeds in the direction of the arrows, entering at intersection a, exiting at d, and passing through b or c or both. These data values are given for the roads connecting the intersections:

From	To	Time	Capacity	Sensitivity
a	b	5.0	10	0.1
a	c	1.0	30	0.9
c	b	2.0	10	0.9
b	d	1.0	30	0.9
c	d	5.0	10	0.1

To be specific, we imagine that the times are in minutes, the capacities are in cars per minute, and the sensitivities are in minutes per (car per minute).

The following AMPL statements can be used to represent networks of this kind:

```
set inters;    # road intersections

param EN symbolic in inters;    # entrance to network
param EX symbolic in inters;    # exit from network

set roads within {i in inters, j in inters: i <> EX and j <> EN};

param time {roads} > 0;
param cap {roads} > 0;
param sens {roads} > 0;
```

(a) What is the shortest path, in minutes, from the entrance to the exit of this network? Construct a shortest path model, along the lines of Figure 15-7, that verifies your answer.

(b) What is the maximum traffic flow from entrance to exit, in cars per minute, that the network can sustain? Construct a maximum flow model, along the lines of Figure 15-6, that verifies your answer.

(c) Question (a) above was concerned only with the speed of traffic, and question (b) only with the volume of traffic. In reality, these quantities are interrelated. As the traffic volume on a road increases from zero, the time required to travel the road also increases.

Travel time increases only moderately when there are just a few cars, but it increases very rapidly as the traffic approaches capacity. Thus a nonlinear function is needed to model this phenomenon. Let us define `T[i,j]`, the travel time on road `(i,j)`, by the following constraints:

```
var X {roads} >= 0;    # cars per minute entering road (i,j)
var T {roads};         # travel time for road (i,j)

subject to Travel_Time {(i,j) in roads}:
   T[i,j] = time[i,j] + (sens[i,j]*X[i,j]) / (1-X[i,j]/cap[i,j]);
```

You can confirm that the travel time on `(i,j)` is close to `time[i,j]` when the road is lightly used, but goes to infinity as the use approaches `cap[i,j]` cars per minute. The magnitude of `sens[i,j]` controls the rate at which travel time increases, as more cars try to enter the road.

Suppose that we want to analyze how the network can best handle some number of cars per minute. The objective is to minimize average travel time from entrance to exit:

```
param through > 0;   # cars per minute using the network

minimize Avg_Time:
   (sum {(i,j) in roads} T[i,j] * X[i,j]) / through;
```

The nonlinear expression `T[i,j] * X[i,j]` is travel minutes on road `(i,j)` times cars per minute entering the road — hence, the number of cars on road `(i,j)`. The summation in the objective thus gives the total cars in the entire system. Dividing by the number of cars per minute getting through, we have the average number of minutes for each car.

Complete the model by adding the following:

– A constraint that total cars per minute in equals total cars per minute out at each intersection, except the entrance and exit.

– A constraint that total cars per minute leaving the entrance equals the total per minute (represented by `through`) that are using the network.

– Appropriate bounds on the variables. (The example in Section 18.4 should suggest what bounds are needed.)

Use AMPL to show that, for the example given above, a throughput of 4.0 cars per minute is optimally managed by sending half the cars along $a \rightarrow b \rightarrow d$ and half along $a \rightarrow c \rightarrow d$, giving an average travel time of about 8.18 minutes.

(d) By trying values of parameter `through` both greater than and less than 4.0, develop a graph of minimum average travel time as a function of throughput. Also, keep a record of which travel routes are used in the optimal solutions for different throughputs, and summarize this information on your graph.

What is the relationship between the information in your graph and the solutions from parts (a) and (b)?

(e) The model in (c) assumes that you can make the cars' drivers take certain routes. For example, in the optimal solution for a throughput of 4.0, no drivers are allowed to ''cut through'' from c to b.

What would happen if instead all drivers could take whatever route they pleased? Observation has shown that, in such a case, the traffic tends to reach a *stable* solution in which no route has a travel time less than the average. The optimal solution for a throughput of 4.0 is not stable, since — as you can verify — the travel time on $a \to c \to b \to d$ is only about 7.86 minutes; some drivers would try to cut through if they were permitted.

To find a stable solution using AMPL, we have to add some data specifying the possible routes from entrance to exit:

```
param choices integer > 0;   # number of routes
set route {1..choices} within roads;
```

Here `route` is an indexed collection of sets; for each `r` in `1..choices`, the expression `route[r]` denotes a different subset of roads that together form a route from EN to EX. For our network, `choices` should be 3, and the `route` sets should be `{(a,b),(b,d)}`, `{(a,c),(c,d)}` and `{(a,c),(c,b),(b,d)}`. Using these data values, the stability conditions may be ensured by one additional collection of constraints, which say that the time to travel any route is no less than the average travel time for all routes:

```
subject to Stability {r in 1..choices}:
    sum {(i,j) in route[r]} T[i,j] >=
        (sum {(i,j) in roads} T[i,j] * X[i,j]) / through;
```

Show that, in the stable solution for a throughput of 4.0, a little more than 5% of the drivers cut through, and the average travel time increases to about 8.27 minutes. Thus traffic would have been faster if the road from b to c had never been built! (This phenomenon is known as Braess's paradox, in honor of a traffic analyst who noticed that when a certain link was added to Munich's road system, traffic seemed to get worse.)

(f) By trying throughput values both greater than and less than 4.0, develop a graph of the stable travel time as a function of throughput. Indicate, on the graph, for which throughputs the stable time is greater than the optimal time.

(g) Suppose now that you have been hired to analyze the possibility of opening an additional winding road, directly from a to d, with travel time 5 minutes, capacity 10, and sensitivity 1.5. Working with the models developed above, write an analysis of the consequences of making this change, as a function of the throughput value.

18-5. Return to the model constructed in (e) of the previous exercise. This exercise asks about reducing the number of variables by substituting some out, as explained in Section 18.2.

(a) Set the option `show_stats` to 1, and solve the problem. From the extra output you get, verify that there are 10 variables.

Next repeat the session with option `substout` set to 1. Verify from the resulting messages that some of the variables are eliminated by substitution. Which of the variables must these be?

(b) Rather than setting `substout`, you can specify that a variable is to be substituted out by placing an appropriate = phrase in its `var` declaration. Modify your model from (a) to use this feature, and verify that the results are the same.

(c) There is a long expression for average travel time that appears twice in this model. Define a new variable `Avg` to stand for this expression. Verify that AMPL also substitutes this variable out when you solve the resulting model, and that the result is the same as before.

18-6. In *Modeling and Optimization with GINO*, Judith Liebman, Leon Lasdon, Linus Schrage and Allan Waren describe the following problem in designing a steel tank to hold ammonia. The decision variables are

T the temperature inside the tank
I the thickness of the insulation

The pressure inside the tank is a function of the temperature,

$$P = e^{-3950/(T+460)+11.86}$$

It is desired to minimize the cost of the tank, which has three components: insulation cost, which depends on the thickness; the cost of the steel vessel, which depends on the pressure; and the cost of a recondensation process for cooling the ammonia, which depends on both temperature and insulation thickness. A combination of engineering and economic considerations has yielded the following formulas:

$$C_I = 400 I^{0.9}$$
$$C_V = 1000 + 22(P - 14.7)^{1.2}$$
$$C_R = 144(80 - T)/I$$

(a) Formulate this problem in AMPL as a two-variable problem, and alternatively as a six-variable problem in which four of the variables can be substituted out. Which formulation would you prefer to work with?

(b) Using your preferred formulation, determine the parameters of the least-cost tank.

(c) Increasing the factor 144 in C_R causes a proportional increase in the recondensation cost. Try several larger values, and describe in general terms how the total cost increases as a function of the increase in this factor.

18-7. A social accounting matrix is a table that shows the flows from each sector in an economy to each other sector. Here is simple five-sector example, with blank entries indicating flows known to be zero:

	LAB	H1	H2	P1	P2	total
LAB		15	3	130	80	220
H1	?					?
H2	?					?
P1		15	130		20	190
P2		25	40	55		105

If the matrix were estimated perfectly, it would be *balanced:* each row sum (the flow out of a sector) would equal the corresponding column sum (the flow in). As a practical matter, however, there are several difficulties:

– Some entries, marked ? above, have no reliable estimates.

– In the estimated table, the row sums do not necessarily equal the column sums.

– We have separate estimates of the total flows into (or out of) each sector, shown to the right of the rows in our table. These do not necessarily equal the sums of the estimated rows or columns.

Nonlinear programming can be used to adjust this matrix by finding the balanced matrix that is closest, in some sense, to the one given.

For a set S of sectors, let $E_T \subseteq S$ be the subset of sectors for which we have estimated total flows, and let $E_A \subseteq S \times S$ contain all sector pairs (i, j) for which there are known estimates. The given data values are:

t_i estimated row/column sums, $i \in E_T$
a_{ij} estimated social accounting matrix entries, $(i, j) \in E_A$

Let $S_A \subseteq S \times S$ contain all row-column pairs (i, j) for which there should be entries in the matrix — this includes entries that contain ? instead of a number. We want to determine adjusted entries A_{ij}, for each $(i, j) \in S_A$, that are truly balanced:

$$\sum_{j \in S:(i,j) \in S_A} A_{ij} = \sum_{j \in S:(j,i) \in S_A} A_{ji}$$

for all $i \in S$. You can think of these equations as the constraints on the variables A_{ij}.

There is no best function for measuring "close", but one popular choice is the sum of squared differences between the estimates and the adjusted values — for both the matrix and the row and column sums — scaled by the estimated values. For convenience, we write the adjusted sums as defined variables:

$$T_i = \sum_{j \in S:(i,j) \in S_A} A_{ij}$$

Then the objective is to minimize

$$\sum_{(i,j) \in E_A} (a_{ij} - A_{ij})^2 / a_{ij} + \sum_{i \in E_T} (t_i - T_i)^2 / t_i$$

Formulate an AMPL model for this problem, and determine an optimal adjusted matrix.

18-8. A network of pipes has the following layout:

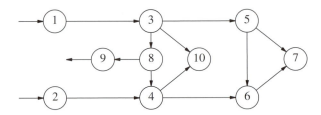

The circles represent joints, and the arrows are pipes. Joints 1 and 2 are sources of flow, and joint 9 is a sink or destination for flow, but flow through a pipe can be in either direction. Associated with each joint i is an amount w_i to be withdrawn from the flow at that joint, and an elevation e_i:

	1	2	3	4	5	6	7	8	9	10
w_i	0	0	200	0	0	200	150	0	0	150
e_i	50	40	20	20	0	0	0	20	20	20

Our decision variables are the flows F_{ij} through the pipes from i to j, with a positive value representing flow in the direction of the arrow, and a negative value representing flow in the opposite direction. Naturally, flow in must equal flow out plus the amount withdrawn at every joint, except for the sources and the sink.

The "head loss" of a pipe is a measure of the energy required to move a flow through it. In our situation, the head loss for the pipe from i to j is proportional to the square of the flow rate:

$$H_{ij} = Kc_{ij}F_{ij}^2 \quad \text{if } F_{ij} > 0,$$

$$H_{ij} = -Kc_{ij}F_{ij}^2 \quad \text{if } F_{ij} < 0,$$

where $K = 4.96407 \times 10^{-6}$ is a conversion constant, and c_{ij} is a factor computed from the diameter, friction, and length of the pipe:

from	to	c_{ij}
1	3	6.36685
2	4	28.8937
3	10	28.8937
3	5	6.36685
3	8	43.3406
4	10	28.8937
4	6	28.8937
5	6	57.7874
5	7	43.3406
6	7	28.8937
8	4	28.8937
8	9	705.251

For two joints i and j at the same elevation, the pressure drop for flow from i to j is equal to the head loss. Both pressure and head loss are measured in feet, so that after correcting for differences in elevation between the joints we have the relation:

$$H_{ij} = (P_i + e_i) - (P_j + e_j)$$

Finally, we wish to maintain the pressure at both the sources and the sink at 200 feet.

(a) Formulate a general AMPL model for this situation, and put together data statements for the data given above.

(b) There is no objective function here, but you can still employ a nonlinear solver to seek a feasible solution. By setting the option show_stats to 1, confirm that the number of variables equals the number of equations, so that there are no ''degrees of freedom'' in the solution. (This does not guarantee that there is just one solution, however.)

Check that your solver finds a solution to the equations, and display the results.

19

Complementarity Problems

A variety of physical and economic phenomena are most naturally modeled by saying that certain pairs of inequality constraints must be *complementary,* in the sense that at least one must hold with equality. These conditions may in principle be accompanied by an objective function, but are more commonly used to construct *complementarity problems* for which a feasible solution is sought. Indeed, optimization may be viewed as a special case of complementarity, since the standard optimality conditions for linear and smooth nonlinear optimization are complementarity problems. Other kinds of complementarity problems do not arise from optimization, however, or cannot be conveniently formulated or solved as optimization problems.

The AMPL operator `complements` permits complementarity conditions to be specified directly in constraint declarations. Complementarity models can thereby be formulated in a natural way, and instances of such models are easily sent to special solvers for complementarity problems.

To motivate the syntax of `complements`, we begin by describing how it would be used to model a few simple economic equilibrium problems, some equivalent to linear programs and some not. We then give a general definition of the `complements` operator for pairs of inequalities and for more general "mixed" complementarity conditions via double inequalities. Where appropriate in these sections, we also comment on an AMPL interface to the PATH solver for "square" mixed complementarity problems. In a final section, we describe how complementarity constraints are accommodated in several of AMPL's existing features, including presolve, constraint-name suffixes, and generic synonyms for constraints.

19.1 Sources of complementarity

Economic equilibria are one of the best-known applications of complementarity conditions. We begin this section by showing how a previous linear programming example in production economics has an equivalent form as a complementarity model, and how

```
set PROD;   # products
set ACT;    # activities

param cost {ACT} > 0;       # cost per unit of each activity
param demand {PROD} >= 0;   # units of demand for each product
param io {PROD,ACT} >= 0;   # units of each product from
                            # 1 unit of each activity

var Level {j in ACT} >= 0;

minimize Total_Cost:  sum {j in ACT} cost[j] * Level[j];

subject to Demand {i in PROD}:
    sum {j in ACT} io[i,j] * Level[j] >= demand[i];
```

Figure 19-1: Production cost minimization model (`econmin.mod`).

bounded variables are handled though an extension to the concept of complementarity. We then describe a further extension to price-dependent demands that is not motivated by optimization or equivalent to any linear program. We conclude by briefly describing other complementarity models and applications.

A complementarity model of production economics

In Section 2.4 we observed that the form of a diet model also applies to a model of production economics. The decision variables may be taken as the levels of production activities, so that the objective is the total production cost,

```
minimize Total_Cost:  sum {j in ACT} cost[j] * Level[j];
```

where `cost[j]` and `Level[j]` are the cost per unit and the level of activity j. The constraints say that the totals of the product outputs must be at least the product demands:

```
subject to Demand {i in PROD}:
    sum {j in ACT} io[i,j] * Level[j] >= demand[i];
```

with `io[i,j]` being the amount of product i produced per unit of activity j, and `demand[i]` being the total quantity of product i demanded. Figures 19-1 and 19-2 show this "economic" model and some data for it.

Minimum-cost production levels are easily computed by a linear programming solver:

```
ampl: model econmin.mod;
ampl: data econ.dat;

ampl: solve;
CPLEX 8.0.0: optimal solution; objective 6808640.553
3 dual simplex iterations (0 in phase I)
```

```
param: ACT:    cost  :=
          P1    2450       P1a    1290
          P2    1850       P2a    3700       P2b    2150
          P3    2200       P3c    2370
          P4    2170  ;

param: PROD:   demand :=
          AA1     70000
          AC1     80000
          BC1     90000
          BC2     70000
          NA2    400000
          NA3    800000  ;

param io (tr):
          AA1   AC1   BC1   BC2    NA2    NA3  :=
     P1    60    20    10    15    938    295
     P1a    8     0    20    20   1180    770
     P2     8    10    15    10    945    440
     P2a   40    40    35    10    278    430
     P2b   15    35    15    15   1182    315
     P3    70    30    15    15    896    400
     P3c   25    40    30    30   1029    370
     P4    60    20    15    10   1397    450  ;
```

Figure 19-2: Data for production models (econ.dat).

```
ampl: display Level;
Level [*] :=
 P1      0
P1a   1555.3
 P2      0
P2a      0
P2b      0
 P3    147.465
P3c   1889.4
 P4      0
```

Recall (from Section 12.5) that there are also dual or marginal values — or "prices" — associated with the constraints:

```
ampl: display Demand.dual;
Demand.dual [*] :=
AA1   16.7051
AC1    5.44585
BC1   57.818
BC2    0
NA2    0
NA3    0
```

In the conventional linear programming interpretation, the price on constraint i gives, within a sufficiently small range, the change in the total cost per unit change in the demand for product i.

Consider now an alternative view of the production economics problem, in which we define variables Price[i] as well as Level[j] and seek an *equilibrium* rather than an optimum solution. There are two requirements that the equilibrium solution must satisfy.

First, for each product, total output must meet demand and the price must be nonnegative, and in addition there must be a *complementarity* between these relationships: where production exceeds demand the price must be zero, or equivalently, where the price is positive the production must equal the demand. This relationship is expressed in AMPL by means of the complements operator:

```
subject to Pri_Compl {i in PROD}:
   Price[i] >= 0 complements
       sum {j in ACT} io[i,j] * Level[j] >= demand[i];
```

When two inequalities are joined by complements, they both must hold, and at least one must hold with equality. Because our example is indexed over the set PROD, it sets up a relationship of this kind for each product.

Second, for each activity, there is another relationship that may at first be less obvious. Consider that, for each unit of activity j, the value of the resulting product i output in terms of the model's prices is Price[i] * io[i,j]. The total value of all outputs from one unit of activity j is thus

```
sum {i in ACT} Price[i] * io[i,j]
```

At equilibrium prices, this total value cannot exceed the activity's cost per unit, cost[j]. Moreover, there is a complementarity between this relationship and the level of activity j: where cost *exceeds* total value the activity must be zero, or equivalently, where the activity is positive the total value must equal the cost. Again this relationship can be expressed in AMPL with the complements operator:

```
subject to Lev_Compl {j in ACT}:
   Level[j] >= 0 complements
       sum {i in PROD} Price[i] * io[i,j] <= cost[j];
```

Here the constraint is indexed over ACT, so that we have a complementarity relationship for each activity.

Putting together the two collections of complementarity constraints, we have the linear *complementarity problem* shown in Figure 19-3. The number of variables and the number of complementarity relationships are equal (to activities plus products), making this a "square" complementarity problem that is amenable to certain solution techniques, though not the same techniques as those for linear programs.

Applying the PATH solver, for example, the complementarity problem can be seen to have the same solution as the related minimum-cost production problem:

```
set PROD;    # products
set ACT;     # activities

param cost {ACT} > 0;       # cost per unit of each activity
param demand {PROD} >= 0;   # units of demand for each product
param io {PROD,ACT} >= 0;   # units of each product from
                            # 1 unit of each activity

var Price {i in PROD};
var Level {j in ACT};

subject to Pri_Compl {i in PROD}:
   Price[i] >= 0 complements
      sum {j in ACT} io[i,j] * Level[j] >= demand[i];

subject to Lev_Compl {j in ACT}:
   Level[j] >= 0 complements
      sum {i in PROD} Price[i] * io[i,j] <= cost[j];
```

Figure 19-3: Production equilibrium model (econ.mod).

```
ampl: model econ.mod;
ampl: data econ.dat;
ampl: option solver path;
ampl: solve;
Path v4.5: Solution found.
7 iterations (0 for crash); 33 pivots.
20 function, 8 gradient evaluations.

ampl: display sum {j in ACT} cost[j] * Level[j];
sum{j in ACT} cost[j]*Level[j] = 6808640
```

Further application of display shows that Level is the same as in the production economics LP and that Price takes the same values that Demand.dual has in the LP.

Complementarity for bounded variables

Suppose now that we extend our models by placing bounds on the activity levels: level_min[j] <= Level[j] <= level_max[j]. The equivalence between the optimization problem and a square complementarity problem can be maintained, provided that the complementarity relationship for the activities is generalized to a ''mixed'' form. Where an activity's cost is greater than its total value (per unit), the activity's level must be *at its lower bound* (much as before). Where an activity's level is *between its bounds,* its cost must equal its total value. And an activity's cost may also be less than its total value, provided that its level is *at its upper bound.* These three relationships are summarized by another form of the complements operator:

```
subject to Lev_Compl {j in ACT}:
   level_min[j] <= Level[j] <= level_max[j] complements
      cost[j] - sum {i in PROD} Price[i] * io[i,j];
```

```
set PROD;    # products
set ACT;     # activities

param cost {ACT} > 0;        # cost per unit of each activity
param demand {PROD} >= 0;    # units of demand for each product
param io {PROD,ACT} >= 0;    # units of each product from
                             # 1 unit of each activity

param level_min {ACT} > 0;  # min allowed level for each activity
param level_max {ACT} > 0;  # max allowed level for each activity

var Price {i in PROD};
var Level {j in ACT};

subject to Pri_Compl {i in PROD}:
   Price[i] >= 0 complements
      sum {j in ACT} io[i,j] * Level[j] >= demand[i];

subject to Lev_Compl {j in ACT}:
   level_min[j] <= Level[j] <= level_max[j] complements
      cost[j] - sum {i in PROD} Price[i] * io[i,j];
```

Figure 19-4: Bounded version of production equilibrium model (econ2.mod).

When a double inequality is joined to an expression by complements, the inequalities must hold, and either the expression must be zero, or the lower inequality must hold with equality and the expression must be nonnegative, or the upper inequality must hold with equality and the expression must be nonpositive.

A bounded version of our complementarity examples is shown in Figure 19-4. The PATH solver can be applied to this model as well:

```
ampl: model econ2.mod;
ampl: data econ2.dat;
ampl: option solver path;
ampl: solve;
Path v4.5: Solution found.
9 iterations (4 for crash); 8 pivots.
22 function, 10 gradient evaluations.

ampl: display level_min, Level, level_max;
:    level_min   Level level_max   :=
P1       240       240    1000
P1a      270      1000    1000
P2       220       220    1000
P2a      260       680    1000
P2b      200       200    1000
P3       260       260    1000
P3c      220      1000    1000
P4       240       240    1000
;
```

The results are the same as for the LP that is derived from our previous example (Figure 19-1) by adding the bounds above to the variables.

```
set PROD;    # products
set ACT;     # activities

param cost {ACT} > 0;        # cost per unit of each activity
param io {PROD,ACT} >= 0;    # units of each product from
                             # 1 unit of each activity

param demzero {PROD} > 0;    # intercept and slope of the demand
param demrate {PROD} >= 0;   # as a function of price

var Price {i in PROD};
var Level {j in ACT};

subject to Pri_Compl {i in PROD}:
   Price[i] >= 0 complements
      sum {j in ACT} io[i,j] * Level[j]
         >= demzero[i] - demrate[i] * Price[i];

subject to Lev_Compl {j in ACT}:
   Level[j] >= 0 complements
      sum {i in PROD} Price[i] * io[i,j] <= cost[j];
```

Figure 19-5: Price-dependent demands (`econn1.mod`).

Complementarity for price-dependent demands

If complementarity problems only arose from linear programs, they would be of very limited interest. The idea of an economic equilibrium can be generalized, however, to problems that have no LP equivalents. Rather than taking the demands to be fixed, for example, it makes sense to view the demand for each product as a decreasing function of its price.

The simplest case is a decreasing linear demand, which could be expressed in AMPL as

```
demzero[i] - demrate[i] * Price[i]
```

where `demzero[i]` and `demrate[i]` are nonnegative parameters. The resulting complementarity problem simply substitutes this expression for `demand[i]`, as seen in Figure 19-5. The complementarity problem remains square, and can still be solved by PATH, but with clearly different results:

```
ampl: model econn1.mod;
ampl: data econn1.dat;

ampl: option solver path;

ampl: solve;
Path v4.5: Solution found.
11 iterations (3 for crash); 11 pivots.
12 function, 12 gradient evaluations.
```

```
ampl: display Level;
Level [*] :=
 P1   240
 P1a  710.156
 P2   220
 P2a  260
 P2b  200
 P3   260
 P3c  939.063
 P4   240
 ;
```

The balance between demands and prices now tends to push down the equilibrium production levels.

Because the `Price[i]` variables appear on both sides of the `complements` operator in this model, there is no equivalent linear program. There does exist an equivalent nonlinear optimization model, but it is not as easy to derive and may be harder to solve as well.

Other complementarity models and applications

This basic example can be extended to considerably more complex models of economic equilibrium. The activity and price variables and their corresponding complementarity constraints can be comprised of several indexed collections each, and both the cost and price functions can be nonlinear. A solver such as PATH handles all of these extensions, so long as the problem remains square in the sense of having equal numbers of variables and complementarity constraints (or being easily converted to such a form as explained in the next section).

More ambitious models may add an objective function and may mix equality, inequality and complementarity constraints in arbitrary numbers. Solution techniques for these so-called MPECs — mathematical programs with equilibrium constraints — are at a relatively experimental stage, however.

Complementarity problems also arise in physical systems, where they can serve as models of equilibrium conditions between forces. A complementarity constraint may represent a discretization of the relationship between two objects, for example. The relationship on one side of the `complements` operator may hold with equality at points where the objects are in contact, while the relationship on the other side holds with equality where they do not touch.

Game theory provides another class of examples. The Nash equilibrium for a bi-matrix game is characterized by complementarity conditions, for example, in which the variables are the probabilities with which the two players make their available moves. For each move, either its probability is zero, or a related equality holds to insure there is nothing to be gained by increasing or decreasing its probability.

Surveys that describe a variety of complementarity problems in detail are cited in the references at the end of this chapter.

19.2 Forms of complementarity constraints

An AMPL complementarity constraint consists of two expressions or constraints separated by the `complements` operator. There are always *two inequalities,* whose position determines how the constraint is interpreted.

If there is one inequality on either side of `complements`, the constraint has the general form

>   *single-inequality* `complements` *single-inequality* ;

where a *single-inequality* is any valid ordinary constraint — linear or nonlinear — containing one `>=` or `<=` operator. A constraint of this type is satisfied if both of the *single-inequality* relations are satisfied, and at least one is satisfied with equality.

If both inequalities are on the same side of the `complements` operator, the constraint has instead one of the forms

>   *double-inequality* `complements` *expression* ;
>   *expression* `complements` *double-inequality* ;

where *double-inequality* is any ordinary AMPL constraint containing two `>=` or two `<=` operators, and *expression* is any numerical expression. Variables may appear nonlinearly in either the *double-inequality* or the *expression* (or both). The conditions for a constraint of this type to be satisfied are as follows:

- if the left side `<=` or the right side `>=` of the *double-inequality* holds with equality, then the *expression* is greater than or equal to 0;

- if the right side `<=` or the left side `>=` of the *double-inequality* holds with equality, then the *expression* is less than or equal to 0;

- if neither side of the *double-inequality* holds with equality, then the *expression* equals 0.

In the special case where the *double-inequality* has the form $0 <=$ *body* $<=$ `Infinity`, these conditions reduce to those for complementarity of a pair of single inequalities.

For completeness, the special case in which the left-hand side equals the right-hand side of the double inequality may be written using one of the forms

>   *equality* `complements` *expression* ;
>   *expression* `complements` *equality* ;

A constraint of this kind is equivalent to an ordinary constraint consisting only of the *equality*; it places no restrictions on the *expression*.

For the use of solvers that require ''square'' complementarity systems, AMPL converts to square any model instance in which the number of variables equals the number of complementarity constraints plus the number of equality constraints. There may be any number of additional inequality constraints, but there must not be any objective. Each equality is trivially turned into a complementarity condition, as observed above; each added inequality is made complementary to a new, otherwise unused variable, preserving the squareness of the problem overall.

19.3 Working with complementarity constraints

All of AMPL's features for ordinary equalities and inequalities extend in a straightforward way to complementarity constraints. This section covers extensions in three areas: expressions for related solution values, effects of presolve and related displays of problem statistics, and generic synonyms for constraints.

Related solution values

AMPL's built-in suffixes for values related to a problem and its solution extend to complementarity constraints, but with two collections of suffixes — of the form *cname*.L*suf* and *cname*.R*suf* — corresponding to the left and right operands of complements, respectively. Thus after econ2.mod (Figure 19-4) has been solved, for example, we can use the following display command to look at values associated with the constraint Lev_Compl:

```
ampl: display Lev_Compl.Llb, Lev_Compl.Lbody,
ampl?           Lev_Compl.Rbody, Lev_Compl.Rslack;
```

: Lev_Compl.Llb	Lev_Compl.Lbody	Lev_Compl.Rbody	Lev_Compl.Rslack :=
P1	240	240	1392.86	Infinity
P1a	270	1000	-824.286	Infinity
P2	220	220	264.286	Infinity
P2a	260	680	5.00222e-12	Infinity
P2b	200	200	564.286	Infinity
P3	260	260	614.286	Infinity
P3c	220	1000	-801.429	Infinity
P4	240	240	584.286	Infinity
```
;
```

Because the right operand of Lev_Compl is an expression, it is treated as a ''constraint'' with infinite lower and upper bounds, and hence infinite slack.

A suffix of the form *cname*.slack is also defined for complementarity constraints. For complementary pairs of single inequalities, it is equal to the lesser of *cname*.Lslack and *cname*.Rslack. Hence it is nonnegative if and only if both inequalities are satisfied and is zero if the complementarity constraint holds exactly. For complementary double inequalities of the form

> *expr* complements *lbound* <= *body* <= *ubound*
> *lbound* <= *body* <= *ubound* complements *expr*

cname.slack is defined to be

$$\texttt{min}(expr, \ body \ - \ lbound) \qquad \text{if } body \ \texttt{<=} \ lbound$$
$$\texttt{min}(-expr, \ ubound \ - \ body) \qquad \text{if } body \ \texttt{>=} \ ubound$$
$$\texttt{-abs}(expr) \qquad \text{otherwise}$$

Hence in this case it is always nonpositive, and is zero when the complementarity constraint is satisfied exactly.

If *cname* for a complementarity constraint appears unsuffixed in an expression, it is interpreted as representing *cname*.`slack`.

Presolve

As explained in Section 14.1, AMPL incorporates a presolve phase that can substantially simplify some linear programs. In the presence of complementarity constraints, several new kinds of simplifications become possible.

As an example, given a constraint of the form

$expr_1$ `>= 0 complements` $expr_2$ `>= 0`

if presolve can deduce that $expr_1$ is strictly positive for all feasible points — in other words, that it has a positive lower bound — it can replace the constraint by $expr_2 \ = \ 0$.

Similarly, in a constraint of the form

lbound `<=` *body* `<=` *ubound* `complements` *expr*

there are various possibilities, including the following:

If presolve can deduce for all feasible points that	Then the constraint can be replaced by
$body \ < \ ubound$	*lbound* `<=` *body* `complements` *expr* `>= 0`
$lbound \ < \ body \ < \ ubound$	$expr \ = \ 0$
$expr \ < \ 0$	$body \ = \ ubound$

Transformations of these kinds are carried out automatically, unless `option presolve 0` is used to turn off the presolve phase. As with ordinary constraints, results are reported in terms of the original model.

By displaying a few predefined parameters:

`_ncons`	the number of ordinary constraints before presolve
`_nccons`	the number of complementarity conditions before presolve
`_sncons`	the number of ordinary constraints after presolve
`_snccons`	the number of complementarity conditions after presolve

or by setting `option show_stats 1`, you can get some information on the number of simplifying transformations that presolve has made:

```
ampl: model econ2.mod; data econ2.dat;
ampl: option solver path;
ampl: option show_stats 1;
ampl: solve;
```

```
Presolve eliminates 16 constraints and 2 variables.
Presolve resolves 2 of 14 complementarity conditions.
Adjusted problem:
12 variables, all linear
12 constraints, all linear; 62 nonzeros
12 complementarity conditions among the constraints:
        12 linear, 0 nonlinear.
0 objectives.

Path v4.5: Solution found.
7 iterations (1 for crash); 30 pivots.
8 function, 8 gradient evaluations.

ampl: display _ncons, _nccons, _sncons, _snccons;
_ncons = 28
_nccons = 14
_sncons = 12
_snccons = 12
```

When first instantiating the problem, AMPL counts each complementarity constraint as two ordinary constraints (the two arguments to `complements`) and also as a complementarity condition. Thus `_nccons` equals the number of complementarity constraints before presolve, and `_ncons` equals twice `_nccons` plus the number of any non-complementarity constraints before presolve. The presolve messages at the beginning of the `show_stats` output indicate how much presolve was able to reduce these numbers.

In this case the reason for the reduction can be seen by comparing each product's demand to the minimum possible output of that product — the amount that results from setting each `Level[j]` to `level_min[j]`:

```
ampl: display {i in PROD}
ampl?    (sum{j in ACT} io[i,j]*level_min[j], demand[i]);
:    sum{j in ACT} io[i,j]*level_min[j]  demand[i]    :=
AA1                69820                    70000
AC1                45800                    80000
BC1                37300                    90000
BC2                29700                    70000
NA2              1854920                    4e+05
NA3               843700                    8e+05
;
```

We see that for products NA2 and NA3, the total output exceeds demand even at the lowest activity levels. Hence in the constraint

```
subject to Pri_Compl {i in PROD}:
   Price[i] >= 0 complements
      sum {j in ACT} io[i,j] * Level[j] >= demand[i];
```

the right-hand argument to `complements` never holds with equality for NA2 or NA3. Presolve thus concludes that `Price["NA2"]` and `Price["NA3"]` can be fixed at zero, removing them from the resulting problem.

Generic synonyms

AMPL's generic synonyms for constraints (Section 12.6) extend to complementarity conditions, mainly through the substitution of ccon for con in the synonym names.

From the modeler's view (before presolve), the ordinary constraint synonyms remain:

_ncons	number of ordinary constraints before presolve
_conname	names of the ordinary constraints before presolve
_con	synonyms for the ordinary constraints before presolve

The complementarity constraint synonyms are:

_nccons	number of complementarity constraints before presolve
_cconname	names of the complementarity constraints before presolve
_ccon	synonyms for the complementarity constraints before presolve

Because each complementarity constraint also gives rise to two ordinary constraints, as explained in the preceding discussion of presolve, there are two entries in _conname corresponding to each entry in _cconname:

```
ampl: display {i in 1..6} (_conname[i], _cconname[i]);
:       _conname[i]           _cconname[i]         :=
1   "Pri_Compl['AA1'].L"   "Pri_Compl['AA1']"
2   "Pri_Compl['AA1'].R"   "Pri_Compl['AC1']"
3   "Pri_Compl['AC1'].L"   "Pri_Compl['BC1']"
4   "Pri_Compl['AC1'].R"   "Pri_Compl['BC2']"
5   "Pri_Compl['BC1'].L"   "Pri_Compl['NA2']"
6   "Pri_Compl['BC1'].R"   "Pri_Compl['NA3']"
;
```

For each complementarity constraint *cname*, the left and right arguments to the complements operator are the ordinary constraints named *cname*.L and *cname*.R. This is confirmed by using the synonym terminology to expand the complementarity constraint Pri_Compl['AA1'] and the corresponding two ordinary constraints from the example above:

```
ampl: expand Pri_Compl['AA1'];
subject to Pri_Compl['AA1']:
        Price['AA1'] >= 0
    complements
        60*Level['P1'] + 8*Level['P1a'] + 8*Level['P2'] +
        40*Level['P2a'] + 15*Level['P2b'] + 70*Level['P3'] +
        25*Level['P3c'] + 60*Level['P4'] >= 70000;

ampl: expand _con[1], _con[2];
subject to Pri_Compl.L['AA1']:
        Price['AA1'] >= 0;
subject to Pri_Compl.R['AA1']:
        60*Level['P1'] + 8*Level['P1a'] + 8*Level['P2'] +
        40*Level['P2a'] + 15*Level['P2b'] + 70*Level['P3'] +
        25*Level['P3c'] + 60*Level['P4'] >= 70000;
```

From the solver's view (after presolve), a more limited collection of synonyms is defined:

_sncons	number of all constraints after presolve
_snccons	number of complementarity constraints after presolve
_sconname	names of all constraints after presolve
_scon	synonyms for all constraints after presolve

Necessarily _snccons is less than or equal to _sncons, with equality only when all constraints are complementarity constraints.

To simplify the problem description that is sent to the solver, AMPL converts every complementarity constraint into one of the following canonical forms:

```
expr complements lbound <= var <= ubound
expr <= 0 complements var <= ubound
expr >= 0 complements lbound <= var
```

where *var* is the name of a *different* variable for each constraint. (Where an expression more complicated than a single variable appears on both sides of complements, this involves the introduction of an auxiliary variable and an equality constraint defining the variable to equal one of the expressions.) By using solexpand in place of expand, you can see the form in which AMPL has sent a complementarity constraint to the solver:

```
ampl: solexpand Pri_Compl['AA1'];
subject to Pri_Compl['AA1']:
    -70000 + 60*Level['P1'] + 8*Level['P1a'] + 8*Level['P2'] +
    40*Level['P2a'] + 15*Level['P2b'] + 70*Level['P3'] +
    25*Level['P3c'] + 60*Level['P4'] >= 0
complements
    0 <= Price['AA1'];
```

A predefined array of integers, _scvar, gives the indices of the complementing variables in the generic variable arrays _var and _varname. This terminology can be used to display a list of names of such variables:

```
ampl: display {i in 1..3} (_sconname[i],_svarname[_scvar[i]]);
:          _sconname[i]        _svarname[_scvar[i]]     :=
1    "Pri_Compl['AA1'].R"     "Price['AA1']"
2    "Pri_Compl['AC1'].R"     "Price['AC1']"
3    "Pri_Compl['BC1'].R"     "Price['BC1']"
;
```

When constraint i is an ordinary equality or inequality, _scvar[i] is 0. The names of complementarity constraints in _sconname are suffixed with .L or .R according to whether the *expr* in the constraint sent to the solver was derived from the left or right argument to complements in the original constraint.

Bibliography

Richard W. Cottle, Jong-Shi Pang, and Richard E. Stone, *The Linear Complementarity Problem*, Academic Press (San Diego, CA, 1992). An encyclopedic account of linear complementarity problems with a nice overview of how these problems arise.

Steven P. Dirkse and Michael C. Ferris, ''MCPLIB: A Collection of Nonlinear Mixed Complementarity Problems.'' Optimization Methods and Software **5**, 4 (1995) pp. 319–345. An extensive survey of nonlinear complementarity, including problem descriptions and mathematical formulations.

Michael C. Ferris and Jong-Shi Pang, ''Engineering and Economic Applications of Complementarity Problems.'' SIAM Review **39**, 4 (1997) pp. 669–713. A variety of complementarity test problems, originally written in the GAMS modeling language but now in many cases translated to AMPL.

Exercises

19-1. The economics example in Section 19.1 used a demand function that was linear in the price. Construct a nonlinear demand function that has each of the characteristics described below. Define a corresponding complementarity problem, using the data from Figure 19-2 as much as possible.

Use a solver such as PATH to compute an equilibrium solution. Compare this solution to those for the constant-demand and linear-demand alternatives shown in Section 19.1.

(a) For price i near zero the demand is near `demzero[i]` and is decreasing at a rate near `demrate[i]`. After price i has increased substantially, however, both the demand and the rate of decrease of the demand approach zero.

(b) For price i near zero the demand is approximately constant at `demzero[i]`, but as price i approaches `demlim[i]` the demand drops quickly to zero.

(c) Demand for i actually rises with price, until it reaches a value `demmax[i]` at a price of `demarg[i]`. Then demand falls with price.

19-2. For each scenario in the previous problem, experiment with different starting points for the `Level` and `Price` values. Determine whether there appears a unique equilibrium point.

19-3. A *bimatrix game* between players A and B is defined by two m by n ''payoff'' matrices, whose elements we denote by a_{ij} and b_{ij}. In one round of the game, player A has a choice of m alternatives and player B a choice of n alternatives. If A plays (chooses) i and B plays j, then A and B win amounts a_{ij} and b_{ij}, respectively; negative winnings are interpreted as losses.

We can allow for ''mixed'' strategies in which A plays i with probability p_i^A and B plays j with probability p_j^B. Then the expected value of player A's winnings is:

$$\sum_{j=1}^{n} a_{ij} \times p_j^B, \text{ if A plays } i$$

and the expected value of player B's winnings is:

$$\sum_{i=1}^{m} b_{ij} \times p_i^A, \text{ if B plays } j$$

A ''pure'' strategy is the special case in which each player has one probability equal to 1 and the rest equal to 0.

A pair of strategies is said to represent a *Nash equilibrium* if neither player can improve his expected payoff by changing only his own strategy.

(a) Show that the requirement for a Nash equilibrium is equivalent to the following complementarity-like conditions:

> for all i such that $p_i^A > 0$, A's expected return when playing i equals A's maximum expected return over all possible plays

> for all j such that $p_j^B > 0$, B's expected return when playing j equals B's maximum expected return over all possible plays

(b) To build a complementarity problem in AMPL whose solution is a Nash equilibrium, the parameters representing the payoff matrices can be defined by the following `param` declarations:

```
param nA > 0;   # actions available to player A
param nB > 0;   # actions available to player B

param payoffA {1..nA, 1..nB};   # payoffs to player A
param payoffB {1..nA, 1..nB};   # payoffs to player B
```

The probabilities that define the mixed strategies are necessarily variables. In addition it is convenient to define a variable to represent the maximum expected payoff for each player:

```
var PlayA {i in 1..nA};   # player A's mixed strategy
var PlayB {j in 1..nB};   # player B's mixed strategy

var MaxExpA;   # maximum expected payoff to player A
var MaxExpB;   # maximum expected payoff to player B
```

Write AMPL declarations for the following constraints:

> - The probabilities in any mixed strategy must be nonnegative.
> - The probabilities in each player's mixed strategy must sum to 1.
> - Player A's expected return when playing any particular i
>   must not exceed A's maximum expected return over all possible plays
> - Player B's expected return when playing any particular j
>   must not exceed B's maximum expected return over all possible plays

(c) Write an AMPL model for a square complementarity system that enforces the constraints in (b) and the conditions in (a).

(d) Test your model by applying it to the "rock-scissors-paper" game in which both players have the payoff matrix

```
 0   1  -1
-1   0   1
 1  -1   0
```

Confirm that an equilibrium is found where each player chooses between all three plays with equal probability.

(e) Show that the game for which both players have the payoff matrix

```
-3   1   3  -1
 2   3  -1  -5
```

has several equilibria, at least one of which uses mixed strategies and one of which uses pure strategies.

Running a solver such as PATH will only return one equilibrium solution. To find more, experiment with changing the initial solution or fixing some of the variables to 0 or 1.

19-4. Two companies have to decide now whether to adopt standard 1 or standard 2 for future introduction in their products. If they decide on the same standard, company A has the greater pay-off because its technology is superior. If they decide on different standards, company B has the greater payoff because its market share is greater. These considerations lead to a bimatrix game whose payoff matrices are

$$A = \begin{matrix} 10 & 3 \\ 2 & 9 \end{matrix} \qquad B = \begin{matrix} 4 & 6 \\ 7 & 5 \end{matrix}$$

(a) Use a solver such as PATH to find a Nash equilibrium. Verify that it is a mixed strategy, with A's probabilities being 1/2 for both standards and B's probabilities being 3/7 and 4/7 for standards 1 and 2, respectively.

Why is a mixed strategy not appropriate for this application?

(b) You can see what happens when company A decides on standard 1 by issuing the following commands:

```
ampl: fix PlayA[1] := 1;
ampl: solve;
presolve, constraint ComplA[1].L:
        all variables eliminated, but upper bound = -1 < 0
```

Explain how AMPL's presolve phase could deduce that the complementarity problem has no feasible solution in this case.

(c) Through further experimentation, show that there are no Nash equilibria for this situation that involve only pure strategies.

20

Integer Linear Programs

Many linear programming problems require certain variables to have whole number, or integer, values. Such a requirement arises naturally when the variables represent entities like packages or people that can not be fractionally divided — at least, not in a meaningful way for the situation being modeled. Integer variables also play a role in formulating equation systems that model logical conditions, as we will show later in this chapter.

In some situations, the optimization techniques described in previous chapters are sufficient to find an integer solution. An integer optimal solution is guaranteed for certain network linear programs, as explained in Section 15.5. Even where there is no guarantee, a linear programming solver may happen to find an integer optimal solution for the particular instances of a model in which you are interested. This happened in the solution of the multicommodity transportation model (Figure 4-1) for the particular data that we specified (Figure 4-2).

Even if you do not obtain an integer solution from the solver, chances are good that you'll get a solution in which most of the variables lie at integer values. Specifically, many solvers are able to return an ''extreme'' solution in which the number of variables not lying at their bounds is at most the number of constraints. If the bounds are integral, all of the variables at their bounds will have integer values; and if the rest of the data is integral, many of the remaining variables may turn out to be integers, too. You may then be able to adjust the relatively few non-integer variables to produce a completely integer solution that is close enough to feasible and optimal for practical purposes.

An example is provided by the scheduling linear program of Figures 16-4 and 16-5. Since the number of variables greatly exceeds the number of constraints, most of the variables end up at their bound of 0 in the optimal solution, and some other variables come out at integer values as well. The remaining variables can be rounded up to a nearby integer solution that is a little more costly but still satisfies the constraints.

Despite all these possibilities, there remain many circumstances in which the restriction to integrality must be enforced explicitly by the solver. Integer programming solvers face a much more difficult problem than their linear programming counterparts, however; they generally require more computer time and memory, and often demand more help from the user in formulation and in choice of options. As a result, the size of problem that you can solve will be more limited for integer programs than for linear ones.

This chapter first describes AMPL declarations of ordinary integer variables, and then introduces the use of zero-one (or binary) variables for modeling logical conditions. A concluding section offers some advice on formulating and solving integer programs effectively.

20.1 Integer variables

By adding the keyword `integer` to the qualifying phrases of a `var` declaration, you restrict the declared variables to integer values.

As an example, in analyzing the diet model in Section 2.3, we arrived at the following optimal solution:

```
ampl: model diet.mod;
ampl: data diet2a.dat;

ampl: solve;
MINOS 5.5: optimal solution found.
13 iterations, objective 118.0594032

ampl: display Buy;
Buy [*] :=
BEEF   5.36061
 CHK   2
FISH   2
 HAM  10
 MCH  10
 MTL  10
 SPG   9.30605
 TUR   2
;
```

If we want the foods to be purchased in integral amounts, we add `integer` to the model's `var` declaration (Figure 2-1) as follows:

```
var Buy {j in FOOD} integer >= f_min[j], <= f_max[j];
```

We can then try to re-solve:

```
ampl: model dieti.mod; data diet2a.dat;
ampl: solve;
MINOS 5.5: ignoring integrality of 8 variables
MINOS 5.5: optimal solution found.
13 iterations, objective 118.0594032
```

As you can see, the MINOS solver does not handle integrality constraints. It has ignored them and returned the same optimal value as before.

To get the integer optimum, we switch to a solver that does accommodate integrality:

```
ampl: option solver cplex;
ampl: solve;
CPLEX 8.0.0: optimal integer solution; objective 119.3
11 MIP simplex iterations
1 branch-and-bound nodes

ampl: display Buy;
Buy [*] :=
BEEF    9
 CHK    2
FISH    2
 HAM    8
 MCH   10
 MTL   10
 SPG    7
 TUR    2
;
```

Comparing this solution to the previous one, we see a few features typical of integer programming. The minimum cost has increased from $118.06 to $119.30; because integrality is an additional constraint on the values of the variables, it can only make the objective less favorable. The amounts of the different foods in the diet have also changed, but in unpredictable ways. The two foods that had fractional amounts in the original optimum, BEEF and SPG, have increased from 5.36061 to 9 and decreased from 9.30605 to 7, respectively; also, HAM has dropped from the upper limit of 10 to 8. Clearly, you cannot always deduce the integer optimum by rounding the non-integer optimum to the closest integer values.

20.2 Zero-one variables and logical conditions

Variables that can take only the values zero and one are a special case of integer variables. By cleverly incorporating these ''zero-one'' or ''binary'' variables into objectives and constraints, integer linear programs can specify a variety of logical conditions that cannot be described in any practical way by linear constraints alone.

To introduce the use of zero-one variables in AMPL, we return to the multicommodity transportation model of Figure 4-1. The decision variables Trans[i,j,p] in this model represent the tons of product p in set PROD to be shipped from originating city i in ORIG to destination city j in DEST. In the small example of data given by Figure 4-2, the products are bands, coils and plate; the origins are GARY, CLEV and PITT, and there are seven destinations.

The cost that this model associates with shipment of product p from i to j is cost[i,j,p] * Trans[i,j,p], regardless of the amount shipped. This ''variable'' cost is typical of purely linear programs, and in this case allows small shipments between many origin-destination pairs. In the following examples we describe ways to use zero-one variables to discourage shipments of small amounts; the same techniques can be adapted to many other logical conditions as well.

To provide a convenient basis for comparison, we focus on the tons shipped from each origin to each destination, summed over all products. The optimal values of these total shipments are determined by a linear programming solver as follows:

```
ampl: model multi.mod;
ampl: data multi.dat;
ampl: solve;
MINOS 5.5: optimal solution found.
41 iterations, objective 199500

ampl: option display_eps .000001;
ampl: option display_transpose -10;

ampl: display {i in ORIG, j in DEST}
ampl?     sum {p in PROD} Trans[i,j,p];
sum{p in PROD} Trans[i,j,p] [*,*]
:       DET    FRA    FRE    LAF    LAN    STL    WIN    :=
CLEV    625    275    325    225    400    550    200
GARY      0      0    625    150      0    625      0
PITT    525    625    225    625    100    625    175
;
```

The quantity 625 appears often in this solution, as a consequence of the multicommodity constraints:

```
subject to Multi {i in ORIG, j in DEST}:
    sum {p in PROD} Trans[i,j,p] <= limit[i,j];
```

In the data for our example, `limit[i,j]` is 625 for all `i` and `j`; its six appearances in the solution correspond to the six routes on which the multicommodity limit constraint is tight. Other routes have positive shipments as low as 100; the four instances of 0 indicate routes that are not used.

Even though all of the shipment amounts happen to be integers in this solution, we would be willing to ship fractional amounts. Thus we will not declare the `Trans` variables to be integer, but will instead extend the model by using zero-one integer variables.

Fixed costs

One way to discourage small shipments is to add a "fixed" cost for each origin-destination route that is actually used. For this purpose we rename the `cost` parameter `vcost`, and declare another parameter `fcost` to represent the fixed assessment for using the route from `i` to `j`:

```
param vcost {ORIG,DEST,PROD} >= 0; # variable cost on routes
param fcost {ORIG,DEST} > 0;       # fixed cost on routes
```

We want `fcost[i,j]` to be added to the objective function if the total shipment of products from `i` to `j` — that is, sum {p in PROD} `Trans[i,j,p]` — is positive; we want nothing to be added if the total shipment is zero. Using AMPL expressions, we could write the objective function most directly as follows:

```
minimize Total_Cost:     # NOT PRACTICAL
    sum {i in ORIG, j in DEST, p in PROD}
       vcost[i,j,p] * Trans[i,j,p]
 + sum {i in ORIG, j in DEST}
       if sum {p in PROD} Trans[i,j,p] > 0 then fcost[i,j];
```

AMPL accepts this objective, but treats it as merely "not linear" in the sense of Chapter 18, so that you are unlikely to get acceptable results trying to minimize it.

As a more practical alternative, we may associate a new variable Use[i,j] with each route from i to j, as follows: Use[i,j] takes the value 1 if

```
sum {p in PROD} Trans[i,j,p]
```

is positive, and is 0 otherwise. Then the fixed cost associated with the route from i to j is fcost[i,j] * Use[i,j], a linear term. To declare these new variables in AMPL, we can say that they are integer with bounds >= 0 and <= 1; equivalently we can use the keyword binary:

```
var Use {ORIG,DEST} binary;
```

The objective function can then be written as a linear expression:

```
minimize Total_Cost:
    sum {i in ORIG, j in DEST, p in PROD}
       vcost[i,j,p] * Trans[i,j,p]
 + sum {i in ORIG, j in DEST} fcost[i,j] * Use[i,j];
```

Since the model has a combination of continuous (non-integer) and integer variables, it yields what is known as a "mixed-integer" program.

To complete the model, we need to add constraints to assure that Trans and Use are related in the intended way. This is the "clever" part of the formulation; we simply modify the Multi constraints cited above so that they incorporate the Use variables:

```
subject to Multi {i in ORIG, j in DEST}:
    sum {p in PROD} Trans[i,j,p] <= limit[i,j] * Use[i,j];
```

If Use[i,j] is 0, this constraint says that

```
sum {p in PROD} Trans[i,j,p] <= 0
```

Since this total of shipments from i to j is a sum of nonnegative variables, it must equal 0. On the other hand, when Use[i,j] is 1, the constraint reduces to

```
sum {p in PROD} Trans[i,j,p] <= limit[i,j]
```

which is the multicommodity limit as before. Although there is nothing in the constraint to directly prevent sum {p in PROD} Trans[i,j,p] from being 0 when Use[i,j] is 1, so long as fcost[i,j] is positive this combination can never occur in an optimal solution. Thus Use[i,j] will be 1 if and only if sum {p in PROD} Trans[i,j,p] is positive, which is what we want. The complete model is shown in Figure 20-1a.

```
set ORIG;     # origins
set DEST;     # destinations
set PROD;     # products

param supply {ORIG,PROD} >= 0;   # amounts available at origins
param demand {DEST,PROD} >= 0;   # amounts required at destinations

   check {p in PROD}:
       sum {i in ORIG} supply[i,p] = sum {j in DEST} demand[j,p];

param limit {ORIG,DEST} >= 0;    # maximum shipments on routes

param vcost {ORIG,DEST,PROD} >= 0; # variable shipment cost on routes
var Trans {ORIG,DEST,PROD} >= 0;   # units to be shipped

param fcost {ORIG,DEST} >= 0;      # fixed cost for using a route
var Use {ORIG,DEST} binary;        # = 1 only for routes used

minimize Total_Cost:
   sum {i in ORIG, j in DEST, p in PROD} vcost[i,j,p] * Trans[i,j,p]
 + sum {i in ORIG, j in DEST} fcost[i,j] * Use[i,j];

subject to Supply {i in ORIG, p in PROD}:
   sum {j in DEST} Trans[i,j,p] = supply[i,p];

subject to Demand {j in DEST, p in PROD}:
   sum {i in ORIG} Trans[i,j,p] = demand[j,p];

subject to Multi {i in ORIG, j in DEST}:
   sum {p in PROD} Trans[i,j,p] <= limit[i,j] * Use[i,j];
```

Figure 20-1a: Multicommodity model with fixed costs (multmip1.mod).

To show how this model might be solved, we add a table of fixed costs to the sample data (Figure 20-1b):

```
param fcost:    FRA  DET  LAN  WIN  STL  FRE  LAF :=
        GARY   3000 1200 1200 1200 2500 3500 2500
        CLEV   2000 1000 1500 1200 2500 3000 2200
        PITT   2000 1200 1500 1500 2500 3500 2200 ;
```

If we apply the same solver as before, the integrality restrictions on the Use variables are ignored:

```
ampl: model multmip1.mod;
ampl: data multmip1.dat;
ampl: solve;
MINOS 5.5: ignoring integrality of 21 variables
MINOS 5.5: optimal solution found.
43 iterations, objective 223504

ampl: option display_eps .000001;
ampl: option display_transpose -10;
```

```
set ORIG := GARY CLEV PITT ;
set DEST := FRA DET LAN WIN STL FRE LAF ;
set PROD := bands coils plate ;

param supply (tr):     GARY    CLEV    PITT :=
               bands    400     700     800
               coils    800    1600    1800
               plate    200     300     300 ;

param demand (tr):
                FRA   DET   LAN   WIN   STL   FRE   LAF :=
        bands   300   300   100    75   650   225   250
        coils   500   750   400   250   950   850   500
        plate   100   100     0    50   200   100   250 ;

param limit default 625 ;

param vcost :=

 [*,*,bands]:  FRA   DET   LAN   WIN   STL   FRE   LAF :=
        GARY    30    10     8    10    11    71     6
        CLEV    22     7    10     7    21    82    13
        PITT    19    11    12    10    25    83    15

 [*,*,coils]:  FRA   DET   LAN   WIN   STL   FRE   LAF :=
        GARY    39    14    11    14    16    82     8
        CLEV    27     9    12     9    26    95    17
        PITT    24    14    17    13    28    99    20

 [*,*,plate]:  FRA   DET   LAN   WIN   STL   FRE   LAF :=
        GARY    41    15    12    16    17    86     8
        CLEV    29     9    13     9    28    99    18
        PITT    26    14    17    13    31   104    20 ;

param fcost:   FRA   DET   LAN   WIN   STL   FRE   LAF :=
       GARY   3000  1200  1200  1200  2500  3500  2500
       CLEV   2000  1000  1500  1200  2500  3000  2200
       PITT   2000  1200  1500  1500  2500  3500  2200 ;
```

Figure 20-1b: Data for Figure 20-1a (`multmip1.dat`).

```
ampl: display sum {i in ORIG, j in DEST, p in PROD}
ampl?     vcost[i,j,p] * Trans[i,j,p];
sum{i in ORIG, j in DEST, p in PROD}
     vcost[i,j,p]*Trans[i,j,p] = 199500

ampl: display Use;
Use [*,*]
:     DET    FRA    FRE    LAF    LAN    STL    WIN      :=
CLEV   1     0.44   0.52   0.36   0.64   0.88   0.32
GARY   0     0      1      0.24   0      1      0
PITT   0.84  1      0.36   1      0.16   1      0.28
;
```

As you can see, the total variable cost is the same as before, and Use assumes a variety of fractional values. This solution tells us nothing new, and there is no simple way to convert it into a good integer solution. An integer programming solver is essential to get any practical results in this situation.

Switching to an appropriate solver, we find that the true optimum with fixed costs is as follows:

```
ampl: option solver cplex; solve;
CPLEX 8.0.0: optimal integer solution; objective 229850
295 MIP simplex iterations
19 branch-and-bound nodes

ampl: display {i in ORIG, j in DEST}
ampl?     sum {p in PROD} Trans[i,j,p];
sum{p in PROD} Trans[i,j,p] [*,*]
:       DET    FRA    FRE    LAF    LAN    STL    WIN      :=
CLEV    625    275      0    425    350    550    375
GARY      0      0    625      0    150    625      0
PITT    525    625    550    575      0    625      0
;

ampl: display Use;
Use [*,*]
:    DET FRA FRE LAF LAN STL WIN      :=
CLEV   1   1   0   1   1   1   1
GARY   0   0   1   0   1   1   0
PITT   1   1   1   1   0   1   0
;
```

Imposing the integer constraints has increased the total cost from $223,504 to $229,850; but the number of unused routes has increased, to seven, as we had hoped.

Zero-or-minimum restrictions

Although the fixed-cost solution uses fewer routes, there are still some on which the amounts shipped are relatively low. As a practical matter, it may be that even the variable costs are not applicable unless some minimum number of tons is shipped. Suppose, therefore, that we declare a parameter minload to represent the minimum number of tons that may be shipped on a route. We could add a constraint to say that the shipments on each route, summed over all products, must be at least minload:

```
subject to Min_Ship {i in ORIG, j in DEST}:      # WRONG
    sum {p in PROD} Trans[i,j,p] >= minload;
```

But this would force the shipments on every route to be at least minload, which is not what we have in mind. We want the tons shipped to be either zero, or at least minload. To say this directly, we might write:

```
subject to Min_Ship {i in ORIG, j in DEST}:      # NOT ALLOWED
    sum {p in PROD} Trans[i,j,p] = 0 or
    sum {p in PROD} Trans[i,j,p] >= minload;
```

But the current version of AMPL does not accept logical operators in constraints.

The desired zero-or-minimum restrictions can be imposed by employing the variables Use[i,j], much as in the previous example:

```
subject to Min_Ship {i in ORIG, j in DEST}:
    sum {p in PROD} Trans[i,j,p] >= minload * Use[i,j];
```

When total shipments from i to j are positive, Use[i,j] is 1, and Min_Ship[i,j] becomes the desired minimum-shipment constraint. On the other hand, when there are no shipments from i to j, Use[i,j] is zero; the constraint reduces to 0 >= 0 and has no effect.

With these new restrictions and a minload of 375, the solution is found to be as follows:

```
ampl: model multmip2.mod;
ampl: data multmip2.dat;

ampl: solve;
CPLEX 8.0.0: optimal integer solution; objective 233150
279 MIP simplex iterations
17 branch-and-bound nodes

ampl: display {i in ORIG, j in DEST}
ampl?     sum {p in PROD} Trans[i,j,p];
sum{p in PROD} Trans[i,j,p] [*,*]
:       DET    FRA    FRE    LAF    LAN    STL    WIN      :=
CLEV    625    425    425      0    500    625      0
GARY      0      0    375    425      0    600      0
PITT    525    475    375    575      0    575    375
;
```

Comparing this to the previous solution, we see that although there are still seven unused routes, they are not the same ones; a substantial rearrangement of the solution has been necessary to meet the minimum-shipment requirement. The total cost has gone up by about 1.4% as a result.

Cardinality restrictions

Despite the constraints we have added so far, origin PITT still serves 6 destinations, while CLEV serves 5 and GARY serves 3. We would like to explicitly add a further restriction that each origin can ship to at most maxserve destinations, where maxserve is a parameter to the model. This can be viewed as a restriction on the size, or cardinality, of a certain set. Indeed, it could in principle be written in the form of an AMPL constraint as follows:

```
subject to Max_Serve {i in ORIG}:     # NOT ALLOWED
    card {j in DEST:
        sum {p in PROD} Trans[i,j,p] > 0} <= maxserve;
```

Such a declaration will be rejected, however, because AMPL currently does not allow constraints to use sets that are defined in terms of variables.

```
set ORIG;    # origins
set DEST;    # destinations
set PROD;    # products

param supply {ORIG,PROD} >= 0;  # amounts available at origins
param demand {DEST,PROD} >= 0;  # amounts required at destinations

    check {p in PROD}:
        sum {i in ORIG} supply[i,p] = sum {j in DEST} demand[j,p];

param limit {ORIG,DEST} >= 0;    # maximum shipments on routes
param minload >= 0;              # minimum nonzero shipment
param maxserve integer > 0;      # maximum destinations served

param vcost {ORIG,DEST,PROD} >= 0; # variable shipment cost on routes
var Trans {ORIG,DEST,PROD} >= 0;   # units to be shipped

param fcost {ORIG,DEST} >= 0;      # fixed cost for using a route
var Use {ORIG,DEST} binary;        # = 1 only for routes used

minimize Total_Cost:
    sum {i in ORIG, j in DEST, p in PROD} vcost[i,j,p] * Trans[i,j,p]
  + sum {i in ORIG, j in DEST} fcost[i,j] * Use[i,j];

subject to Supply {i in ORIG, p in PROD}:
    sum {j in DEST} Trans[i,j,p] = supply[i,p];

subject to Max_Serve {i in ORIG}:
    sum {j in DEST} Use[i,j] <= maxserve;

subject to Demand {j in DEST, p in PROD}:
    sum {i in ORIG} Trans[i,j,p] = demand[j,p];

subject to Multi {i in ORIG, j in DEST}:
    sum {p in PROD} Trans[i,j,p] <= limit[i,j] * Use[i,j];

subject to Min_Ship {i in ORIG, j in DEST}:
    sum {p in PROD} Trans[i,j,p] >= minload * Use[i,j];
```

Figure 20-2a: Multicommodity model with further restrictions (multmip3.mod).

Zero-one variables again offer a convenient alternative. Since the variables Use[i,j] are 1 precisely for those destinations j served by origin i, and are zero otherwise, we can write sum {j in DEST} Use[i,j] for the number of destinations served by i. The desired constraint becomes:

```
        subject to Max_Serve {i in ORIG}:
            sum {j in DEST} Use[i,j] <= maxserve;
```

Adding this constraint to the previous model, and setting maxserve to 5, we arrive at the mixed integer model shown in Figure 20-2a, with data shown in Figure 20-2b. It is optimized as follows:

```
set ORIG := GARY CLEV PITT ;
set DEST := FRA DET LAN WIN STL FRE LAF ;
set PROD := bands coils plate ;

param supply (tr):    GARY      CLEV      PITT :=
             bands     400       700       800
             coils     800      1600      1800
             plate     200       300       300 ;

param demand (tr):
                FRA   DET   LAN   WIN   STL   FRE   LAF :=
        bands   300   300   100    75   650   225   250
        coils   500   750   400   250   950   850   500
        plate   100   100     0    50   200   100   250 ;

param limit default 625 ;

param vcost :=

  [*,*,bands]:  FRA   DET   LAN   WIN   STL   FRE   LAF :=
        GARY     30    10     8    10    11    71     6
        CLEV     22     7    10     7    21    82    13
        PITT     19    11    12    10    25    83    15

  [*,*,coils]:  FRA   DET   LAN   WIN   STL   FRE   LAF :=
        GARY     39    14    11    14    16    82     8
        CLEV     27     9    12     9    26    95    17
        PITT     24    14    17    13    28    99    20

  [*,*,plate]:  FRA   DET   LAN   WIN   STL   FRE   LAF :=
        GARY     41    15    12    16    17    86     8
        CLEV     29     9    13     9    28    99    18
        PITT     26    14    17    13    31   104    20 ;

param fcost:    FRA   DET   LAN   WIN   STL   FRE   LAF :=
        GARY   3000  1200  1200  1200  2500  3500  2500
        CLEV   2000  1000  1500  1200  2500  3000  2200
        PITT   2000  1200  1500  1500  2500  3500  2200 ;

param minload := 375 ;

param maxserve := 5 ;
```

Figure 20-2b: Data for Figure 20-2a (multmip3.dat).

```
ampl: model multmip3.mod;
ampl: data multmip3.dat;

ampl: solve;
CPLEX 8.0.0: optimal integer solution; objective 235625
392 MIP simplex iterations
36 branch-and-bound nodes
```

```
ampl: display {i in ORIG, j in DEST}
ampl?     sum {p in PROD} Trans[i,j,p];
sum{p in PROD} Trans[i,j,p] [*,*]
 :       DET   FRA   FRE   LAF   LAN   STL   WIN       :=
CLEV    625   375   550     0   500   550     0
GARY      0     0     0   400     0   625   375
PITT    525   525   625   600     0   625     0
;
```

At the cost of a further 1.1% increase in the objective, rearrangements have been made so that GARY can serve WIN, bringing the number of destinations served by PITT down to five.

Notice that this section's three integer solutions have served WIN from each of the three different origins — a good example of how solutions to integer programs can jump around in response to small changes in the restrictions.

20.3 Practical considerations in integer programming

As a rule, any integer program is much harder to solve than a linear program of the same size and structure. A rough idea of the difference in the previous examples is given by the number of iterations reported by the solvers; it is 41 for solving the original linear multicommodity transportation problem, but ranges from about 280 to 400 for the mixed integer versions. The computation times vary similarly; the linear programs are solved almost instantly, while the mixed integer ones are noticeably slower.

As the size of the problem increases, the difficulty of an integer program also grows more quickly than the difficulty of a comparable linear program. Thus the practical limits to the size of a solvable integer program will be much more restrictive. Indeed, AMPL can easily generate integer programs that are too difficult for your computer to solve in a reasonable amount of time or memory. Before you make a commitment to use a model with integer variables, therefore, you should consider whether an alternative continuous linear or network formulation might give adequate answers. If you must use an integer formulation, try it on collections of data that increase gradually in size, so that you can get an idea of the computer resources required.

If you do encounter an integer program that is difficult to solve, you should consider reformulations that can make it easier. An integer programming solver attempts to investigate all of the different possible combinations of values of the integer variables; although it employs a sophisticated strategy that rules out the vast majority of combinations as infeasible or suboptimal, the number of combinations remaining to be checked can still be huge. Thus you should try to use as few integer variables as possible. For those variables that are not zero-one, lower and upper bounds should be made as tight as possible to reduce the number of combinations that might be investigated.

Solvers derive valuable information about the solutions to an integer program by fixing or restricting some of the integer variables, then solving the linear programming ''relaxation'' that results when the remaining integer restrictions are dropped. You may be able to assist in this strategy by reformulating the model so that the solution of a relaxation will be closer to the solution of the associated integer problem. In the multicommodity transportation model, for example, if Use[i,j] is 0 then each of the individual variables Trans[i,j,p] must be 0, while if Use[i,j] is 1 then Trans[i,j,p] cannot be larger than either supply[i,p] or demand[j,p]. This suggests that we add the following constraints:

```
subject to Avail {i in ORIG, j in DEST, p in PROD}:
    Trans[i,j,p] <= min(supply[i,p],demand[j,p]) * Use[i,j];
```

Although these constraints do not rule out any previously admissible integer solutions, they tend to force Use[i,j] to be closer to 1 for any solution of the relaxation that uses the route from i to j. As a result, the relaxation is more accurate, and may help the solver to find the optimum integer solution more quickly; this advantage may outweigh the extra cost of handling more constraints. Tradeoffs of this kind are most often observed for problems substantially larger than the examples in this section, however.

As this example suggests, the choice of a formulation is much more critical in integer than in linear programming. For large problems, solution times can change dramatically in response to simple reformulations. The effect of a reformulation can be hard to predict, however; it depends on the structure of your model and data, and on the details of the strategy used by your solver. Generally you will need to do some experimentation to see what works best.

You can also try to help a solver by changing some of the default settings that determine how it initially processes a problem and how it searches for integer solutions. Many of these settings can be manipulated from within AMPL, as explained in the separate instructions for using particular solvers.

Finally, a solver may provide options for stopping prematurely, returning an integer solution that has been determined to bring the objective value to within a small percentage of optimality. In some cases, such a solution is found relatively early in the solution process; you may save a great deal of computation time by not insisting that the solver go on to find a provably optimal integer solution.

In summary, experimentation is usually necessary to solve integer or mixed-integer linear programs. Your chances of success are greatest if you approach the use of integer variables with caution, and if you are willing to keep trying alternative formulations, settings and strategies until you get an acceptable result.

Bibliography

Ellis L. Johnson, Michael M. Kostreva and Uwe H. Suhl, ''Solving 0-1 Integer Programming Problems Arising from Large-Scale Planning Models.'' Operations Research **33** (1985) pp. 803–819. A case study in which preprocessing, reformulation, and algorithmic strategies were brought to bear on the solution of a difficult class of integer linear programs.

George L. Nemhauser and Laurence A. Wolsey, *Integer and Combinatorial Optimization.* John Wiley & Sons (New York, NY, 1988). A survey of integer programming problems, theory and algorithms.

Alexander Schrijver, *Theory of Linear and Integer Programming.* John Wiley & Sons (New York, NY, 1986). A guide to the fundamentals of the subject, with a particularly thorough collection of references.

Laurence A. Wolsey, *Integer Programming.* Wiley-Interscience (New York, NY, 1998). A practical, intermediate-level guide for readers familiar with linear programming.

Exercises

20-1. Exercise 1-1 optimizes an advertising campaign that permits arbitrary combinations of various media. Suppose that instead you must allocate a $1 million advertising campaign among several media, but for each, your only choice is whether to use that medium or not. The following table shows, for each medium, the cost and audience in thousands if you use that medium, and the creative time in three categories that will have to be committed if you use that medium. The final column shows your limits on cost and on hours of creative time.

	TV	Magazine	Radio	Newspaper	Mail	Phone	Limits
Audience	300	200	100	150	100	50	
Cost	600	250	100	120	200	250	1000
Writers	120	50	20	40	30	5	200
Artists	120	80	0	60	40	0	300
Others	20	20	20	20	20	200	200

Your goal is to maximize the audience, subject to the limits.

(a) Formulate an AMPL model for this situation, using a zero-one integer variable for each medium.

(b) Use an integer programming solver to find the optimal solution.

Also solve the problem with the integrality restrictions relaxed and compare the resulting solution. Could you have guessed the integer optimum from looking at the non-integer one?

20-2. Return to the multicommodity transportation problem of Figures 4-1 and 4-2. Use appropriate integer variables to impose each of the restrictions described below. (Treat each part separately; don't try to put all the restrictions in one model.) Solve the resulting integer program, and comment in general terms on how the solution changes as a result of the restrictions. Also solve the corresponding linear program with the integrality restrictions relaxed, and compare the LP solution to the integer one.

(a) Require the amount shipped on each origin-destination link to be a multiple of 100 tons. To accommodate this restriction, allow demand to be satisfied only to the nearest 100 — for example, since demand for bands at FRE is 225 tons, allow either 200 or 300 tons to be shipped to FRE.

(b) Require each destination except STL to be served by at most two origins.

(c) Require the number of origin-destination links used to be as small as possible, regardless of cost.

(d) Require each origin that supplies product p to destination j to ship either nothing, or at least the lesser of demand[j,p] and 150 tons.

20-3. Employee scheduling problems are a common source of integer programs, because it may not make sense to schedule a fraction of a person.

(a) Solve the problem of Exercise 4-4(b) with the requirement that the variables Y_{st}, representing the numbers of crews employed on shift s in period t, must all be integer. Confirm that the optimal integer solution just "rounds up" the linear programming solution to the next highest integer.

(b) Similarly attempt to find an integer solution for the problem of Exercise 4-4(c), where inventories have been added to the formulation. Compare the difficulty of this problem to the one in (a). Is the optimal integer solution the same as the rounded-up linear programming solution in this case?

(c) Solve the scheduling problem of Figures 16-4 and 16-5 with the requirement that an integer number of employees be assigned to each shift. Show that this solution is better than the one obtained by rounding up.

(d) Suppose that the number of supervisors required is 1/10 the number of employees, rounded to the nearest whole number. Solve the scheduling problem again for supervisors. Does the integer program appear to be harder or easier to solve in this case? Is the improvement over the rounded-up solution more or less pronounced?

20-4. The so-called knapsack problem arises in many contexts. In its simplest form, you start with a set of objects, each having a known weight and a known value. The problem is to decide which items to put into your knapsack. You want these items to have as great a total value as possible, but their weight cannot exceed a certain preset limit.

(a) The data for the simple knapsack problem could be written in AMPL as follows:

```
set OBJECTS;
param weight {OBJECTS} > 0;
param value {OBJECTS} > 0;
```

Using these declarations, formulate an AMPL model for the knapsack problem.

Use your model to solve the knapsack problem for objects of the following weights and values, subject to a weight limit of 100:

object	a	b	c	d	e	f	g	h	i	j
weight	55	50	40	35	30	30	15	15	10	5
value	1000	800	750	700	600	550	250	200	200	150

(b) Suppose instead that you want to fill several identical knapsacks, as specified by this extra parameter:

```
param knapsacks > 0 integer;    # number of knapsacks
```

Formulate an AMPL model for this situation. Don't forget to add a constraint that each object can only go into one knapsack!

Using the data from (a), solve for 2 knapsacks of weight limit 50. How does the solution differ from your previous one?

(c) Superficially, the preceding knapsack problem resembles an assignment problem; we have a collection of objects and a collection of knapsacks, and we want to make an optimal assignment from members of the former to members of the latter. What is the essential difference between the kinds of assignment problems described in Section 15.2, and the knapsack problem described in (b)?

(d) Modify the formulation from (a) so that it accommodates a volume limit for the knapsack as well as a weight limit. Solve again using the following volumes:

```
object    a    b    c    d    e    f    g    h    i    j
volume    3    3    3    2    2    2    2    1    1    1
```

How do the total weight, volume and value of this solution compare to those of the solution you found in (a)?

(e) How can the media selection problem of Exercise 20-1 be viewed as a knapsack problem like the one in (d)?

(f) Suppose that you can put up to 3 of each object in the knapsack, instead of just 1. Revise the model of (a) to accommodate this possibility, and re-solve with the same data. How does the optimal solution change?

(g) Review the roll-cutting problem described in Exercise 2-6. Given a supply of wide rolls, orders for narrower rolls, and a collection of cutting patterns, we seek a combination of patterns that fills all orders using the least material.

When this problem is solved, the algorithm returns a "dual value" corresponding to each ordered roll width. It is possible to interpret the dual value as the saving in wide rolls that you might achieve for each extra narrow roll that you obtain; for example, a value of 0.5 for a 50" roll would indicate that you might save 5 wide rolls if you could obtain 10 extra 50" rolls.

It is natural to ask: Of all the patterns of narrow rolls that fit within one wide roll, which has the greatest total (dual) value? Explain how this can be regarded as a knapsack problem.

For the problem of Exercise 2-6(a), the wide rolls are 110"; in the solution using the six patterns given, the dual values for the ordered widths are:

20"	0.0
40"	0.5
50"	1.0
55"	0.0
75"	1.0

What is the maximum-value pattern? Show that it is not one of the ones given, and that adding it allows you to get a better solution.

20-5. Recall the multiperiod model of production that was introduced in Section 4.2. Add zero-one variables and appropriate constraints to the formulation from Figure 4-4 to impose each of the restrictions described below. (Treat each part separately.) Solve with the data of Figure 4-5, and confirm that the solution properly obeys the restrictions.

(a) Require for each product p and week t, that either none of the product is made that week, or at least 2500 tons are made.

(b) Require that only one product be made in any one week.

(c) Require that each product be made in at most two weeks out of any three-week period. Assume that only bands were made in "week −1" and that both bands and coils were made in "week 0".

A

AMPL Reference Manual

AMPL is a language for algebraic modeling and mathematical programming: a computer-readable language for expressing optimization problems such as linear programming in algebraic notation. This appendix summarizes the features of AMPL, with particular emphasis on technical details not fully covered in the preceding chapters. Nevertheless, not every feature, construct, and option is listed; the AMPL web site `www.ampl.com` contains the most up to date and complete information.

The following notational conventions are used. Literal text is printed in `constant width font`, while syntactic categories are printed in *italic font*. Phrases or subphrases enclosed in slanted square brackets *[* and *]* are optional, as are constructs with the subscript *opt*.

A.1 Lexical rules

AMPL models involve variables, constraints, and objectives, expressed with the help of sets and parameters. These are called *model entities*. Each model entity has an alphanumeric name: a string of one or more Unicode UTF-8 letters, digits, and underscores, in a pattern that cannot be mistaken for a numeric constant. Upper-case letters are distinct from lower-case letters.

Numeric constants are written in standard scientific notation: an optional sign, a sequence of digits that may contain a decimal point, and an optional exponent field that begins with one of the letters d, D, e or E, as in `1.23D-45`. All arithmetic in AMPL is in the same precision (double precision on most machines), so all exponent notations are synonymous.

Literals are strings delimited either by single quotes `'` or by double quotes `"`; the delimiting character must be doubled if it appears within the literal, as in `'x''y'`, which is a literal containing the three characters `x'y`. Newline characters may appear within a literal only if preceded by `\`. The choice of delimiter is arbitrary; `'abc'` and `"abc"` denote the same literal.

Literals are distinct from numeric constants: `1` and `'1'` are unrelated.

Input is free form; white space (any sequence of space, tab or newline characters) may appear between any tokens. Each statement ends with a semicolon.

Comments begin with # and extend to the end of the current line, or are delimited by `/*` and `*/`, in which case they may extend across several lines and do not nest. Comments may appear anywhere in declarations, commands, and data.

These words are reserved, and may not be used in other contexts:

```
Current          complements      integer          solve_result_num
IN               contains         less             suffix
INOUT            default          logical          sum
Infinity         dimen            max              symbolic
Initial          div              min              table
LOCAL            else             option           then
OUT              environ          setof            union
all              exists           shell_exitcode   until
binary           forall           solve_exitcode   while
by               if               solve_message    within
check            in               solve_result
```

Words beginning with underscore are also reserved. The other keywords, function names, etc., are predefined, but their meanings can be redefined. For example, the word `prod` is predefined as a product operator analogous to `sum`, but it can be redefined in a declaration, such as

```
set prod;   # products
```

Once a word has been redefined, the original meaning is inaccessible.

AMPL provides synonyms for several keywords and operators; the preferred forms are on the left.

```
^                **
=                ==
<>               !=
and              &&
not              !
or               ||
prod             product
```

A.2 Set members

A set contains zero or more elements or members, each of which is an ordered list of one or more components. Each member of a set must be distinct. All members must have the same number of components; this common number is called the set's dimension.

A literal set is written as a comma-separated list of members, between braces { and }. If the set is one-dimensional, the members are simply numeric constants or literal strings, or any expressions that evaluate to numbers or strings:

```
{"a","b","c"}
{1,2,3,4,5,6,7,8,9}
{t,t+1,t+2}
```

For a multidimensional set, each member must be written as a parenthesized comma-separated list of the above:

```
{("a",2),("a",3),("b",5)}
{(1,2,3),(1,2,4),(1,2,5),(1,3,7),(1,4,6)}
```

The value of a numeric member is the result of rounding its decimal representation to a floating-point number. Numeric members that appear different but round to the same floating-point number, such as 1 and 0.01E2, are considered the same.

A.3 Indexing expressions and subscripts

Most entities in AMPL can be defined in collections indexed over a set; individual items are selected by appending a bracketed subscript to the name of the entity. The range of possible subscripts is indicated by an *indexing expression* in the entity's declaration. Reduction operators, such as sum, also use indexing expressions to specify sets over which operations are iterated.

A subscript is a list of symbolic or numeric expressions, separated by commas and enclosed in square brackets, as in supply[i] and cost[j,p[k]+1,"O+"]. Each subscripting expression must evaluate to a number or a literal. The resulting value or sequence of values must give a member of a relevant one-dimensional or multidimensional indexing set.

An indexing expression is a comma-separated list of set expressions, followed optionally by a colon and a logical ''such that'' expression, all enclosed in braces:

> *indexing*:
> > { *sexpr-list* }
> > { *sexpr-list* : *lexpr* }
>
> *sexpr-list*:
> > *sexpr*
> > *dummy-member* in *sexpr*
> > *sexpr-list* , *sexpr*

Each set expression may be preceded by a dummy member and the keyword in. A dummy member for a one-dimensional set is an unbound name, that is, a name not currently defined. A dummy member for a multidimensional set is a comma-separated list, enclosed in parentheses, of expressions or unbound names; the list must include at least one unbound name.

A dummy member introduces one or more dummy indices (the unbound names in its components), whose *scopes*, or ranges of definition, begin just after the following *sexpr*; an index's scope runs through the rest of the indexing expression, to the end of the declaration using the indexing expression, or to the end of the operand that uses the indexing expression. When a dummy member has one or more expression components, the dummy indices in the dummy member range over a *slice* of the set, i.e., they assume all values for which the dummy member is in the set.

```
{A}                        # a set
{A, B}                     # all pairs, one from A, one from B
{i in A, j in B}           # the same
{i in A, B}                # the same
{i in A, C[i]}             # all pairs, one from A, one from C[i]
{i in A, (j,k) in D}       # 1 from A and 1 (itself a pair) from D
{i in A: p[i] > 0}         # all i in A such that p[i] is positive
{i in A, j in C[i]: i <= j} # i and j must be numeric
{i in A, (i,j) in D: i <= j} # all pairs with i in A and i,j in D
                           # (same value of i) and i <= j
```

The optional : *lexpr* in an indexing expression selects only the members that satisfy the logical expression and excludes the others. The *lexpr* typically involves one or more dummy indices of the indexing expression.

A.4 Expressions

Various items can be combined in AMPL's arithmetic and logical expressions. An expression that may not contain variables is denoted *cexpr* and is sometimes called a ''constant expression'',

Precedence	Name	Type	Remarks
1	if-then-else	A, S	A: if no else, then "else 0" assumed
			S: "else *sexpr*" required
2	or ||	L	
3	exists forall	L	logical reduction operators
4	and &&	L	
5	< <= = == <> != >= >	L	
6	in not in	L	membership in set
6	within not within	L	S within T means set S ⊆ set T
7	not !	L	logical negation
8	union diff symdiff	S	symdiff ≡ symmetric difference
9	inter	S	set intersection
10	cross	S	cross or Cartesian product
11	setof .. by	S	set constructors
12	+ - less	A	a less $b \equiv \max(a - b, 0)$
13	sum prod min max	A	arithmetic reduction operators
14	* / div mod	A	div ≡ truncated quotient of integers
15	+ -	A	unary plus, unary minus
16	^ **	A	exponentiation

Operators are listed in increasing precedence order. Exponentiation and if-then-else are right-associative; the other operators are left-associative. The 'Type' column indicates result types: A for arithmetic, L for logical, S for set.

Table A-1: Arithmetic, logical and set operators.

even though it may involve dummy indices. A logical expression, denoted *lexpr*, may not contain variables when it is part of a *cexpr*. Set expressions are denoted *sexpr*.

Table A-1 summarizes the arithmetic, logical and set operators; the type column indicates whether the operator produces an arithmetic value (A), a logical value (L), or a set value (S).

Arithmetic expressions are formed from the usual arithmetic operators, built-in functions, and arithmetic reduction operators like sum:

expr:
    number
    variable
    expr arith-op expr   *arith-op* is + - less * / mod div ^ **
    unary-op expr   *unary-op* is + -
    built-in (*exprlist*)
    if *lexpr* then *expr* [else *expr*]
    reduction-op indexing expr   *reduction-op* is sum prod max min
    (*expr*)

Built-in functions are listed in Table A-2.

The arithmetic reduction operators are used in expressions like

```
sum {i in Prod} cost[i] * Make[i]
```

The scope of the *indexing* expression extends to the end of the *expr*. If the operation is over an empty set, the result is the identity value for the operation: 0 for sum, 1 for prod, Infinity for min, and -Infinity for max.

Logical expressions appear where a value of "true" or "false" is required: in check statements, the "such that" parts of indexing expressions (following the colon), and in if *lexpr* then ... else ... expressions. Numeric values that appear in any of these contexts are implicitly coerced to logical values: 0 is interpreted as false, and all other numeric values as true.

lexpr:
    expr
    expr compare-op expr   *compare-op* is < <= = == != <> > >=
    lexpr logic-op lexpr   *logic-op* is or || and &&
    not *lexpr*
    member in *sexpr*
    member not in *sexpr*
    sexpr within *sexpr*
    sexpr not within *sexpr*
    opname indexing lexpr   *opname* is exists or forall
    (*lexpr*)

The in operator tests set membership. Its left operand is a potential set member, i.e., an expression or comma-separated list of expressions enclosed in parentheses, with the number of expressions equal to the dimension of the right operand, which must be a set expression. The within operator tests whether one set is contained in another. The two set operands must have the same dimension.

The logical reduction operators exists and forall are the iterated counterparts of or and and. When applied over an empty set, exists returns false and forall returns true.

Set expressions yield sets.

sexpr:
    { [*member* [, *member*...]] }
    sexpr set-op sexpr   *set-op* is union diff symdiff inter cross
    opname indexing sexpr   *opname* is union or inter
    expr .. *expr* [by *expr*]
    setof *indexing member*
    if *lexpr* then *sexpr* else *sexpr*
    (*sexpr*)
    interval
    infinite-set
    indexing

Components of members can be arbitrary constant expressions. Section A.6.3 describes *interval*s and *infinite-set*s.

When used as binary operators, union and inter denote the binary set operations of union and intersection. These keywords may also be used as reduction operators.

The .. operator constructs sets. The default by clause is by 1. In general, $e_1 .. e_2$ by e_3 means the numbers

$$e_1, e_1 + e_3, e_1 + 2e_3, \ldots, e_1 + \left\lfloor \frac{e_2 - e_1}{e_3} \right\rfloor e_3$$

rounded to set members. (The notation $\lfloor x \rfloor$ denotes the floor of x, that is, the largest integer $\leq x$.)

The setof operator is a set construction operator; *member* is either an expression or a comma-separated list of expressions enclosed in parentheses. The resulting set consists of all the *member*s obtained by iterating over the *indexing* expression; the dimension of the resulting expression is the number of components in *member*.

abs (x)	absolute value $\lvert x \rvert$
acos (x)	inverse cosine, $\cos^{-1}(x)$
acosh (x)	inverse hyperbolic cosine, $\cosh^{-1}(x)$
alias (v)	alias of model entity v
asin (x)	inverse sine, $\sin^{-1}(x)$
asinh (x)	inverse hyperbolic sine, $\sinh^{-1}(x)$
atan (x)	inverse tangent, $\tan^{-1}(x)$
atan2 (y, x)	inverse tangent, $\tan^{-1}(y/x)$
atanh (x)	inverse hyperbolic tangent, $\tanh^{-1}(x)$
ceil (x)	ceiling of x (next higher integer)
ctime $(\ )$	current time as a string
ctime (t)	time t as a string
cos (x)	cosine
exp (x)	e^x
floor (x)	floor of x (next lower integer)
log (x)	$\log_e(x)$
log10 (x)	$\log_{10}(x)$
max $(x, y, \ldots)$	maximum (2 or more arguments)
min $(x, y, \ldots)$	minimum (2 or more arguments)
precision (x, n)	x rounded to n significant decimal digits
round (x, n)	x rounded to n digits past decimal point
round (x)	x rounded to an integer
sin (x)	sine
sinh (x)	hyperbolic sine
sqrt (x)	square root
tan (x)	tangent
tanh (x)	hyperbolic tangent
time $(\ )$	current time in seconds
trunc (x, n)	x truncated to n digits past decimal point
trunc (x)	x truncated to an integer

Table A-2: Built-in arithmetic functions.

```
ampl: set y = setof {i in 1..5} (i,i^2);
ampl: display y;
set y := (1,1) (2,4) (3,9) (4,16) (5,25);
```

A.4.1 Built-in functions

The built-in arithmetic functions are listed in Table A-2. The function alias takes as its argument the name of a model entity and returns its *alias*, a literal value described in Section A.5. The functions round(x, n) and trunc(x, n) convert x to a decimal string and round or truncate it to n places past the decimal point (or to $-n$ places before the decimal point if $n < 0$); similarly, precision(x, n) rounds x to n significant decimal digits. For round and trunc, a missing n is taken as 0, thus providing the usual rounding or truncation to an integer.

Several built-in random number generation functions are available, as listed in Table A-3. All are based on a uniform random number generator with a very long period. An initial seed n can be specified with the -sn command-line argument (A.23) or option randseed, while -s or

`Beta(a, b)`	$density(x) = x^{a-1}(1-x)^{b-1}/(\Gamma(a)\Gamma(b)/\Gamma(a+b))$, x in $[0, 1]$
`Cauchy()`	$density(x) = 1/(\pi(1+x^2))$
`Exponential()`	$density(x) = e^{-x}$, $x > 0$
`Gamma(a)`	$density(x) = x^{a-1}e^{-x} / \Gamma(a)$, $x \geq 0$, $a > 0$
`Irand224()`	integer uniform on $[0, 2^{24})$
`Normal(μ, σ)`	normal distribution with mean μ, variance σ
`Normal01()`	normal distribution with mean 0, variance 1
`Poisson(μ)`	$probability(k) = e^{-\mu}\mu^k/k!$, $k = 0, 1, \ldots$
`Uniform(m, n)`	uniform on $[m, n)$
`Uniform01()`	uniform on $[0, 1)$

Table A-3: Built-in random number generation functions.

`option randseed ' '` instructs AMPL to choose and print a seed. Giving no `-s` argument is the same as specifying `-s1`.

`Irand224()` returns an integer in the range $[0, 2^{24})$. Given the same seed, an expression of the form `floor(m*Irand224()/n)` will yield the same value on most computers when m and n are integer expressions of reasonable magnitude, i.e., $|n| < 2^{k-24}$ and $|m| < 2^k$, for machines that correctly compute k-bit floating-point integer products and quotients; $k \geq 47$ for most machines of current interest.

Functions that operate on sets are described in Section A.6.

A.4.2 Strings and regular expressions

In almost all contexts in a model or command where a literal string could be used, it is also possible to use a string expression, enclosed in parentheses. Strings are created by concatenation and from the built-in string and regular expression functions listed in Table A-4.

The string concatenation operator `&` concatenates its arguments into a single string; it has precedence below all arithmetic operators. Numeric operands are converted to full-precision decimal strings as though by `printf` format `%.g` (A.16).

`s & t`	concatenate strings s and t
`num(s)`	convert string s to number; error if stripping leading and trailing white space does not yield a valid decimal number
`num0(s)`	strip leading white space, and interpret as much as possible of s as a number, but do not raise error
`ichar(s)`	Unicode value of the first character in string s
`char(n)`	string representation of character n; inverse of `ichar`
`length(s)`	length of string s
`substr(s, m, n)`	n character substring of s starting at position m; if n omitted, rest of string
`sprintf(f, exprlist_{opt})`	format arguments according to format string fmt
`match(s, re)`	starting position of regular expression re in s, or 0 if not found
`sub(s, re, repl)`	substitute $repl$ for the first occurrence of regular expression re in s
`gsub(s, re, repl)`	substitute $repl$ for all occurrences of regular expression re in s

Table A-4: Built-in string and regular expression functions.

There is no implicit conversion of strings to numbers, but `num(s)` and `num0(s)` perform explicit conversions. Both ignore leading and trailing white space; num complains if what remains is not a valid number, whereas num0 converts as much as it can, returning 0 if it sees no numeric prefix.

The `match`, `sub`, and `gsub` functions accept strings representing regular expressions as their second arguments. AMPL regular expressions are similar to standard Unix regular expressions. Aside from certain *metacharacters,* any literal character *c* is a regular expression that matches an occurrence of itself in the target string. The metacharacter ''`.`'' is a regular expression that matches any character. A list of characters enclosed in brackets is a regular expression that matches any character in the list, and lists may be abbreviated: `[a-z0-9]` matches any lower case letter or digit. A list of characters that begins with the character ^ and is enclosed in brackets is a regular expression that matches any character *not* in the list: `[^0-9]` matches any non-digit. If *r* is a regular expression, then *r* `*` matches 0 or more occurrences of *r*, *r* `+` matches 1 or more occurrences, and *r* `?` matches 0 or 1 occurrence. `^`*r* matches *r* only if *r* occurs at the beginning of the string, and *r* `$` matches *r* only at the end of the string. Parentheses are used for grouping and | means ''or''; $r_1 \mid r_2$ matches r_1 or r_2. The special meaning of a metacharacter can be turned off by preceding it by a backslash.

In the replacement pattern (third argument) for `sub` and `gsub`, & stands for the whole matched string, as does `\0`, while `\1`, `\2`, ..., `\9` stand for the string matched by the first, second, ..., ninth parenthesized expression in the pattern.

Options (A.14.1) are named string values, some of which affect various AMPL commands (A.14). The current value of option *opname* is denoted $opname.

A.4.3 Piecewise-linear terms

In variable, constraint and objective declarations, piecewise-linear terms have one of the following forms:

```
<< breakpoints ; slopes >> var
<< breakpoints ; slopes >> (cexpr)
<< breakpoints ; slopes >> (var, cexpr)
<< breakpoints ; slopes >> (cexpr, cexpr)
```

where *breakpoints* is a list of breakpoints and *slopes* a list of slopes. Each such list is a comma-separated list of *cexpr*'s, each optionally preceded by an indexing expression (whose scope covers just the *cexpr*). The indexing expression must specify a set that is manifestly ordered (see A.6.2), or it can be of the special form

```
{if lexpr}
```

which causes the *expr* to be omitted if the *lexpr* is false. In commands, the more general forms

```
<< breakpoints ; slopes >> (expr)
<< breakpoints ; slopes >> (expr, expr)
```

are also allowed, and variables may appear in expressions in the *breakpoints* and *slopes*.

After the lists of slopes and breakpoints are extended (by indexing over any indexing expressions), the number of slopes must be one more than the number of breakpoints, and the breakpoints must be in non-decreasing numeric order. (There is no requirement on the order of the slopes.) AMPL interprets the result as the piecewise-linear function $f(x)$ defined as follows. Let s_j, $1 \le j \le n$, and b_i, $1 \le i \le n - 1$, denote the slopes and breakpoints, respectively, and let

$b_0 = -\infty$ and $b_n = +\infty$. Then $f(0)=0$, and for $b_{i-1} \leq x \leq b_i$, f has slope s_i, i.e., $f'(x) = s_i$. For the forms having just one argument (either a variable *var* or a constant expression *expr*), the result is $f(var)$ or $f(expr)$. The form with two operands is interpreted as $f(var) - f(expr)$. This adds a constant that makes the result vanish when the *var* equals the *expr*.

When piecewise-linear terms appear in an otherwise linear constraint or objective, AMPL collects two or more piecewise-linear terms involving the same variable into a single term.

A.5 Declarations of model entities

Declarations of model entities have the following common form:

> *entity name alias*_{opt} *indexing*_{opt} *body*_{opt} ;

where *name* is an alphanumeric name that has not previously been assigned to an entity by a declaration, *alias* is an optional literal, *indexing* is an optional indexing expression, and *entity* is one of the keywords

```
set
param
var
arc
minimize
maximize
subject to
node
```

In addition, several other constructs are technically entity declarations, but are described later; these include `environ`, `problem`, `suffix` and `table`.

The *entity* may be omitted, in which case `subject to` is assumed. The *body* of various declarations consists of other, mostly optional, phrases that follow the initial part. Each declaration ends with a semicolon.

Declarations may appear in any order, except that each name must be declared before it is used.

As with piecewise-linear terms, a special form of indexing expression is allowed for variable, constraint, and objective declarations:

> {if *lexpr*}

If the logical expression *lexpr* is true, then a simple (unsubscripted) entity results; otherwise the entity is excluded from the model, and subsequent attempts to reference it cause an error message. For example, this declaration includes the variable `Test` in the model if the parameter `testing` has been given a value greater than 100:

```
param testing;
var Test {if testing > 100} >= 0;
```

A.6 Set declarations

A set declaration has the form

set declaration:
    set *name* *alias*_{opt} *indexing*_{opt} *attributes*_{opt} ;

in which *attributes* is a list of attributes optionally separated by commas:

attribute:
    dimen *n*
    within *sexpr*
    = *sexpr*
    default *sexpr*

The dimension of the set is either the constant positive integer *n*, or the dimension of *sexpr*, or 1 by default. The phrase within *sexpr* requires the set being declared to be a subset of *sexpr*. Several within phrases may be given. The = phrase specifies a value for the set; it implies that the set will not be given a value later in a data section (A.12.1) or a command such as let (A.18.9). The default phrase specifies a default value for the set, to be used if no value is given in a data section. The = and default phrases are mutually exclusive. If neither is given and the set is not defined by a data statement, references to the set during model generation cause an error message. For historical reasons, := is currently a synonym for = in declarations of sets and parameters, but this use of := is deprecated.

The *sexpr* in a = or default phrase can be {}, the empty set, which then has the dimension implied by any dimen or within phrases in the declaration, or 1 if none is present. In other contexts, {} denotes the empty set.

Recursive definitions of indexed sets are allowed, so long as the assigned values can be computed in a sequence that only references previously computed values. For example,

```
set nodes;
set arcs within nodes cross nodes;

param max_iter = card(nodes)-1; # card(s) = number of elements in s

set step {s in 1..max_iter} dimen 2 =
    if s == 1
        then arcs
        else step[s-1] union setof {k in nodes,
                (i,k) in step[s-1], (k,j) in step[s-1]} (i,j);

set reach = step[max_iter];
```

computes in set reach the transitive closure of the graph represented by nodes and arcs.

A.6.1 Cardinality and arity functions

The function card operates on any finite set: card(*sexpr*) returns the number of members in *sexpr*. If *sexpr* is an indexing expression, the parentheses may be omitted. For example,

```
card({i in A: x[i] >= 4})
```

may also be written

```
card {i in A: x[i] >= 4}
```

The function arity returns the arity of its set argument; the function indexarity returns the arity of its argument's indexing set.

A.6.2   Ordered sets

A named one-dimensional set may have an order associated with it. Declarations for ordered sets include one of the phrases

```
ordered [ by [ reversed ] sexpr ]
circular [ by [ reversed ] sexpr ]
```

The keyword `circular` indicates that the set is ordered and wraps around, i.e., its first member is the successor of its last, and its last member is the predecessor of its first.

Sets of dimension two or higher may not currently be `ordered` or `circular`.

If set S is `ordered by` T or `circular by` T, then set T must itself be an ordered set that contains S, and S inherits the order of T. If the ordering phrase is `by reversed` T, then S still inherits its order from T, but in reverse order.

If S is `ordered` or `circular` and no ordering `by` *sexpr* is specified, then one of two cases applies. If the members of S are explicitly specified by a = {*member-list*} expression in the model or by a list of members in a data section, S adopts the ordering of this list. If S is otherwise computed from an assigned or default *sexpr* in the model, AMPL will retain the order of manifestly ordered sets (explained below) and is otherwise free to pick an arbitrary order.

Functions of ordered sets are summarized in Table A-5.

If S is an expression for an ordered set of n members, e is the jth member of S, and k is an integer expression, then next (e,S,k) is member $j + k$ of S if $1 \leq j + k \leq n$, and an error otherwise. If S is `circular`, then next (e,S,k) is member $1 + ((j + k - 1)$ mod $n)$ of S. The function nextw (next with wraparound) is the same as next, except that it treats all ordered sets as `circular`; prev $(e,S,k) \equiv$ next $(e,S,-k)$, and prevw $(e,S,k) \equiv$ nextw $(e,S,-k)$.

Several abbreviations are allowed. If k is not given, it is taken as 1. If both k and S are omitted, then e must be a dummy index in the scope of an indexing expression that runs over S, for example as in {e in S}.

Five other functions apply to ordered sets: first (S) returns the first member of S, last (S) the last, member (j,S) returns the jth member of S, ord (e,S) and ord0 (e,S) the ordinal position of member e in S. If e is not in S, then ord0 returns 0, whereas ord complains of an error. Again, ord(e) = ord(e,S) and ord0(e) = ord0(e,S) if e is a dummy index in the scope of an indexing expression for S.

Some sets are manifestly ordered, such as arithmetic progressions, intervals, subsets of ordered sets, `if` expressions whose `then` and `else` clauses are both ordered sets, and set differences (but not symmetric differences) whose left operands are ordered.

A.6.3   Intervals and other infinite sets

For convenience in specifying ordered sets and prototypes for imported functions (A.22), there are several kinds of infinite sets. Iterating over infinite sets is forbidden, but one can check membership in them and use them to specify set orderings. The most natural infinite set is the interval, which may be closed, open, or half-open and which follows standard mathematical notation. There are intervals of real (floating-point) numbers and of integers, introduced by the keywords `interval` and `integer` respectively:

next (e, S, k)	member k positions after member e in set S
next (e, S)	same, with $k = 1$
next (e)	next member of set for which e is dummy index
nextw (e, S, k)	member k positions after member e in set S, wrapping around
nextw (e, S)	wrapping version of next(e, S)
nextw (e)	wrapping version of next(e)
prev (e, S, k)	member k positions before member e in set S
prev (e, S)	same, with $k = 1$
prev (e)	previous member of set for which e is dummy index
prevw (e, S, k)	member k positions before member e in set S, wrapping around
prevw (e, S)	wrapping version of prev(e, S)
prevw (e)	wrapping version of prev(e)
first (S)	first member of S
last (S)	last member of S
member (j, S)	jth member of S; $1 \le j \le$ card(S), j integer
ord (e, S)	ordinal position of member e in S
ord (e)	ordinal position of member e in set for which it is dummy index
ord0 (e, S)	ordinal position of member e in S; 0 if not present
ord0 (e)	same as ord(e)
card (S)	number of members in set S
arity (S)	arity of S if S is a set, else 0; for use with _SETS
indexarity (E)	arity of entity E's indexing set
	card, arity, and indexarity also apply to unordered sets

Table A-5: Functions of ordered sets.

interval:
```
interval [a, b] ≡ {x: a ≤ x ≤ b},
interval (a, b] ≡ {x: a < x ≤ b},
interval [a, b) ≡ {x: a ≤ x < b},
interval (a, b) ≡ {x: a < x < b},
integer  [a, b] ≡ {x: a ≤ x ≤ b and x ∈ I},
integer  (a, b] ≡ {x: a < x ≤ b and x ∈ I},
integer  [a, b) ≡ {x: a ≤ x < b and x ∈ I},
integer  (a, b) ≡ {x: a < x < b and x ∈ I}
```

where a and b denote arbitrary arithmetic expressions, and I denotes the set of integers. In function prototypes (A.22) and the declaration phrases

```
in interval
within interval
ordered by [ reversed ] interval
circular by [ reversed ] interval
```

the keyword interval may be omitted.

The predefined infinite sets Reals and Integers are the sets of all floating-point numbers and integers, respectively, in numeric order. The predefined infinite sets ASCII, EBCDIC, and Display all represent the universal set of strings and numbers from which members of any one-dimensional set are drawn; ASCII is ordered by the ASCII collating sequence, EBCDIC by the

EBCDIC collating sequence, and `Display` by the ordering used in the `display` command (Section A.16). Numbers precede literals in `Display` and are ordered numerically; literals are sorted by the ASCII collating sequence.

A.7 Parameter declarations

Parameter declarations have a list of optional attributes, optionally separated by commas:

> *parameter declaration*:
>   `param` *name alias* $_{opt}$ *indexing* $_{opt}$ *attributes* $_{opt}$ `;`

The *attributes* may be any of the following:

> *attribute*:
>   `binary`
>   `integer`
>   `symbolic`
>   *relop expr*
>   `in` *sexpr*
>   `=` *expr*
>   `default` *expr*
>
> *relop*:
>   `<   <=  =  ==  !=  <>  >  >=`

The keyword `integer` restricts the parameter to be an integer; `binary` restricts it to 0 or 1. If `symbolic` is specified, then the parameter may assume any literal or numeric value (rounded as for set membership), and the attributes involving `<`, `<=`, `>=` and `>` are disallowed; otherwise the parameter is numeric and can only assume a numeric value.

The attributes involving comparison operators specify that the parameter must obey the given relation. The `in` attribute specifies a check that the parameter lies in the given set.

The `=` and `default` attributes are analogous to the corresponding ones in set declarations, and are mutually exclusive.

Recursive definitions of indexed parameters are allowed, so long as the assigned values can be computed in a sequence that only references previously computed values. For example,

```
param comb 'n choose k' {n in 0..N, k in 0..n}
   = if k = 0 or k = n then 1 else comb[n-1,k-1] + comb[n-1,k];
```

computes the number of ways of choosing *n* things *k* at a time.

In a recursive definition of a symbolic parameter, the keyword `symbolic` must precede all references to the parameter.

A.7.1 Check statements

Check statements supply assertions to help verify that correct data have been read or generated; they have the syntax

> `check [` *indexing* $_{opt}$ `:  ]` *lexpr* `;`

Each `check` statement is evaluated when one of the commands `solve`, `write`, `solution`, or `check` is executed.

A.7.2 Infinity

`Infinity` is a predefined parameter; it is the threshold above which upper bounds are considered absent (i.e., infinite), and `-Infinity` is the threshold below which lower bounds are considered absent. Thus given

```
set A;
param Ub{A} default  Infinity;
param Lb{A} default -Infinity;
var V {i in A} >= Lb[i], <= Ub[i];
```

components of V for which no `Lb` value is given in a data section are unbounded below and components for which no `Ub` value is given are unbounded above. One can similarly arrange for optional lower and upper constraint bounds. On computers with IEEE arithmetic (most modern systems) `Infinity` is the IEEE ∞ value.

A.8 Variable declarations

Variable declarations begin with the keyword `var`:

> *variable declaration*:
>   var *name alias*_{opt} *indexing*_{opt} *attributes*_{opt} ;

Optional attributes of variable declarations may be separated by commas; these attributes include

> *attribute*:
>   binary
>   integer
>   symbolic
>   >= *expr*
>   <= *expr*
>   := *expr*
>   default *expr*
>   = *expr*
>   coeff *indexing*_{opt} *constraint expr*
>   cover *indexing*_{opt} *constraint*
>   obj *indexing*_{opt} *objective expr*
>   in *sexpr*
>   suffix *sufname expr*

As with parameters, `integer` restricts the variable to integer values, and `binary` restricts it to 0 or 1. The `>=` and `<=` phrases specify bounds, and the `:=` phrase an initial value. The `default` phrase specifies a default for initial values that may be provided in a data section (A.12.2); `default` and `:=` are mutually exclusive. The = *expr* phrase is allowed only if none of the previous attributes appears; it makes the variable a *defined variable* (A.8.1). Each `suffix` *sufname expr* phrase specifies an initial value for a previously declared suffix *sufname*.

If `symbolic` is specified, in *sexpr* must also appear and attributes requiring a numeric value, such as `>=` *expr*, are excluded. If in *sexpr* appears without `symbolic`, the set expression *sexpr* must be the union of intervals and discrete sets of numbers. Either way, in *sexpr* restricts the variable to lie in *sexpr*.

The `coeff` and `obj` phrases are for columnwise coefficient generation; they specify coefficients to be placed in the named constraint or objective, which must have been previously declared using the placeholder `to_come` (see A.9 and A.10). The scope of *indexing* is limited to the phrase, and may have the special form

> `{if` *lexpr*`}`

which contributes the coefficient only if the *lexpr* is true. A `cover` phrase is equivalent to a `coeff` phrase in which the *expr* is 1.

Arcs are special network variables, declared with the keyword `arc` instead of `var`. They may contribute coefficients to `node` constraints (A.9) via optional attribute phrases of the form

> `from` *indexing*$_{opt}$ `node` *expr*$_{opt}$
> `to` *indexing*$_{opt}$ `node` *expr*$_{opt}$

These phrases are analogous in syntax to the `coeff` phrase, except that the final *expr* is optional; omitting it is the same as specifying 1.

A.8.1 Defined variables

In some nonlinear models, it is convenient to define named values that contribute somehow to the constraints or objectives. For example,

```
set A;
var v {A};
var w {A};
subject to C {i in A}: w[i] = vexpr;
```

where v*expr* is an expression involving the variables v.

As things stand, the members of C are constraints, and we have turned an unconstrained problem into a constrained one, which may not be a good idea. Setting option `substout` to 1 instructs AMPL to eliminate the collection of constraints C. AMPL does so by telling solvers that the constraints define the variables on their left-hand sides, so that, in effect, these defined variables become named common subexpressions.

When option `substout` is 1, a constraint such as C that provides a definition for a defined variable is called a *defining constraint*. AMPL decides which variables are defined variables by scanning the constraints once, in order of their appearance in the model. A variable is eligible to become a defined variable only if its declaration imposes no constraints on it, such as integrality or bounds. Once a variable has appeared on the right-hand side of a defining constraint, it is no longer eligible to be a defined variable — without this restriction, AMPL might have to solve implicit equations. Once a variable has been recognized as a defined variable, its subsequent appearance on the left-hand side of what would otherwise be a defining constraint results in the constraint being treated as an ordinary constraint.

Some solvers give special treatment to linear variables because their higher derivatives vanish. For such solvers, it may be helpful to treat linear defined variables specially. Otherwise, variables involved in the right-hand side of the equation for a defined variable appear to solvers as nonlinear variables, even if they are used only linearly in the right-hand side. By doing Gaussian elimination rather than conveying linear variable definitions explicitly, AMPL can arrange for solvers to see such right-hand variables as linear variables. This often causes fill-in, i.e., makes the problem less sparse, but it may give the solvers a more accurate view of the problem. When option `linelim` has its default value 1, AMPL treats linear defined variables in this special way; when option `linelim` is 0, AMPL treats all defined variables alike.

A variable declaration may have a phrase of the form = *expr*, where *expr* is an expression involving variables. Such a phrase indicates that the variable is to be defined with the value *expr*. Such defining declarations allow some models to be written more compactly.

Recognizing defined variables is not always a good idea — it leads to a problem in fewer variables, but one that may be more nonlinear and that may be more expensive to solve because of loss of sparsity. By using defining constraints (instead of using defining variable declarations) and translating and solving a problem both with $substout = 0 and with $substout = 1, one can see whether recognizing defined variables is worthwhile. On the other hand, if recognizing a defined variable seems clearly worthwhile, defining it in its declaration is often more convenient than providing a separate defining constraint; in particular, if all defined variables are defined in their declarations, one need not worry about $substout.

One restriction on defining declarations is that subscripted variables must be defined before they are used.

A.9 Constraint declarations

The form of a constraint declaration is

> *constraint declaration* :
>> [subject to] *name alias*$_{opt}$ *indexing*$_{opt}$
>>> [:= *initial_dual*] [default *initial_dual*]
>>> [: *constraint expression*] [*suffix-initializations*] ;

The keyword subject to in a constraint declaration may be omitted but is usually best retained for clarity. The optional := *initial_dual* specifies an initial guess for the dual variable (Lagrange multiplier) associated with the constraint. Again, default and := clauses are mutually exclusive, and default is for initial values not given in a data section. Constraint declarations must specify a constraint in one of the following forms:

> *constraint expression* :
>> *expr* <= *expr*
>> *expr* = *expr*
>> *expr* >= *expr*
>> *cexpr* <= *expr* <= *cexpr*
>> *cexpr* >= *expr* >= *cexpr*

To enable columnwise coefficient generation for the constraint, one of the *expr*s may have one of the following forms:

> to_come + *expr*
> *expr* + to_come
> to_come

Terms for this constraint that are specified in a var declaration (A.8) are placed at the location of to_come.

Nodes are special constraints that may send flow to or receive flow from arcs. Their declarations begin with the keyword node instead of subject to. Pure transshipment nodes do not have a constraint body; they must have ''flow in'' equal to ''flow out''. Other nodes are sources or sinks; they specify constraints in one of the forms above, except that they may not mention to_come, and exactly one *expr* must have one of the following forms:

```
net_in + expr
net_out + expr
expr + net_in
expr + net_out
net_in
net_out
```

The keyword `net_in` is replaced by an expression representing the net flow into the node: the terms contributed to the node constraint by `to` phrases of `arc` declarations (A.8), minus the terms contributed by `from` phrases. The treatment of `net_out` is analogous; it is the negative of `net_in`.

The optional *suffix-initialization* phrases each have the form

> *suffix-initialization* :
>   `suffix` *sufname expr*

optionally preceded by a comma, where *sufname* is a previously declared suffix.

A.9.1 Complementarity constraints

For expressing complementarity constraints, in addition to the forms above, constraint declarations may have the form

> *name alias*$_{opt}$ *indexing*$_{opt}$: *constr*$_1$ `complements` *constr*$_2$;

in which *constr*$_1$ and *constr*$_2$ consist of 1, 2, or 3 expressions separated by the operators `<=`, `>=` or `=`. In *constr*$_1$ and *constr*$_2$ together, there must be a total of two explicit inequality operators, with `=` counting as two. A complementarity constraint is satisfied if both *constr*$_1$ and *constr*$_2$ hold and at least one inequality is tight, i.e., satisfied as an equality. If one of *constr*$_1$ or *constr*$_2$ involves two inequalities, then the constraint must have one of the forms

> *expr*$_1$ `<=` *expr*$_2$ `<=` *expr*$_3$ `complements` *expr*$_4$
> *expr*$_3$ `>=` *expr*$_2$ `>=` *expr*$_1$ `complements` *expr*$_4$
> *expr*$_4$ `complements` *expr*$_1$ `<=` *expr*$_2$ `<=` *expr*$_3$
> *expr*$_4$ `complements` *expr*$_3$ `>=` *expr*$_2$ `>=` *expr*$_1$

In all of these cases, the constraint requires the inequalities to hold, with

> *expr*$_4$ $\geq$ 0 `if` *expr*$_1$ `=` *expr*$_2$
> *expr*$_4$ $\leq$ 0 `if` *expr*$_2$ `=` *expr*$_3$
> *expr*$_4$ `=` 0 `if` *expr*$_1$ `<` *expr*$_2$ `<` *expr*$_3$

For expressing mathematical programs with equilibrium constraints, complementarity constraints may coexist with other constraints and objectives.

Solvers see complementarity constraints in the standard form

> *expr* `complements` *lower bound* `<=` *variable* `<=` *upper bound*

A synonym (A.19.4), `_scvar{i in 1.._sncons}`, indicates which variable, if any, complements each constraint the solver sees. If `_scvar[i]` `in` `1.._snvars`, then variable `_svar[_scvar[i]]` complements constraint `_scon[i]`; otherwise `_scvar[i] == 0`, and `_con[i]` is an ordinary constraint. The synonyms `_cconname{1.._nccons}` are the names of the complementarity constraints as the modeler sees them.

A.10 Objective declarations

The declaration of an objective is one of

> *objective declaration* :
> maximize *name* *alias*~opt~ *indexing*~opt~ [: *expression*] [*suffix-initializations*] ;
> minimize *name* *alias*~opt~ *indexing*~opt~ [: *expression*] [*suffix-initializations*] ;

and may specify an expression in one of the following forms:

> *expression* :
> *expr*
> to_come + *expr*
> *expr* + to_come
> to_come

The to_come forms permit columnwise coefficient generation, as for constraints (A.9). Specifying none of the above expressions is the same as specifying ": to_come". *Suffix-initializations* may appear as in constraint declarations.

If there are multiple objectives, the one sent to a solver can be set by the objective command; see section A.18.6. By default, all objectives are sent.

A.11 Suffix notation for auxiliary values

Variables, constraints, objectives and problems have a variety of associated auxiliary values. For example, variables have bounds and reduced costs, and constraints have dual values and slacks. Such values are accessed as *name* . *suffix*, where *name* is a simple or subscripted variable, constraint, objective or problem name, and . *suffix* is one of the possibilities listed in Tables A-6, A-7, and A-8.

For a constraint, the .body, .lb, and .ub values correspond to a modified form of the constraint. If the constraint involves a single inequality, subtract the right-hand side from the left, then move any constant terms to the right-hand side; if the constraint involves a double inequality, similarly subtract any constant terms in the middle from all three expressions (left, middle, right). Then the constraint has the form $lb \leq body \leq ub$, where *lb* and *ub* are (possibly infinite) constants.

The following rules determine lower and upper dual values (*c* . ldual and *c* . udual) for a constraint *c*. The solver returns a single dual value, *c* . dual, which might apply either to $body \geq lb$ or to $body \leq ub$. For an equality constraint ($lb = ub$), AMPL uses the sign of *c* . dual to decide. For a minimization problem, *c* . dual > 0 implies that the same optimum would be found if the constraint were $body \geq lb$, so AMPL sets *c* . ldual = *c* . dual and *c* . udual = 0; similarly, *c* . dual < 0 implies that *c* . ldual = 0 and *c* . udual = *c* . dual. For a maximization problem, the inequalities are reversed.

For inequality constraints ($lb < ub$), AMPL uses nearness to bound to decide whether *c* . ldual or *c* . udual equals *c* . dual. If $body - lb < ub - body$, then *c* . ldual = *c* . dual and *c* . udual = 0; otherwise, *c* . ldual = 0 and *c* . udual = *c* . dual.

Model declarations may reference any of the suffixed values described in Tables A-6, A-7 and A-8. This is most often useful in new declarations that are introduced after one model has already been translated and solved. In particular, the suffixes .val and .dual are provided so that new constraints can refer to current optimal values of the primal and dual variables and of the objective.

`.astatus`	AMPL status (A.11.2)
`.init`	current initial guess
`.init0`	initial initial guess (set by `:=`, data, or `default`)
`.lb`	current lower bound
`.lb0`	initial lower bound
`.lb1`	weaker lower bound from presolve
`.lb2`	stronger lower bound from presolve
`.lrc`	lower reduced cost (for `var >= lb`)
`.lslack`	lower slack (`val - lb`)
`.rc`	reduced cost
`.relax`	ignore integrality restriction if positive
`.slack`	min(`lslack`, `uslack`)
`.sstatus`	solver status (A.11.2)
`.status`	status (A.11.2)
`.ub`	current upper bound
`.ub0`	initial upper bound
`.ub1`	weaker upper bound from presolve
`.ub2`	stronger upper bound from presolve
`.urc`	upper reduced cost (for `var <= ub`)
`.uslack`	upper slack (`ub - val`)
`.val`	current value of variable

Table A-6: Dot suffixes for variables.

For a complementarity constraint, suffix notations like *constraint* `.lb`, *constraint* `.body`, etc., are extended so that *constraint* `.L`*suffix* and *constraint* `.R`*suffix* correspond to *constr*$_1$ `.` *suffix* and *constr*$_2$ `.` *suffix*, respectively, and *complementarity-constraint* `.slack` (or the unadorned name) stands for a measure of the extent to which the complementarity constraint is satisfied: if *constr*$_1$ and *constr*$_2$ each involve one inequality, the new measure is

$$\min(constr_1 . \texttt{slack}, \ constr_2 . \texttt{slack}),$$

which is positive if both are satisfied as strict inequalities, 0 if the complementarity constraint is satisfied exactly, and negative if at least one of *constr*$_1$ or *constr*$_2$ is violated. For constraints of the form *expr*$_1$ `<=` *expr*$_2$ `<=` *expr*$_3$ `complements` *expr*$_4$, the `.slack` value is

```
min(expr₂-expr₁,   expr₄) if expr₁ >= expr₂
min(expr₃-expr₂,  -expr₄) if expr₃ <= expr₂
-abs(expr₄) if expr₁ < expr₂ < expr₃
```

so in all cases, the `.slack` value is 0 if the complementarity constraint holds exactly and is negative if one of the requisite inequalities is violated.

A.11.1 Suffix declarations

Suffix declarations introduce new suffixes, which may be assigned values in subsequent declarations, `let` commands and function invocations (with `OUT` arguments, A.22). Suffix declarations begin with the keyword `suffix`:

.astatus	AMPL status (A.11.2)
.body	current value of constraint *body*
.dinit	current initial guess for dual variable
.dinit0	initial initial guess for dual variable (set by :=, data, or default)
.dual	current dual variable
.lb	lower bound
.lbs	lb for solver (adjusted for fixed variables)
.ldual	lower dual value (for body >= lb)
.lslack	lower slack (body - lb)
.slack	min(lslack, uslack)
.sstatus	solver status (A.11.2)
.status	status (A.11.2)
.ub	upper bound
.ubs	ub for solver (adjusted for fixed variables)
.udual	upper dual value (for body <= ub)
.uslack	upper slack (ub - body)

Table A-7: Dot suffixes for constraints.

.val	current value of objective

Table A-8: Dot suffix for objectives.

suffix declaration :
> `suffix` *name* *alias* _{opt} *attributes* _{opt} ;

Optional attributes of suffix declarations may be separated by commas; these attributes include

attribute :
> `binary`
> `integer`
> `symbolic`
> `>=` *expr*
> `<=` *expr*
> *direction*

direction :
> `IN`
> `OUT`
> `INOUT`
> `LOCAL`

At most one *direction* may be specified; AMPL assumes `INOUT` if no *direction* is given. These directions are from a solver's perspective: `IN` suffix values are input to the solver; `OUT` suffix values are assigned by the solver; `INOUT` values are both `IN` and `OUT`; and `LOCAL` values are not seen by the solver.

Symbolic suffixes are declared with the `symbolic` attribute; appending _num to the name of a symbolic suffix gives the name of an associated numeric suffix; solvers see the associated numeric value. If *symsuf* is a symbolic suffix, option *symsuf* _table connects *symsuf* with *symsuf* _num as follows. Each line of $*symsuf* _table should begin with a numeric limit value, followed by a string value and optional comments, all separated by white space. The numeric limit values must increase with each line. The string value with the greatest numeric limit value less than or equal to the .*sufname*_num value is the associated string value. If the .*sufname*_num value is less than the limit value in the first line of $*symsuf*_table, then the .*symsuf*_num value is used as the .*symsuf* value.

A.11.2 Statuses

Some solvers maintain a *basis* and distinguish among *basic* and various kinds of *nonbasic* variables and constraints. AMPL's built-in symbolic suffix .sstatus permits solvers to return basis information to AMPL and, in a subsequent solve (A.18.1), to be given the previously optimal basis, which sometimes leads to solving the updated problem faster.

AMPL's drop/restore status (A.18.6) of constraints and its fix/unfix status (A.18.7) of variables is reflected in the built-in symbolic suffix .astatus. The built-in symbolic suffix .status is derived from .astatus and .sstatus: if the variable or constraint, say x, is in the current problem, x.status = x.sstatus; otherwise x.status = x.astatus. AMPL assigns x.astatus_num = 0 if x is in the current problem, so the rule for determining .status is

> x.status = if x.astatus_num == 0 then x.sstatus else x.astatus.

When option astatus_table has its default value, x.astatus = 'in' when x.astatus_num = 0.

A.12 Standard data format

AMPL supports a standard format to describe the sets and parameter values that it combines with a model to yield a specific optimization problem.

A data section consists of a sequence of tokens (literals and strings of printing characters) separated by white space (spaces, tabs, and newlines). Tokens include keywords, literals, numbers, and the delimiters ( ) [ ] : , ; := *. A statement is a sequence of tokens terminated by a semicolon. Comments may appear as in declarations. In all cases, arrangement of data into neat rows and columns is for human readability; AMPL ignores formatting.

A data section begins with a data command and ends with end-of-input or with a command that returns to model mode (A.14).

In a data section, model entities may be assigned values in any convenient order, independent of their order of declaration.

A.12.1 Set data

Statements defining sets consist of the keyword set, the set name, an optional :=, and the members. A one-dimensional set is most easily specified by listing its members, optionally separated by commas. The single or double quotes of a literal string may be omitted if the string is alphanumeric but does not specify a number.

An *object* in a data section may be a number or a character string. As in a model, a character string may be specified by surrounding it with quotes (' or "). Since so many strings appear in data, however, AMPL allows data statements to drop the quotes around any string that consists only of characters that may occur in a name or number, unless quotes are needed to distinguish a string from a number.

The general form of a set data statement is

> *set-data-statement*:
>   set *set-name* := *set-spec* *set-spec* ... ;
>
> *set-spec*:
>   *set-template*_{opt} *member-list*
>   *set-template*_{opt} *member-table* *member-table* ...

The *set-name* must be the name of an individually declared set, or the subscripted name of a set from an indexed collection. The optional template has the form

> *set-template*:
>   (*templ-item*, *templ-item*, ...)
>
> *templ-item*:
>   *object*
>   *
>   :

where the number of *templ-item*s must equal the dimension of the named set. If no template is given, a template of all *'s is assumed.

There are two forms of *set-spec*, list format and table format. The list format of *set-spec* is

> *member-list*:
>   *member-item* *member-item* ...
>
> *member-item*:
>   *object* *object* ...

The number of *object*s in a *member-item* must match the number of *'s in the preceding template, which may not have : as a *templ-item*; the *object*s are substituted for the *'s, from left to right, to produce a member that is added to the set being specified. In the special case that the template contains no *'s, the *member-list* should be empty, while the template itself specifies one member to be added.

The table format of *set-spec* looks like this:

> *member-table*:
>   (tr)_{opt} *t-header* *t-header*   ...   :=
>   *row-label*   ±   ±   ±
>   *row-label*   ±   ±   ±
>   ...
>
> *t-header*:
>   :   *object*   *object*   ...
>
> *row-label*:
>   *object*   *object*   ...

There must be at least one *t-header*, at least one *object* in each *row-label*, and as many *t-header*'s and *row-label*'s as *'s and :'s in the preceding template. If the preceding template involves any :'s, there must be as many :'s as *t-headers*; otherwise if the optional (tr) appears, the initial *'s are treated as :'s, and if (tr) does not appear, the final *'s are treated as :'s. Each table entry shown as ± must be either a + or a − symbol. Each − entry is ignored, while each + entry's *row-label*s are substituted for the template's *'s in sequence, and the *object*s in the *t-headers* corresponding to the + are substituted for the :'s to produce a set member.

To define a compound set, one can list all members. Each member is a parenthesized, comma-separated list of components, and successive members have an optional comma between them. Alternatively, one can describe the members of a two-dimensional set by a table or sequence of tables. In such a table, the row labels are for the first subscript, the columns for the second; ''+'' stands for a pair that occurs in the set, and ''−'' for a pair that does not. The colon introduces a table, and is mandatory in this context. If (`tr`) precedes the colon, the table is transposed, interchanging the roles of rows and columns.

In general, a `set` statement involves a sequence of 1D and 2D `set` tables. 1D tables start with either a new template (after which a `:=` is optional) or with `:=` alone, in which case the previous template is retained. The default (initial) template is (`*,...,*`), that is, as many `*`'s as the set's dimension. 2D tables start with an optional new template, followed by `:` or (`tr`) and an optional colon, followed by a list of column labels and a `:=`. Templates containing no `*`'s stand for themselves. The effect of (`tr`) persists until a new template appears.

For indexed sets, each component set must be given in a separate data statement. It is not necessary to specify subset members in the same order as in their parent set.

A.12.2 *Parameter data*

There are two forms of the statement that specifies parameter data or variable initial values. The first form is analogous to the set data statement:

> *param-data-statement*:
>> `param` *param-name* *param-default*$_{opt}$ `:=` *param-spec* *param-spec* ... `;`
>
> *param-spec*:
>> *param-template*$_{opt}$ *value-list*
>> *param-template*$_{opt}$ *value-table* *value-table* ...

with the addition of an optional *param-default* that will be described below. The *param-name* is usually the name of a parameter declared in the model, but may also be the name of a variable or constraint; the keyword `var` may be used instead of `param` to make the distinction clear.

The `param` statement's templates have the same content as in the set data statement, but are given in brackets (like subscripts) rather than parentheses:

> *param-template*:
>> `[` *templ-item*, *templ-item*, ... `]`
>
> *templ-item*:
>> *object*
>> `*`
>> `:`

The *value-list* is like the previously defined *member-list*, except that it also specifies a parameter or variable value:

> *value-list*:
>> *value-item* *value-item* ...
>
> *value-item*:
>> *object* *object* ... *entry*

The objects are substituted for `*`'s in the template to define a set member, and the parameter or variable indexed by this set member is assigned the value associated with the *entry* (see below).

The *value-table* is like the previously defined *member-table*, except that its *entry*s are values rather than + or −:

value-table:
    (tr) *opt* :   *t-header*   *t-header*   *t-header*   ...   :=
    row-label   *entry*   *entry*   *entry*
    row-label   *entry*   *entry*   *entry*
    ...

t-header:
    :   *object*   *object*   ...

row-label:
    object   *object*   ...

entry:
    number
    string
    default-symbol

As in set statements, the notation (tr) means *transpose*; it implies a 2D table, and a : after it is optional. It remains in effect until a new template appears.

A table may be given in several chunks of columns.

Each *entry*'s *row-label* and *t-header* entries are substituted for *'s and :'s in the template to define a set member, and the parameter or variable indexed by this set member is assigned the value specified by the *entry*. The *entry* may be a number for variables and for parameters that take numerical values, or a string for variables and parameters declared with the attribute symbolic. An *entry* that is the default symbol (see below) is ignored.

The second form of parameter data statement provides for the definition of multiple parameters, and also optionally the set over which they are indexed:

param-data-alternate:
    param *param-default opt* :
        param-name param-name ... := *value-item value-item* ... ;

    param *param-default opt* : *set-name*:
        param-name param-name ... := *value-item value-item* ... ;

The named parameters must all have the same dimension. If the optional *set-name* is specified, its membership is also defined by this statement. Each *value-item* consists of an optional template followed by a list of objects and a list of values:

value-item:
    template opt *object* ... *entry entry* ...

An initial template of all *'s (as many as the common dimension of the named parameters) is assumed, and a template remains in effect until a new one appears. The objects must be equal in number to the number of *'s in the current template; when substituted for the *'s in the current template, they define a set member. If a set is being defined, this member is added to it. The parameters indexed by this member are assigned the values associated with the subsequent *entry*s, which obey the same rules as the table *entry*s previously described. Values are assigned in the order in which the parameters' names appeared at the beginning of the statement; the number of *entry*s must equal the number of named parameters.

A param data statement's optional default phrase has the form

> *param-default*:
>   default *number*

If this phrase is present, any parameter named but not explicitly assigned a value in the statement is given the value of *number*.

A data item may be specified as ``.`` rather than an explicit value. This is normally taken as a missing value, and a reference to it in the model would cause an error message. If there is a default value, however, each missing value is determined from that default. A default value may be specified either through a default phrase in a parameter's declaration in the model (A.7), or from an optional phrase

>   default *r*

that follows the parameter's name in the data statement. In the latter case, *r* must be a numeric constant.

Default-value symbols may appear in both 1D and 2D tables. The *default-symbol* is initially a dot (.). A stack of default-value symbols is maintained, with the current symbol at the top. The defaultsym statement (which is recognized only in a data section) pushes a new symbol onto the stack, and nodefaultsym pushes a ``no symbol`` indicator onto the stack. The statement

>   defaultsym;

(without a symbol) pops the stack.

Parameters having three or more indices may be given values by a sequence of 1D and 2D tables, one or more for each slice of nondefault values.

In summary, a param statement defining one indexed parameter starts with the keyword param and the name of the parameter, followed by an optional default value and an optional :=. Then comes a sequence of 1D and 2D param tables, which are similar to 1D and 2D set tables, except that templates involve square brackets rather than parentheses, 2D tables contain numbers (or, for a symbolic parameter, literals) rather than +'s and -'s, and 1D tables corresponding to a template of k *'s contain $k + 1$ rather than k columns, the last being a column of numbers or default symbols (.'s). A special form, the keyword param, an optional default value, and a single (untransposed) 2D table, defines several parameters indexed by a common set, and another special form, param followed by the parameter name, an optional :=, and a numeric value, defines a scalar parameter.

Variable and constraint names may appear in data sections anywhere that a parameter name may appear, to specify initial values for variables and for the dual variables associated with constraints. The rules for default values are the same as for parameters. The keyword var is a synonym for param in data statements.

A.13 Database access and tables

AMPL's table facility permits obtaining data from and returning data to an external medium, such as a database, file, or spreadsheet. A table declaration establishes connections between columns of an external relational table and sets, parameters, variables and expressions in AMPL. The read table and write table commands use these connections to read data values into AMPL from tables and write them back. AMPL uses table *handlers* to implement these connections. Built-in table handlers permit reading and writing ``.tab`` and ``.bit`` files to save and restore values and experiment with AMPL's table facilities; to access databases and spreadsheets, at least one other handler must be installed or loaded (A.22). The built-in set _HANDLERS names the currently available handlers, and the symbolic parameter _handler_lib{_HANDLERS} tells which shared library each handler came from.

Table declarations have the form

> *table-declaration*:
>   table *table-name indexing*$_{opt}$ *in-out*$_{opt}$ *string-list*$_{opt}$:
>   *key-spec*, *data-spec*, *data-spec*, ... ;

in-out is one of IN, OUT, or INOUT; IN means into AMPL, OUT means out of AMPL, and INOUT means both. INOUT is assumed if *in-out* is not given. The optional *string-list* gives the names of drivers, files, and driver-specific options that are used to access external data; the contents depend on the handler used for the table and perhaps on the operating system.

The *key-spec* in a table declaration specifies key columns that uniquely identify the data to be accessed:

> *set-io*$_{opt}$ [*key-col-spec*, *key-col-spec*, ...]

The optional *set-io* phrase has the form

> *set-name arrow*

in which *arrow* is one of <-, ->, or <->; it points in the direction that the information is moved, from the key columns to the AMPL set (by read table), from the set to the columns (by write table), or both, depending on *in-out*. Each *key-col-spec* names a column in the external table and associates a dummy index with it. A *key-col-spec* of the form

> *key-name*

uses *key-name* for both purposes, and a *key-col-spec* of the form

> *key-name ~ data-col-name*

introduces *key-name* as the dummy index and *data-col-name* as the name of the column in the external medium; *data-col-name* may be a name, quoted string, or parenthesized string expression.

Each *data-spec* names a data column in the external table. In the simplest case, the external name and the AMPL name are the same. If not, however, an external name can be associated with an internal name with the syntax

> *data-spec:*
>   *param-name ~ data-col-name*

Each *data-spec* optionally ends with one of IN, OUT, or INOUT, which overrides the default table direction and indicates whether read table should read the column into AMPL (IN or INOUT), and whether write table should write the column to the external medium (OUT or INOUT).

Special syntax permits use of an indexing expression to describe one or more columns of data:

> *indexing expr-col-desc*
> *indexing* (*expr-col-desc* , *expr-col-desc* , ...)

in which *expr-col-desc* has the form

> *expr* [~ *colname*] *in-out*$_{opt}$

Another special syntax permits iterating data columns:

> *indexing* < *data-spec* , *data-spec* , ... >

The latter may not be nested, but may contain the former.

After a table declaration, data access is done with

```
read table table-name subscript_opt ;
write table table-name subscript_opt ;
```

which refer back to the information given in the `table` declaration.

A.14 Command language overview

AMPL recognizes the commands listed in Table A-9. Commands are not part of a model, but cause AMPL to act as described below.

The command environment recognizes two modes. In *model mode*, where AMPL starts upon invocation, it recognizes model declarations (A.5) and all of the commands described below. Upon seeing a `data` statement, it switches to *data mode*, where only data-mode statements (A.12) are recognized. It returns to model mode upon encountering a keyword that cannot begin a data-mode statement, or the end of the file. Commands other than `data`, `end`, `include`, `quit` and `shell` also cause AMPL to enter model mode.

A phrase of the form

```
include filename
```

causes the indicated file to be interpolated. Here, and in subsequent contexts where a *filename* appears, if *filename* involves semicolons, quotes, white space, or non-printing characters, it must be given as a literal, i.e., `'filename'` or `"filename"`. In contexts other than `include`, *filename* may also be a parenthesized string expression (A.4.2). `include` commands may be nested; they are recognized in both model and data mode. The sequences

```
model; include filename
data; include filename
```

may be abbreviated

```
model filename
data filename
```

The `commands` command is analogous to `include`, but is a statement and must be terminated by a semicolon. When a `data` or `commands` command appears in a compound command (i.e., the body of a loop or the `then` or `else` part of an `if` command (A.20.1), or simply in a sequence of commands enclosed in braces), it is executed when the flow of control reaches it, instead of when the compound command is being read. In this case, if the `data` or `commands` command does not specify a file, AMPL reads commands or data from the current input file until it encounters either an `end` command or the end of the current file.

For `include` phrases as well as `model`, `data`, and `commands` commands, files with simple names, e.g., not involving a slash (/), are sought in directories (folders) specified by option `ampl_include` (A.14.1): each nonblank line of `$ampl_include` specifies a directory; if `$ampl_include` is empty or entirely blank, files are sought in the current directory.

The option `insertprompt` (default `'<%d>'`) specifies an insertion prompt that immediately precedes the usual prompt for input from the standard input. If present, `%d` is the current insert

`call`	invoke imported function
`cd`	change current directory
`check`	perform all `check` commands
`close`	close file
`commands`	read and interpret commands from a file
`data`	switch to data mode; optionally include file contents
`delete`	delete model entities
`display`	print model entities and expressions; also `csvdisplay` and `_display`
`drop`	drop a constraint or objective
`end`	end input from current input file
`environ`	set environment for a problem instance
`exit`	exit AMPL with status value
`expand`	show expansion of model entities
`fix`	freeze a variable at its current value
`include`	include file contents
`let`	change data values
`load`	load dynamic function library
`model`	switch to model mode; optionally include file contents
`objective`	select an objective to be optimized
`option`	set or display option values
`print`	print model entities and expressions unformatted
`printf`	print model entities and expressions formatted
`problem`	define or switch to a named problem
`purge`	remove model entities
`quit`	terminate AMPL
`read`	take input from a file
`read table`	take input from a data table
`redeclare`	change declaration of entity
`reload`	reload dynamic function library
`remove`	remove file
`reset`	reset specified entities to their initial state
`restore`	undo a `drop` command
`shell`	temporary escape to operating system to run commands
`show`	show names of model entities
`solexpand`	show expansion as seen by solver
`solution`	import variable values as if from a solver
`solve`	send current instance to a solver and retrieve solution
`update`	allow updating data
`unfix`	undo a `fix` command
`unload`	unload dynamic function library
`write`	write out a problem instance
`write table`	write data to a data table
`xref`	show dependencies among entities

Table A-9: Commands.

level, i.e., nesting of `data` and `commands` commands specifying files and appearing within a compound command.

A.14.1 Options and environment variables

AMPL maintains the values of a variety of *options* that influence the behavior of commands and solvers. Options resemble the ''environment variables'' of the Windows and Unix operating systems; in fact AMPL inherits its initial options from the environments of these systems. AMPL supplies its own defaults for many options, however, if they are not inherited in this way.

The `option` command provides a way to examine and change options. It has one of the forms

> option *redirection*_{opt} ;
> option *opname* [*evalue*] [, *opname* [*evalue*] ...] *redirection*_{opt} ;

The first form prints all options that have been changed or whose default may be provided by AMPL. In the second form, if an *evalue* is present, it is assigned to *opname*; otherwise the value (a character string) currently associated with *opname* is printed. An *opname* is an option name optionally preceded by an environment name (A.18.8) and a period. The option name also may be a name-pattern, which is a name containing one or more `*`'s. In a name-pattern, a `*` stands for an arbitrary sequence, possibly empty, of name characters, and thus may match multiple names; for example

> option *col*;

lists all options whose names contain the string ''col''. Specific environment or option names may also be given by parenthesized string expressions.

An *evalue* is a white-space-separated sequence of one or more literals, numbers, parenthesized string expressions, and references to options of the form $ *opname* or $ $ *opname*, in which *opname* contains no `*`'s; in general, $ *opname* means the current value of option *opname*, and $ $ *opname* means the default value, i.e., the value inherited from the operating system, if any, or provided by AMPL. The quotes around a literal may be omitted if what remains is a name or number. The displayed option values are in a format that could be read as an `option` command.

A.15 Redirection of input and output

An optional *redirection* phrase can be used with a variety of AMPL commands to capture their output in a file for subsequent processing. It applies to all forms of `display` and `print` and also to most other commands that can produce output, such as `solve`, `objective`, `fix`, `drop`, `restore`, and `expand`.

A redirection has one of the forms

> > *filename*
> >> *filename*
> < *filename* (for `read` command)

in which *filename* may have any of the forms that can appear in `data` and `commands` commands (A.14). The file is opened the first time a command specifies *filename* in a redirection; the first form of *redirection* causes the file to be overwritten upon first being opened, while the second form causes output to be appended to the current contents. The form < *filename* is used only for input from the `read` command (A.17). Once open, *filename* remains open until a `reset` or unless explicitly closed by a `close` command:

> close *filenames*_{opt} ;

As long as *filename* remains open, output forms of *redirection* causes output to be appended to the file's current contents. A `close` command without a filename closes all open files and pipes. A `close` command may specify a comma-separated list of filenames. The variant

> `remove` *filename* `;`

closes and deletes *filename*.

A.16 Printing and display commands

The `display`, `print`, and `printf` commands print arbitrary expressions. They have the forms

> `display [` *indexing*: `]` *disparglist redirection*$_{opt}$ `;`
> `print   [` *indexing*: `]` *arglist redirection*$_{opt}$ `;`
> `printf  [` *indexing*: `]` *fmt* , *arglist redirection*$_{opt}$ `;`

If *indexing* is present, its scope extends to the end of the command, and it causes one execution of the command for each member in the *indexing* set. The format string *fmt* is like a `printf` format string in the C programming language and is explained more fully below.

An *arglist* is a (possibly empty, comma-separated) list of expressions and *iterated-arglist*s; an *iterated-arglist* has one of the forms

> *indexing expr*
> *indexing* (*nonempty-arglist*)

where *expr* is an arbitrary expression. The *expr*s can also involve simple or subscripted variable, constraint, and objective names; a constraint name represents the constraint's current dual value. A *disparglist* is described below.

The optional *redirection* (A.15) causes output to be sent to a file instead of appearing on the standard output.

The `print` command prints the items in its *arglist* on one line, separated by spaces and terminated by a newline; the separator may be changed with option `print_separator`. Literals are quoted only if they would have to be quoted in data mode. By default, numeric expressions are printed to full precision, but this can be changed with option `print_precision` or option `print_round`, as described below.

The `printf` command prints the items in its *arglist* as dictated by its format string *fmt*. It behaves like the `printf` function in C. Most characters in the format string are printed verbatim. Conversion specifications are an exception. They start with a `%` and end with a format letter, as summarized in Table A-10. Between the `%` and the format letter there may be any of `-`, for left-justification; `+`, which forces a sign; `0`, to pad with leading zeros; a minimum field width; a period; and a precision giving the maximum number of characters to be printed from a string or digits to be printed after the decimal point for `%f` and `%e` or significant digits for `%g` or minimum number of digits for `%d`. Field widths and precisions are either decimal numbers or a `*`, which is replaced by the value of the next item in the *arglist*. Each conversion specification consumes one or (when `*`'s are involved) more items from the *arglist* and formats the last item it consumes. With `%g`, a precision of 0 (`%.0g` or `%.g`) specifies the shortest decimal string that rounds to the value being formatted. The standard C escape sequences are allowed: `\a` (alert or bell), `\b` (backspace), `\f` (form-feed), `\n` (newline), `\r` (carriage return), `\t` (horizontal tab), `\v` (vertical tab), `\x`*d* and `\x`*dd*,

`%d`	signed decimal notation
`%i`	signed decimal notation (same as `%d`)
`%u`	unsigned decimal notation
`%o`	unsigned octal notation, without leading 0
`%x`	unsigned hexadecimal, using `abcdef`, without leading `0x`
`%X`	unsigned hexadecimal, using `ABCDEF`, without leading `0X`
`%c`	single character
`%s`	string
`%q`	quote string appropriately for data values
`%Q`	always quote string
`%f`	double-precision floating-point
`%e`	double-precision floating-point, exponential notation using `e`
`%E`	double-precision floating-point, exponential notation using `E`
`%g`	double-precision floating-point, using `%f` or `%e`
`%G`	double-precision floating-point, using `%f` or `%E`
`%%`	literal `%`

Table A-10: Conversion specifications in `printf` formats.

where *d* denotes a hexadecimal digit, and *d*, *dd* and *ddd*, where *d* denotes an octal digit. Format `%q` prints a string value with data-section quoting rules; format `%Q` always quotes the string.

The `sprintf` function (A.4.2) formats its argument list according to a format string that uses the same conversion specifications.

The `display` command formats various entities in tables or lists, as appropriate. Its *disparg-list* is similar to an *arglist* for `print` or `printf`, except that an item to be displayed can also be a set expression or the unsubscripted name of an indexed parameter, variable, constraint, or set; furthermore iterated *arglist*s cannot be nested, i.e., they are restricted to the forms

> *indexing expr*
> *indexing* (*exprlist*)

where *exprlist* is a nonempty, comma-separated list of expressions. The `display` command prints scalar expressions and sets individually, and partitions indexed entities into groups having the same number of subscripts, then prints each group in its own table or sequence of tables.

By default, the `display` command rounds numeric expressions to six significant figures, but this can be changed with the options `display_precision` or `display_round`, as described below.

Several options whose names end with `_precision` control the precision with which floating-point numbers are converted to printable values; positive values imply rounding to that many significant figures, and 0 or other values imply rounding to the shortest decimal string that, when properly rounded to a machine number, would yield the number in question. If set to integral values, `$display_round` and `$print_round` override `$display_precision` and `$print_precision`, respectively, and similarly for the analogous options in Table A-11. For example, `$display_round` *n* causes the `display` command to round numeric values to *n* places past the decimal point (or to −*n* places before the decimal point if *n* < 0). A negative precision with `%f` formats as for the `print` command with `print_round` negative. Options that affect printing include those shown in Table A-11.

`csvdisplay_precision`	precision for `_display` and `csvdisplay` (0 is full precision)
`csvdisplay_round`	rounding for `_display` and `csvdisplay` (' ' is full precision)
`display_1col`	maximum elements for a 1D table to be displayed one element per line
`display_eps`	display absolute numeric values < `$display_eps` as zero
`display_max_2d_cols`	if > 0, maximum data columns in a 2D display
`display_precision`	precision for `display` command when `$display_round` is not numeric
`display_round`	places past decimal for `display` command to round
`display_transpose`	transpose tables if rows − columns < `$display_transpose`
`display_width`	maximum line length for `print` and `display` commands
`expand_precision`	precision for `expand` command when `$expand_round` is not numeric
`expand_round`	places past decimal for `expand` command to round
`gutter_width`	separation between columns for `display` command
`objective_precision`	precision for objective value displayed by solver
`omit_zero_cols`	if nonzero, omit all-zero columns from displays
`omit_zero_rows`	if nonzero, omit all-zero rows from displays
`output_precision`	precision used in nonlinear expression (`.nl`) files
`print_precision`	precision for `print` command when `$print_round` is not numeric
`print_round`	places past decimal for `print` command to round
`print_separator`	separator for values printed by `print` command
`solution_precision`	precision for `solve` or `solution` command when `$solution_round` is not numeric
`solution_round`	places past decimal for `solve` or `solution` command to round

Table A-11: Options that control printing.

Commands `_display` and `csvdisplay` are variants that emit tables in a more regular format than does `display`: each line of a table starts with s subscripts and ends with k items, all separated by commas. `_display` and `csvdisplay` differ in the table headers they emit. The header for `_display` consists of a line starting with `_display` and followed by three integers s, k, and n (the number of table lines that follow the header), each preceded by a space. If `$csvdisplay_header` is 1, `csvdisplay` precedes the data values by a header line listing the k indices and n expressions by name. If `$csvdisplay_header` is 0, this header line is omitted.

A.17 Reading data

The `read` command provides a way of reading unformatted data into AMPL parameters and other components, with syntax similar to the `print` command:

> read [*indexing* :] *arglist* *redirection*$_{opt}$;

As with `print`, the optional *indexing* causes the `read` command to be executed separately for each member of the specified indexing set.

The `read` command treats its input as an unformatted series of data values, separated by white space. The *arglist* is a comma-separated list of arguments that specifies a series of components to which these values are assigned. As with `print`, the *arglist* is a comma-separated list of *args*, which may be any of *arg*:

> *component-ref*
> *indexing-expr component-ref*
> *indexing-expr (arglist)*

The *component-ref* must be a reference to a possibly suffixed parameter, variable, or constraint, or a suffixed problem name; it is meaningless to read a value into a set member or any more general expression. All indexing must be explicit, for example read {j in DEST} demand[j] rather than read demand. Values are assigned to *args* in the order that they are read, so later arguments can use values read earlier in the same read command.

If no *redirection* is specified, values are read from the current input stream. Thus if the read command was issued at an AMPL prompt, one types the values at subsequent prompts until all of the *arglist* entries have been assigned values. The prompt changes from ampl? back to ampl: when all the needed input has been read. If instead read is inside a script file that is read with include or commands, then the input is read from the same file, beginning directly after the ; that ends the read command.

Most often the input to read lies in a separate file, specified by the optional *redirection*; its form is <*filename*, where *filename* is a string or parenthesized string expression that identifies a file. Multiple read's can access the same file, in which case each read starts reading the file where the previous one left off. To force reading to start at the beginning again, close *filename* before re-reading.

If a script is to contain a read command that reads values typed interactively, the source of the values must be redirected to the standard input; specifying a – (minus sign) as the *filename* does so. This is most often used to read interactive responses from a user.

A.18 Modeling commands

A.18.1 The solve command

The solve command has the form

> solve *redirection*_{opt} ;

It causes AMPL to write the current translated problem to temporary files in directory $TMPDIR (unless the current optimization problem has not changed since a previous write command), to invoke a solver, and to attempt to read optimal primal and dual variables from the solver. If this succeeds, the optimal variable values become available for printing and other uses. The optional *redirection* is for the solver's standard output.

The current value of the solver option determines the solver that is invoked. Appending '_oopt' to $solver gives the name of an option which, if defined with a nonempty string, determines (by the first letter of the string) the style of temporary problem files written; otherwise, AMPL uses its generic binary output format (style b). For example, if $solver is supersol, then $supersol_oopt, if nonempty, determines the output style. The command-line option '-o?' (A.23) shows a summary of the currently supported output styles.

AMPL passes two command-line arguments to the solver: the *stub* of the temporary files, and the literal string -AMPL. AMPL expects the solver to write dual and primal variable values to file *stub*.sol, preceded by commentary that, if appropriate, reports the objective value to

$objective_precision significant digits. In reading the solution, AMPL rounds the primal variables to $solution_round places past the decimal point if $solution_round is an integer, or to $solution_precision significant figures if $solution_precision is a positive integer; the defaults for these options imply that no rounding is performed.

A variable always has a current value. A variable declaration or data section can specify the initial value, which is otherwise 0. The option reset_initial_guesses controls the initial guess conveyed to the solver. If option reset_initial_guesses has its default value of 0, then the current variable values are conveyed as the initial guess. Setting option reset_initial_guesses to 1 causes the original initial values to be sent. Thus $reset_initial_guesses affects the starting guess for a second solve command, as well as for an initial solve command that follows a solution command (described below).

A constraint always has an associated current dual variable value (Lagrange multiplier). The initial dual value is 0 unless otherwise given in a data section or specified in the constraint's declaration by a := *initial_dual* or a default *initial_dual* phrase. Whether a dual initial guess is conveyed to solvers is governed by the option dual_initial_guesses. Its default value of 1 causes AMPL to pass the current dual variables (if $reset_initial_guesses is 0) or the original initial dual variables to the solver; if $dual_initial_guesses is set to 0, AMPL will omit initial values for the dual variables.

AMPL's presolve phase computes two sets of bounds on variables. The first set reflects any sharpening of the declared bounds implied by eliminated constraints. The other set incorporates sharpenings of the first set that presolve deduces from constraints it cannot eliminate from the problem. The problem has the same solutions with either set of bounds, but solvers that use active-set strategies (such as the simplex method) may have more trouble with degeneracies with the sharpened bounds. Solvers often run faster with the first set, but sometimes run faster with the second. By default, AMPL passes the first set of bounds, but if option var_bounds is 2, AMPL passes the second set. The .lb and .ub suffixes for variables always reflect the current setting of $var_bounds; .lb1 and .ub1 are for the first set, .lb2 and .ub2 for the second set.

If the output style is m, AMPL writes a conventional MPS file, and the value of option integer_markers determines whether integer variables are indicated in the MPS file by 'INTORG' and 'INTEND' 'MARKER' lines. By default, $integer_markers is 1, causing these lines to be written; specifying

```
option integer_markers 0;
```

causes AMPL to omit the 'MARKER' lines.

The option relax_integrality causes integer and binary attributes of variables to be ignored in solve and write commands. It is also possible to control this by setting the .relax suffix of a variable (A.11).

By default, values of suffixes of type IN or INOUT (A.11.1) are sent to the solver, and updated values for suffixes of type OUT or INOUT are obtained from the solver, but the sending and receiving of suffix values can be controlled by setting option send_suffixes suitably: if $send_suffixes is 1 or 3, suffix values are sent to the solver; and if $send_suffixes is 2 or 3, then updated suffix values are requested from the solver.

Whether .sstatus values (A.11.2) are sent to the solver is determined by options send_suffixes and send_statuses; setting $send_statuses to 0 causes .sstatus values not to be sent when $send_suffixes permits sending other suffixes.

Most solvers supply a value for AMPL's built-in symbolic parameter solve_message. AMPL prints the updated solve_message by default, but setting option solver_msg to 0 suppresses this printing. Most solvers also supply a numeric return code solve_result_num, which has a corresponding symbolic value solve_result that is derived from solve_result_num and $solve_result_table analogously to symbolic suffix values (A.11.1).

By default AMPL permutes variables and constraints so solvers see the nonlinear ones first. Some solvers require this, but with other solvers, occasionally it is useful to suppress these permutations by setting option nl_permute suitably. It is the sum of

1	to permute constraints
2	to permute variables
4	to permute objectives

and its default value is 3.

When complementarity constraints are present, the system of constraints is considered *square* if the number of "inequality complements inequality" constraints plus the number of equations equals the number of variables. Some complementarity solvers require square systems, so by default AMPL warns about nonsquare systems. This can be changed by adjusting option compl_warn, which is the sum of

1	warn about nonsquare complementarity systems
2	warn and regard nonsquare complementarity systems as infeasible
4	disregard explicit matchings of variables to equations

A.18.2 The solution command

The solution command has the form

    solution *filename* ;

This causes AMPL to read primal and dual variable values from *filename*, as though written by a solver during execution of a solve command.

A.18.3 The write command

The write command has the form

    write *outopt-value*~opt~ ;

in which the optional *outopt-value* must adhere to the quoting rules for a *filename*. If *outopt-value* is present, write sets $outopt to *outopt-value*. Whether or not *outopt-value* is present, write then writes the translated problem as $outopt dictates: the first letter of $outopt gives the output style (A.18.1), and the rest is used as a "stub" to form the names of the files that are created. For example, if $outopt is "b/tmp/diet", the write command will create file /tmp/diet.nl, and if $auxfiles so dictates (A.18.4), auxiliary files /tmp/diet.row, /tmp/diet.col, and so forth. The solve command's rules for initial guesses, bounds, suffixes, etc., apply.

A.18.4 Auxiliary files

The solve and write commands may cause AMPL to write auxiliary files. For the solve command, appending _auxfiles to $solver gives the name of an option that governs the auxiliary files written; for the write command, $auxfiles plays this role. The auxiliary files

Key	File	Description
a	*stub*.adj	constant added to objective values
c	*stub*.col	AMPL names of variables the solver sees
e	*stub*.env	environment (written by a solve command)
f	*stub*.fix	variables eliminated from the problem because their values are known
p	*stub*.spc	MINOS ''specs'' file for output style m
r	*stub*.row	AMPL names of constraints and objectives the solver sees
s	*stub*.slc	constraints eliminated from the problem
u	*stub*.unv	unused variables

Table A-12: Auxiliary files.

shown in Table A-12 are written only if the governing option's value contains the indicated key letter. If a key letter is capitalized, the corresponding auxiliary file is written only if the problem is nonlinear.

A.18.5 Changing a model: `delete`, `purge`, `redeclare`

The command

```
delete namelist ;
```

deletes each *name* in *namelist*, restoring any previous meaning *name* had, provided no other entities depend on *name*, i.e., if xref *name* (A.19.2) reports no dependents.

The command

```
purge namelist ;
```

deletes each *name* and all its direct and indirect dependents.

The statement

```
redeclare entity-declaration ;
```

replaces any existing declaration of the specified entity with the given one, provided either that the entity has no dependents, or that the new declaration does not change the character of the entity (its kind, such as set or param, and its number of subscripts). A redeclare can be applied to statements beginning with any of the following:

```
arc       function   minimize   param     set          suffix    var
check     maximize   node       problem   subject to   table
```

Redeclarations that would cause circular dependencies are rejected.

The command

```
delete check n;
```

deletes the *n*th check, while

```
redeclare check n indexing_opt : ... ;
```

redeclares the *n*th check.

A.18.6   The drop, restore and objective commands

These commands have the forms

> drop *indexing_{opt} constr-or-obj-name redirection_{opt}* ;
> restore *indexing_{opt} constr-or-obj-name redirection_{opt}* ;
> objective *objective-name redirection_{opt}* ;

where *constr-or-obj-name* is the possibly subscripted name of a constraint or objective. The drop command instructs AMPL not to transmit the indicated entity (in write and solve commands); the restore command cancels the effect of a corresponding drop command. If *constr-or-obj-name* is not subscripted but is the name of an indexed collection of constraints or objectives, drop and restore affect all members of the collection.

The objective command arranges that only the named objective is passed to the solver. Issuing an objective command is equivalent to dropping all objectives, then restoring the named objective.

A.18.7   The fix and unfix commands

These commands have the forms

> fix *indexing_{opt} varname* [:= *expr*] *redirection_{opt}* ;
> unfix *indexing_{opt} varname* [:= *expr*] *redirection_{opt}* ;

where *varname* is the possibly subscripted name of a variable. The fix command instructs AMPL to treat the indicated variable (in write and solve commands) as though fixed at its current value, i.e., as a constant; the unfix command cancels the effect of a corresponding fix command. If *varname* is not subscripted but is the name of an indexed collection of variables, fix and unfix affect all members of the collection.

An optional := *expr* may appear before the terminating semicolon, in which case the expression is assigned to the variable being fixed or unfixed, as though assigned by let (A.18.9).

A.18.8   Named problems and environments

The problem declaration/command has three functions: declaring a new problem, making a previously declared problem current, and printing the name of the current problem (in the form of a problem command establishing the current problem).

> problem *name indexing_{opt} environ_{opt} suffix-initializations_{opt}* : *itemlist* ;

declares a new problem and specifies the variables, constraints, and objectives that are in it. Other variables appearing in the specified constraints and objectives are fixed (but can be unfixed by the unfix command). The new problem becomes the current problem. Initially the current problem is Initial. The *itemlist* in a problem declaration is a comma-separated list of possibly subscripted names of variables, constraints, and objectives, each optionally preceded by indexing. *Suffix-initializations* are analogous to those in constraint declarations, except that they appear before the colon.

The command

> problem *name* ;

makes *name* (a previously declared problem) current, and

       `problem` *redirection*$_{opt}$ `;`

prints the current problem name. Drop/restore and fix/unfix commands apply only to the current problem. Variable values, like parameters, are global; just the fixed/unfixed status of a variable depends on the problem. Similarly, the drop/restore status of a constraint depends on the problem (as do reduced costs). The current problem does not restrict the `let` command (A.18.9).

    When a problem is declared, it can optionally specify an environment associated with the problem: the *environ* phrase has the form

       `environ` *envname*

to specify that the problem's initial environment is *envname*, which must bear a subscript if the environment is indexed. Otherwise a new unindexed environment with the same name as the problem is created, and it inherits the then current environment (set of option values).

    In `option` commands, unadorned (conventional) option names refer to options in the current environment, and the notation *envname*`.`*opname* refers to $opname in environment *envname*. The declaration

       `environ` *envname* *indexing*$_{opt}$ `;`

declares environment *envname* (or indexed set of environments, if *indexing* is present). If there is no *indexing*, *envname* becomes the current environment for the current problem.

    For previously declared environments, the command

       `environ` *envname* `;`

makes the indicated environment current, and the command

       `environ` *indexing*$_{opt}$ *envname* `:=` *envname*$_1$ `;`

copies environment *envname*$_1$ to *envname*, where *envname* and *envname*$_1$ must be subscripted if declared with *indexing*s. The initial environment is called `Initial`.

A.18.9 Modifying data: `reset, update, let`

    The `reset` command has several forms.

       `reset ;`

causes AMPL to forget all model declarations and data and to close all files opened by redirection, while retaining the current option settings.

       `reset options ;`

causes AMPL to restore all options to their initial state. It ignores the current `$OPTIONS_IN` and `$OPTIONS_INOUT`; the files they name can be included manually, if desired.

       `reset data ;`

causes AMPL to forget all assignments read in data mode and allows reading data for a different problem instance.

       `reset data` *name-list* `;`

causes AMPL to forget any data read for the entities in *name-list*; commas between names are optional.

A `reset data` command forces recomputation of all = expressions, and `reset data p`, even when p is declared with a = expression, forces recomputation of random functions in the = expression (and of any user-defined functions in the expression).

Problems (including the current one) are adjusted when their indexing expressions change, except that previous explicit drop/restore and fix/unfix commands remain in effect. The `reset problem` command cancels this treatment of previous explicit drop, restore, fix, and unfix commands, and brings the problem to its declared drop/fix state. This command has the forms

> ` reset problem ; `   *applies to the current problem*
> ` reset problem `*probname* *[subscript]*_{opt} `;`

If the latter form mentions the current problem, it has the same effect as the first form. `reset problem` does not affect the problem's environment.

> ` update data ; `

permits all data to be updated once in subsequent data sections: current values may be overwritten, but no values are discarded.

> ` update data `*name-list* `;`

grants update permission only to the entities in *name-list*.

The `let` command

> ` let ` *indexing*_{opt} *name* `:=` *expr* `;`

changes the value of the possibly indexed set, parameter or variable *name* to the value of the expression. If *name* is a set or parameter, its declaration must not specify a = phrase.

The command

> ` let ` *indexing*_{opt} *name*`.`*suffix* `:=` *expr* `;`

assigns the corresponding suffix value, if permitted. Some suffix values are derived and cannot be assigned; attempting to do so causes an error message.

A.19 Examining models

A.19.1 *The* `show` *command*

The command

> ` show ` *namelist*_{opt} *redirection*_{opt} `;`

lists all model entities if *namelist* is not present. It shows each *name*'s declaration if it has one, or else lists model entities of the kind indicated by the first letters of each *name*:

```
ch... ==> checks            c... ==> constraints
e...  ==> environments      f... ==> functions
o...  ==> objectives
pr... ==> problems          p... ==> parameters
su... ==> suffixes          s... ==> sets
t...  ==> tables            v... ==> variables
```

A.19.2   The xref command

The command xref shows entities that depend directly or indirectly on specified entities:

>   xref *itemlist redirection*_{opt} ;

A.19.3   The expand command

The expand command prints generated constraints and objectives:

>   expand *indexing*_{opt} *itemlist redirection*_{opt} ;

The *itemlist* can assume the same forms allowed in problem declarations. If it is empty, all non-dropped constraints and objectives are expanded. The variant

>   solexpand *indexing*_{opt} *itemlist redirection*_{opt} ;

shows how constraints and objectives appear to the solver. It omits constraints and variables eliminated by presolve unless they are explicitly specified in the *itemlist*.

Both the expand and solexpand commands permit variables to appear in the *itemlist*; for each, the commands show the linear coefficients of the variable in the relevant (non-dropped and, for solexpand, not eliminated by presolve) constraints and objectives, and indicates "+ nonlinear" when the variable also participates nonlinearly in a constraint or objective.

The options expand_precision and expand_round control printing of numbers by expand. By default they are printed to 6 significant figures.

A.19.4   Generic names

AMPL provides a number of generic names that can be used to access model entities without using model-specific names. Some of these names are described in Table A-13; the complete current list is on the AMPL web site.

These synonyms and sets can be used in display and other commands. They present the modeler's view (before presolve). Similar automatically updated entities with _ changed to _s (i.e., _snvars, _svarnames, _svar, etc.) give the solver's view, i.e., the view after presolve. There are exceptions, however, due to the way complementarity constraints are handled (A.9.1): none of _cvar, _scconname, or _snccons exists.

A.19.5   The check command

The command

>   check;

causes all check statements to be evaluated.

A.20 Scripts and control flow statements

AMPL provides statements similar to control flow statements in conventional programming languages, which make it possible to write a program of statements to be executed automatically.

`_nvars`	number of variables in current model
`_ncons`	number of constraints in current model
`_nobjs`	number of objectives in current model
`_varname{1.._nvars}`	names of variables in current model
`_conname{1.._ncons}`	names of constraints in current model
`_objname{1.._nobjs}`	names of objectives in current model
`_var{1.._nvars}`	synonyms for variables in current model
`_con{1.._ncons}`	synonyms for constraints in current model
`_obj{1.._nobjs}`	synonyms for objectives in current model
`_PARS`	set of all declared parameter names
`_SETS`	set of all declared set names
`_VARS`	set of all declared variable names
`_CONS`	set of all declared constraint names
`_OBJS`	set of all declared objective names
`_PROBS`	set of all declared problem names
`_ENVS`	set of all declared environment names
`_FUNCS`	set of all declared user-defined functions
`_nccons`	number of complementarity constraints before presolve
`_cconname{1.._nccons}`	names of complementarity constraints
`_scvar{1.._sncons}`	if `_scvar[i] > 0`, `_svar[scvar[i]]` complements `_scon[i]`
`_snbvars`	number of binary (0,1) variables
`_snccons`	number of complementarity constraints after presolve
`_snivars`	number of general integer variables (excluding binaries)
`_snlcc`	number of linear complementarity constraints
`_snlnc`	number of linear network constraints
`_snnlcc`	number of nonlinear compl. constrs.: `_snccons=_snlcc+_snnlcc`
`_snnlcons`	number of nonlinear constraints
`_snnlnc`	number of nonlinear network constraints
`_snnlobjs`	number of nonlinear objectives
`_snnlv`	number of nonlinear variables
`_snzcons`	number of constraint Jacobian matrix nonzeros
`_snzobjs`	number of objective gradient nonzeros

Table A-13: Generic synonyms and sets.

A.20.1 The `for`, `repeat` and `if-then-else` statements

Several commands permit conditional execution of and looping over lists of AMPL commands:

```
if lexpr then cmd
if lexpr then cmd else cmd
for loopname_opt indexing cmd
repeat loopname_opt opt-while { cmds } opt-while ;
break loopname_opt ;
continue loopname_opt ;
```

In these statements, *cmd* is either a single, possibly empty, command that ends with a semicolon or a sequence of zero or more commands enclosed in braces. *lexpr* is a logical expression. *loopname* is an optional loop name (which must be unbound before the syntactic start of the loop), which goes out of scope after syntactic end of the loop. If present, an *opt-while* condition has one of the forms

```
while lexpr
until lexpr
```

If *loopname* is specified, `break` and `continue` apply to the named enclosing loop; otherwise they apply to the immediately enclosing loop. A `break` terminates the loop, and `continue` causes its next iteration to begin (if permitted by the optional initial and final *opt-while* clauses of a `repeat` loop, or by the *indexing* of a `for` loop). Dummy indexes from *indexing* may appear in *cmd* in a `for` loop. The entire index set of a `for` loop is instantiated before starting execution of the loop, so the set of dummy indices for which the loop body is executed will be unaffected by assignments in the loop body.

Variants of `break`,

```
break commands ;
break all ;
```

terminate, respectively, the current `commands` command or all `commands` commands, if any, and otherwise act as a `quit` command.

Loops and if-then-else structures are treasured up until syntactically complete. Because `else` clauses are optional, AMPL must look ahead one token to check for their presence. At the outermost level, one must thus issue a null command (just a semicolon) or some other command or declaration to execute an outermost "`else`-less" `if` statement. (In this regard, end-of-file implies an implicit null statement.)

A semicolon is taken to appear at the end of command files that end with a compound command with optional final parts missing:

```
repeat ... { ... }    # no final condition or semicolon
if ... then { ... }   # no else clause
```

AMPL has three pairs of prompts whose text can be changed through option settings. The default settings are:

```
option cmdprompt1 '%s ampl: ';
option cmdprompt2 '%s ampl? ';
option dataprompt1 'ampl data: ';
option dataprompt2 'ampl data? ';
option prompt1 'ampl: ';
option prompt2 'ampl? ';
```

`prompt1` appears when a new statement is expected, and `prompt2` when the previous input line is not yet a complete command (for example, if the semicolon at the end is missing).

In data mode, the values of `dataprompt1` and `dataprompt2` are used instead. When a new line is begun in the middle of an `if`, `for` or `repeat` statement, the values of `cmdprompt1` and `cmdprompt2` are used, with `%s` replaced by the appropriate command name; for example:

```
ampl: for {t in time} {
for{...} { ? ampl: if t <= 6
for{...} { ? ampl?   then let cmin[t] := 3;
if ... then {...} ? ampl: else let cmin[t] := 4;
for{...} { ? ampl: };
ampl:
```

A.20.2 Stepping through commands

It is possible to step through commands in an AMPL script one command at a time. Single-step mode is enabled by

```
option single_step n ;
```

where n is a positive integer; it specifies that if the insert level is at most n, AMPL should behave as though `commands -;` were inserted before each command: it should read commands from the standard input until `end` or other end of file signal (control-D on Unix, control-Z on Windows). Some special commands may appear in this mode:

```
step  n_opt      execute the next command, or n commands
skip  n_opt      skip the next command, or n commands
next  n_opt      if the next command is an if-then-else or looping command, execute the entire
                 compound command, or n commands, before stopping again
                 (unless the compound command itself specifies commands -;)
cont             execute until the end of all currently nested compound commands
                 at the current insert level
```

A.21 Computational environment

AMPL runs in an operating system environment, most often as a standalone program, but sometimes behind the scenes in a graphical user interface or a larger system. Its behavior is influenced by values from the external environment, and it can set values that become part of that environment. The parameter `_pid` gives the process ID of the AMPL process (a number unique among processes running on the system).

A.21.1 The `shell` command

The `shell` command provides a temporary escape to the operating system, if such is permitted, to run commands.

```
shell 'command-line' redirection_opt ;
shell redirection_opt ;
```

The first version runs *command-line*, which is contained in a literal string. In the second version, AMPL invokes an operating-system shell, and control returns to AMPL when that shell terminates. Before invoking the shell, AMPL writes a list of current options and their values to the file (if any) named by option `shell_env_file`. The name of the shell program is determined by option `SHELL`.

A.21.2 The `cd` command

The `cd` command reports or changes AMPL's working directory.

```
cd ;
cd new-directory ;
```

The parameter `_cd` is set to this value.

`_ampl_elapsed_time`	elapsed seconds since the start of the AMPL process
`_ampl_system_time`	system CPU seconds used by the AMPL process itself
`_ampl_user_time`	user CPU seconds used by the AMPL process itself
`_ampl_time`	`_ampl_system_time + _ampl_user_time`
`_shell_elapsed_time`	elapsed seconds for most recent shell command
`_shell_system_time`	system CPU seconds used by most recent shell command
`_shell_user_time`	user CPU seconds used by most recent shell command
`_shell_time`	`_shell_system_time + _shell_user_time`
`_solve_elapsed_time`	elapsed seconds for most recent solve command
`_solve_system_time`	system CPU seconds used by most recent solve command
`_solve_user_time`	user CPU seconds used by most recent solve command
`_solve_time`	`_solve_system_time + _solve_user_time`
`_total_shell_elapsed_time`	elapsed seconds used by all shell commands
`_total_shell_system_time`	system CPU seconds used by all shell commands
`_total_shell_user_time`	user CPU seconds used by all shell commands
`_total_shell_time`	`_total_shell_system_time+_total_shell_user_time`
`_total_solve_elapsed_time`	elapsed seconds used by all solve commands
`_total_solve_system_time`	system CPU seconds used by all solve commands
`_total_solve_user_time`	user CPU seconds used by all solve commands
`_total_solve_time`	`_total_solve_system_time+_total_solve_user_time`

Table A-14: Built-in timing parameters.

A.21.3 The `quit`, `exit` and `end` commands

The `quit` command causes AMPL to stop without writing any files implied by `$outopt`, and the `end` command causes AMPL to behave as though it has reached the end of the current input file, without reverting to model mode. At the top level of command interpretation, either command terminates an AMPL session. The command `exit` is a synonym for `quit`, but it can return a status to the surrounding environment:

> `exit` *expression*$_{opt}$ `;`

A.21.4 Built-in timing parameters

AMPL has built-in parameters that record various CPU and elapsed times, as shown in Table A-14. Most current operating systems keep separate track of two kinds of CPU time: *system* time spent by the operating time on behalf of a process, e.g., for reading and writing files, and *user* time spent by the process itself outside of the operating system. Usually the system time is much smaller than the user time; when not, finding out why not sometimes suggests ways to improve performance. Because seeing separate system and user times can be helpful when performance seems poor, AMPL provides built-in parameters for both sorts of times, as well as for their sums. AMPL runs both solvers and shell commands as separate processes, so it provides separate parameters to record the times taken by each sort of process, as well as for the AMPL process itself.

A.21.5 Logging

If option `log_file` is a nonempty string, it is taken as the name of a file to which AMPL copies everything it reads from the standard input. If option `log_model` is 1, then commands and declarations read from other files are also copied to the log file, and if `log_data` is 1, then data sections read from other files are copied to the log file as well.

A.22 Imported functions

Sometimes it is convenient to express models with the help of functions that are not built into AMPL. AMPL has facilities for importing functions and optionally checking the consistency of their argument lists. **Note:** The practical details of using imported functions are highly system-dependent. This section is concerned only with syntax; specific information will be found in system-specific documentation, e.g., on the AMPL web site.

An imported function may need to be evaluated to translate the problem; for instance, if it plays a role in determining the contents of a set, AMPL must be able to evaluate the function. In this case the function must be linked, perhaps dynamically, with AMPL. On the other hand, if an imported function's only role is in computing the value of a constraint or objective, AMPL never needs to evaluate the function and can simply pass references to it on to a (nonlinear) solver.

Imported functions must be declared in a `function` declaration before they are referenced. This statement has the form

> function *name* *alias*_{opt} (*domain-spec*)_{opt} *type*_{opt} [pipe *litseq*_{opt} [format *fmt*]] ;

in which *name* is the name of the function, and *domain-spec* amounts to a function prototype:

> *domain-spec*:
>   *domain-list*
>   ...
>   *nonempty-domain-list* , ...

A *domain-list* is a (possibly empty, comma-separated) list of set expressions, asterisks (*'s), direction words (IN, OUT, or INOUT), direction words followed by set expressions, and *iterated-domain-list*s:

> *iterated-domain-list*:
>   *indexing* (*nonempty domain-list*)

An *iterated-domain-list* is equivalent to one repetition of its *domain-list* for each member in the *indexing* set, and the domain of dummy variables appearing in the *indexing* extends over that *domain-list*. The direction words indicate which way information flows: into the function (IN), out of the function (OUT), or both, with IN the default. In a function invocation, OUT arguments are assigned values specified by the function at the end of the command invoking the function.

Omitting the optional (*domain-spec*) in the `function` declaration is the same as specifying (...). The function must be invoked with at least or exactly as many arguments as there are sets in the *domain-spec* (after *iterated-domain-list*s have been expanded), depending on whether or not the *domain-spec* ends with ... . AMPL checks that each argument corresponding to a set in the *domain-list* lies in that set. A * by itself in a *domain-list* signifies no domain checking for the corresponding argument.

A function whose return value is not of interest can be invoked with a `call` command:

> call *funcname* (*arglist*) ;

Type can be `symbolic` or `random` or both; `symbolic` means the function returns a literal (string) value rather than a numeric value, and `random` indicates that the "function" may return different values for the same arguments, i.e., AMPL should assume that each invocation of the function returns a different value.

The commands

```
load     libnames opt ;
unload   libnames opt ;
reload   libnames opt ;
```

load, unload, or reload shared libraries (from which functions and table handlers are imported); *libnames* is a comma-separated list of library names. When at least one *libname* is mentioned in the load and unload commands, $AMPLFUNC is modified to reflect the full pathnames of the currently loaded libraries. The reload command first unloads its arguments, then loads them. This can change the order of loaded libraries and affect the visibility of imported functions: the first name wins. With no arguments, load loads all the libraries currently in $AMPLFUNC; unload unloads all currently loaded libraries, and reload reloads them (which is useful if some have been recompiled).

The keyword pipe indicates that this is a *pipe function*, which means AMPL should start a separate process for evaluating the function. Each time a function value is needed, AMPL writes a line of arguments to the function process, then reads a line containing the function value from the process. (Of course, this is only possible on systems that allow multiple processes.) A *litseq* is a sequence of one or more adjacent literals or parenthesized string expressions, which AMPL concatenates and passes to the operating system (i.e., to $SHELL) as the description of the process to be invoked. In the absence of a *litseq*, AMPL passes a single literal, whose value is the name of the function. If the optional format *fmt* is present, *fmt* must be a format, suitable for printf, that tells AMPL how to format each line it sends to the function process. If no *fmt* is specified, AMPL uses spaces to separate the arguments it passes to the pipe function.

For example:

```
ampl: function mean2 pipe "awk '{print ($1+$2)/2}'";
ampl: display mean2(1,2) + 1;
mean2(1, 2) + 1 = 2.5
```

The function mean2 is expected (by default) to return numeric values; AMPL will complain if it returns a string that does not represent a number.

The following functions are symbolic, to illustrate formatting and the passing of arguments.

```
ampl: function f1 symbolic pipe "awk '" '{printf "x%s\n", $1}' "'";
ampl: function g1 symbolic pipe 'awk ''{printf "XX%s\n", $1}''';
ampl: function cat symbolic pipe format ">>%s<<\n";

ampl: display f1(2/3);
f1(2/3) = x0.66666666666666667

ampl: display g1('abc');
g1('abc') = XXabc

ampl: display cat('some words');
cat('some words') = ">>'some words'<<"
```

The declaration of f1 specifies a *litseq* of 3 literals, while g1 specifies one literal; cat, having an empty *litseq*, is treated as though its *litseq* were 'cat'. The literals in each *litseq* are stripped of the quotes that enclose them, have one of each adjacent pair of these quotes removed, and have (*backslash*, *newline*) pairs changed to a single newline character; the results are concatenated to produce the string passed to the operating system as the description of the process to be started. Thus for the four pipe functions above, the system sees the commands

```
awk '{print ($1+$2)/2}'
awk '{printf "x%s\n", $1}'
awk '{printf "XX%s\n", $1}'
cat
```

respectively. Function `cat` illustrates the optional `format` *fmt* phrase. If *fmt* results in a string that does not end in a newline, AMPL appends a newline character. If no *fmt* is given, each numeric argument is converted to the shortest decimal string that rounds to the argument value.

Caution: The line returned by a pipe function must be a complete line, i.e., must end with a *newline* character, and the pipe function process must flush its buffers to prevent deadlock. (Pipe functions do not work with most standard Unix programs, because they don't flush output at the end of each line.)

Imported functions may be invoked with conventional functional notation, as illustrated above. In addition, iterated arguments are allowed. More precisely, if f is an imported function, an invocation of f has the form f (*arglist*) in which *arglist* is as for the `print` and `printf` commands — a possibly empty, comma-separated list of expressions and *iterated-arglist*s:

```
ampl: function mean pipe 'awk ''{x = 0\
        for(i = 1; i <= NF; i++) x += $i\
        printf "%.17g\n", x/NF}''';
ampl: display mean({i in 1..100} i);
mean({i in 1 .. 100} (i)) = 50.5

ampl: display mean({i in 1..50}(i,i+50));
mean({i in 1 .. 50} (i, i + 50)) = 50.5

ampl: display mean({i in 0..90 by 10}({j in 1..10} i + j));
mean({i in 0 .. 90 by 10} ({j in 1 .. 10} (i + j))) = 50.5
```

The command

```
reset function name_opt ;
```

closes all pipe functions, causing them to be restarted if invoked again. If an function is named explicitly, only that function is closed.

A.23 AMPL invocation

AMPL is most often invoked as a separate command in some operating system environment. When AMPL begins execution, the declarations, commands, and data sections described above (A.14) can be entered interactively. Depending on the operating system where AMPL is run, the invocation may be accompanied by one or more *command-line arguments* that set various properties and options and specify files to be read. These can be examined by typing the command

```
ampl '-?'
```

The initialization of some options may be determined by command-line arguments. The '-?' argument produces a listing of these options and their command-line equivalents.

Sometimes it is convenient to have option settings remembered across AMPL sessions. Under operating systems from which AMPL can inherit environment variables as described above, the options OPTIONS_IN, OPTIONS_INOUT, and OPTIONS_OUT provide one way to do this. If $OPTIONS_IN is nonempty in the inherited environment, it names a file (meant to contain option commands) that AMPL reads before processing command-line arguments or entering its command environment. OPTIONS_INOUT is similar to OPTIONS_IN; AMPL reads file $OPTIONS_INOUT (if nonempty) after $OPTIONS_IN. At the end of execution, if $OPTIONS_INOUT is nonempty, AMPL writes the current option settings to file $OPTIONS_INOUT. If nonempty, $OPTIONS_OUT is treated like $OPTIONS_INOUT at the end of execution.

The command-line argument -v prints the version of the AMPL command being used; this is also available as option version.

The command-line option -R (recognized only as the first command-line option and not mentioned in the -? listing of options) puts AMPL into a restricted "server mode," in which it declines to execute cd and shell commands, forbids changes to options TMPDIR, ampl_include, and PATH (or the search path for the operating system being used), disallows pipe functions, and restricts names in option solver and file redirections to be alphanumeric (so they can only write to the current directory, which, on Unix systems at least, cannot be changed). By invoking AMPL from a shell script that suitably adjusts current directory and environment variables before it invokes ampl -R, one can control the directory in which AMPL operates and the initial environment that it sees.

On systems where imported function libraries can be used, the command-line option -i *libs* specifies libraries of imported functions (A.22) and table handlers (A.13) that AMPL should load initially. If -i *libs* does not appear, AMPL assumes -i$AMPLFUNC. Here *libs* is a string, perhaps extending over several lines, with the name of one library or directory per line. For a directory, AMPL looks for library amplfunc.dll in that directory. If *libs* is empty and amplfunc.dll appears in the current directory, AMPL loads amplfunc.dll initially. If library ampltabl.dll is installed in what the operating system considers to be a standard place, AMPL also tries to load this library, which can provide "standard" database handlers and functions.

Index